Energy for Rural Development

Renewable Resources and Alternative Technologies for Developing Countries

Report of an Ad Hoc Panel of the Advisory
Committee on Technology Innovation
Board on Science and Technology for
International Development
Commission on International Relations

NATIONAL ACADEMY OF SCIENCES
Washington, D.C. 1976

This publication has been prepared by the Panel on Renewable Energy Resources of the Advisory Committee on Technology Innovation of the Board on Science and Technology for International Development, Commission on International Relations, National Academy of Sciences–National Research Council, for the Office of Science and Technology, Bureau for Technical Assistance, Agency for International Development, Washington, D.C., under Contract AID/csd-2584, Task Order No. 1.

NOTICE: The project that is the subject of this report was approved by the Governing Board of the National Research Council, whose members are drawn from the Councils of the National Academy of Sciences, the National Academy of Engineering, and the Institute of Medicine. The members of the Committee responsible for the report were chosen for their special competences and with regard for appropriate balance.

This report has been reviewed by a group other than the authors according to procedures approved by a Report Review Committee consisting of members of the National Academy of Sciences, the National Academy of Engineering, and the Institute of Medicine.

Second printing May 1977
Third printing January 1980

A French-language edition of this report was published in 1978 and single copies are available on request from:

Office of Energy
Bureau for Development Support
Agency for International Development
Washington, D.C. 20523

ISBN: 978-1-60322-037-8

PANEL ON RENEWABLE ENERGY RESOURCES

WILLIAM L. HUGHES, Head, School of Electrical Engineering, Oklahoma State University, Stillwater, Oklahoma 74075, USA, *Chairman*

CLANTON C. BLACK, JR., Chairman, Department of Botany, University of Georgia, Athens, Georgia 30601, USA

JAMES A. CRUTCHFIELD, JR., Department of Economics, University of Washington, Seattle, Washington 98105, USA

JESSE C. DENTON, Dean, School of Energy, American Technological University, Killeen, Texas 76541, USA

JOHN A. DUFFIE, Director, Solar Energy Laboratory, University of Wisconsin, Madison, Wisconsin 53706, USA

JOSEPH J. ERMENC, Thayer School of Engineering, Dartmouth College, Hanover, New Hampshire 03755, USA

THOMAS A. LAWAND, Director of Field Operations, Brace Research Institute, Macdonald College, McGill University, Ste. Anne de Bellevue, Quebec, Canada

GEORGE O. G. LÖF, Solar Energy Applications Laboratory, Engineering Research Center, Colorado State University, Denver, Colorado 80521, USA

DALE N. MOSS, Department of Agronomy and Plant Genetics, University of Minnesota, St. Paul, Minnesota 55101, USA

ROBERT W. REX, President, Republic Geothermal, Inc., 11848 E. Washington Boulevard, Whittier, California 90606, USA

GEOFFREY R. ROBSON, Chief, Minerals Section, Energy and Mineral Development Branch, Centre for Natural Resources, Energy and Transport, United Nations, New York 10017, USA

DANIEL I-c. WANG, Department of Nutrition and Food Science, Massachusetts Institute of Technology, Cambridge, Massachusetts 02139, USA

JEROME M. WEINGART, International Institute for Applied Systems Analysis, 2361 Laxenburg, Austria

NORMAN L. BROWN, Board on Science and Technology for International Development, Commission on International Relations, National Academy of Sciences–National Research Council, 2101 Constitution Avenue, Washington, D.C. 20418, USA, *Staff Study Director*

JULIEN ENGEL, Board on Science and Technology for International Development, Commission on International Relations, National Academy of Sciences–National Research Council, *Head, Special Studies*

Preface

Five years ago, the Advisory Committee on Technology Innovation (ACTI) was established by the Board on Science and Technology for International Development of the National Research Council. Its objective was threefold:

- To screen technologies in use in industrialized nations;
- To examine technologies available but not in use; and
- To determine which of these might be successfully applied in developing nations.

Further, ACTI's aim was to determine the direction of research and development of these technologies and encourage any new approaches necessary to adapt them.

In line with these purposes, ACTI began a study of low-power sources of electrical energy for developing countries. The original intent was to concentrate on sources that could provide power of the order of 100 watts for use primarily, but not exclusively, with small communications devices (including village television receivers). A preliminary engineering survey was conducted in 1973 to assess the then-current availability and cost of suitable devices. While the survey recognized the potential applications of solar devices and wind-driven generators, it assumed the continued availability of fossil fuels (e.g., kerosene) in most rural areas of developing countries. This assumption accounted for the emphasis on the economic and mechanical advantages of small, inexpensive, internal-combustion engines.

It soon became apparent, however, that fuel for conventional energy devices was becoming more expensive and, in many cases, less available. Therefore (and at the request of the Agency for International Development), ACTI broadened the scope of its study to consider other energy sources potentially available to developing countries for low-cost exploitation. A panel was appointed to examine energy technologies applicable at the village or rural level—technologies with power capabilities in the range of 10-100 kilowatts maximum. This report is the result of that examination. Each technology was to be discussed in terms of current or short-term (i.e., within 5 years) and intermediate-term (5-10 years) availability. The panel also was asked to identify the specific research and development efforts needed to make intermediate-term applications feasible in those areas in which such advances offered realistic promise.

With nuclear energy excluded from consideration, the technologies considered by the panel fall into two categories: direct and indirect uses of solar energy. Direct uses of solar energy include cookers, stills for potable water, heating and cooling of buildings, refrigeration, heating of water, crop drying, salt production, and photovoltaics. Indirect uses include photosynthesis, microbial conversion of plant materials to fuel, wind devices, and water devices. Finally, falling into neither of these major categories, there are discussions of geothermal energy and the general problem of energy storage. There were some early doubts about including geothermal sources in this study because of the magnitude of their power capabilities (megawatts) and the cost of exploration and exploitation. The panel felt, nevertheless, that special circumstances could make geothermal wells a practical source of energy for some rural areas in some developing countries. The benefit to those regions in which the technology is applicable was considered sufficient to justify its inclusion.

To maximize its usefulness to two distinct audiences, this report is divided into two sections. The first contains a series of nontechnical summaries in which each technology is described and its potential application outlined. This section is directed to the decision maker and planner who must evaluate technical proposals on the basis of a country's needs and prevailing constraints. The second portion of the report—the technical section—is directed to the specialist who may be familiar with the basic principles involved in given technologies but not with local problems, recent developments, or available information sources.

Technologies available within the next 5 years are variously referred to in the text as having "current," "short-term," or "near-term" availability. These are characterized by devices that either are on the market as manufactured commercial articles or have passed the stages of development and testing and can reasonably be expected to be on the market within 5 years. Technologies likely to become available within 5-10 years are designated "intermediate-term" technologies, or technologies "soon to be available." For these, devices for their exploitation are only just emerging from the experimental stage and are being tested as prototypes, but have not yet reached the stage where commercial manufacture is imminent.

The panel recognizes that this report is but one step toward the adoption of the technologies discussed, and that application of these technologies to rural development will entail many individual efforts. With this in mind, the panel members have agreed to respond individually to questions that may arise that cannot be answered by information sources listed in the report—within realistic limits imposed by time and geography. Assistance of a similar nature and, especially, on practical matters of construction and operation, may be obtained from other sources, such as VITA* (Volunteers in Technical Assistance), an organization with long experience in practical applications of technology in the field.

*Inquiries may be addressed to: VITA, 3706 Rhode Island Avenue, Mt.Rainier, Maryland 20822, USA.

LIST OF SYMBOLS

a	acre	km	kilometer
AC	alternating current	kW	kilowatt
amp-hr	ampere-hour	kWe	kilowatt electrical
Btu	British thermal unit	kWh	kilowatt-hour
bu	bushel	l	liter
°C	degrees Celsius	lb	pound
cm	centimeter	m	meter
DC	direct current	μ	micron
°F	degrees Fahrenheit	mA	milliampere
ft	foot	mi	mile
g	gram	MHz	megahertz
gal	gallon (U.S.)	mph	miles per hour
Gcal	gigacalorie	m/sec	meters per second
ha	hectare	MW	megawatt
hl	hectoliter	mV	millivolt
hp	horsepower	Ω-cm	ohm-centimeter
hr	hour	psig	pounds per square inch, gauge
Hz	Hertz	W	watt
in.	inch	Wh	watt-hour
kcal	kilocalorie	Whe	watt-hour electrical
kg	kilogram		

Acknowledgments

It is only appropriate that the members of the steering group who originally recommended this study and suggested its form be thanked for their efforts. In addition to Jesse C. Denton and Clanton C. Black, Jr., who were also members of the study panel, the group included Charles A. Rosen (who served as chairman), Irma M. Adelman, Charles A. Berg, Lloyd O. Herwig, and Thomas F. Malone. The panel members found their suggestions most useful.

Thanks are also due to Kudret Selçuk, Marshal F. Merriam, William T. Beale, W. P. Teagen, and Sohrab K. Ghandhi for their assistance with the sections on crop drying, windpower, Stirling engines, Rankine engines, and photovoltaics, respectively. The editorial assistance of Warren Kornberg is acknowledged with thanks, and the major contribution of F. R. Ruskin, who prepared the manuscript for publication, is particularly appreciated.

Finally, special thanks are due the panel members and the National Academy of Sciences Staff Project Director, Norman L. Brown, for their efforts. The panel members labored long and hard to meet their objectives; Dr. Brown labored even longer and harder in editing and providing interfaces among the specialty groups in each of the technologies examined. Coherence in this report should be attributed to his untiring efforts and insistence.

WILLIAM L. HUGHES, *Chairman*
Ad Hoc Panel on Alternative Energy
Sources for Developing Countries

Contents

TECHNICAL SECTION

APPENDIXES

Introduction

Energy plays a larger role in man's struggle with the vagaries of nature than merely sustaining life. In *The Economic History of World Population*, Cipolla has said that the more successfully man can use his own energy output to control and put to use other forms of energy, "the more he acquires control over his environment and achieves goals other than those strictly related to animal existence." He then adds what is certainly obvious but does not suffer from repetition—that fundamental to the utilization of nonmuscular energy is the problem of transforming it into the needed form "at a selected time and place and at convenient cost."[1]

That is the theme of this report. It is not the first such study; earlier, related efforts by competent groups and individuals have contributed in many ways to the thinking of the members of this panel.[2] This report focuses, however, as few others have done, on those small-scale energy technologies, not based on conventional fuels, that seem to be candidates for rural and village use in developing countries. It also examines the ways in which their candidacy may be affected by technological and economic constraints, present and future. In undertaking this assignment, the panel recognized from the start that the nature of its charge (contained in the key terms "renewable resources," "small-scale technologies," "rural environment") required that it consider only those technologies that could not be expected to have direct impact on the energy economy of a country. Rather, the technologies would be expected to contribute to the improvement of the quality of rural and village life in situations where conventional fuels and power systems have not yet penetrated or are too expensive to become a significant factor in the foreseeable future. Indirectly, however, the energy economy of some countries is likely to improve as adoption of these small-scale technologies becomes more widespread (if only to slow the growth of dependence on conventional power systems).

This report provides a summary of the state of the art of alternative (or "appropriate" or "soft") technologies frequently suggested as solutions to rural or individual-family energy needs. Moreover, it informs both the technologist and the planner where to go for more detailed information and what kinds of research and development are needed before a particular device or process is ready for use. Thus, it should be of considerable assistance in evaluating the potential of each energy source in each candidate situation. Indeed, that was the original purpose of the study. It is not, and could not be, a detailed "how to do it" book: that would not be a book but a substantial library.

REFERENCES

1. Cipolla, Carlo M. 1962. Sources of energy. Conversion of energy. In *The Economic history of world population*, pp. 35-36. Baltimore, Maryland: Penguin Books.
2. See, for example:

General Electric Co. March 1963. *Small-scale power supplies for rural communities in developing countries: Report to U.S. Agency for International Development.* Syracuse, New York: General Electric Co.

Johnson, A. L. 1974. Non-conventional sources of energy for developing countries. Paper prepared for the United Nations Advisory Committee on the Application of Science and Technology to Development. U.N. Document No. E/AC.52/XX/CRP.T/Annex. New York: United Nations.

United Nations. Economic and Social Affairs. 1967. *Small-scale power generation. A study for pioneer electrification work.* U.N. Document No. ST/ECA/94. New York: United Nations.

Little, Arthur D., Inc. 1974. *An overview of alternative energy sources for LDCs.* Report to U.S. Agency for International Development, Technical Assistance Bureau, Office of Science and Technology. Cambridge, Mass.: Arthur D. Little, Inc.

Among the more specialized surveys are:

Golding, E. W. n.d. *Windmills for water lifting and generation of electricity on the farm.* Agricultural Engineering Branch, Land and Water Development Division, Informal Working Bulletin No. 17. Rome: Food and Agriculture Organization.

Heronemeus, W. E. 1974. *A survey of the possible uses of windpower in Thailand and the Philippines.* Prepared for the U.S. Agency for International Development, Office of Science and Technology. Amherst, Mass.: University of Massachusetts, Department of Civil Engineering.

Lustig, H. 15 February 1974. *Solar energy: The state of the art and the art of the states.* UNESCO Document No. SC/WS/575. Paris: United Nations Educational, Scientific, and Cultural Organization.

An interesting collection of nontechnical articles reviewing the status of solar energy utilization and implications for developing countries appeared in the *UNESCO Courier* 1974 (January).

Energy Technology and Economic Development — An Overview

The relationship between energy and economic development is as crucial in the less-developed countries (LDCs) as it was and continues to be in the industrialized nations. The process of economic growth is traceable in large part to the substitution of energy for muscle in the performance of every type of agricultural, industrial, and domestic task. Moreover, many of the pesticides, herbicides, and fertilizers on which successful agriculture in industrialized nations traditionally depends are also derived from energy (fossil fuel) sources. It is hardly surprising, then, that prospects for growth in critical sectors of the less-developed economies are linked, at least in part, to the development and exploitation of energy resources available to them.

No one familiar with the problems of accelerating economic development has any illusions about the speed with which innovation can be achieved and made economically viable; learning to use energy in new forms and new ways may require years–perhaps even a generation or more–to break down the economic, cultural, social, and institutional barriers that impede technical change in developing countries. But although the development of reasonably priced, scaled-down, and diversified sources of energy, adaptable to a range of needs and a range of physical and cultural environments, is necessary, it is not sufficient for accelerated economic progress.

Decisions that have high payoff potential often center on the economics of energy supply and demand in the developing countries; as in any country, the first step in the appraisal of economic feasibility must be a thorough assessment of present and potential technical capabilities and of the human and material resources required to realize these capabilities. The technical sections of this report present assessments of the state of the art of a number of conventional and unconventional energy technologies. These are accompanied, in most cases, by an evaluation of their adaptability to the diverse circumstances of underdeveloped economies.

Two additional steps are required, however, before sensible research policies and longer-term programs for the exploitation of energy technologies can be formulated. First, it is essential to link energy programs with other elements of factor supply in the nations involved. (See the discussion of the second proposal in the chapter on Conclusions and Proposals.) Analysis of the economics of development of any of the technologies cannot, by itself, produce all the necessary information, nor can the manufacture of any of the devices discussed in this report generate an information base sufficient for the evaluation of energy programs. Using either approach alone would underestimate the interrelationships of new types of energy production with other components of the economy, particularly the natural-resource systems and the reservoirs of scarce capital and skilled labor that must also be taken into account. Further, available or potentially available foreign exchange—based upon many complex political and economic functions—must be carefully assessed. The availability of foreign exchange is particularly important, since the "shadow prices" of imports in countries with severe exchange restrictions are usually far above nominal money prices, and except for the most rudimentary energy development, dependence on foreign sources of materials and technology almost inevitably will be involved.

Second, and less obvious but often critical, are the costs associated with dissemination of new technology and step-by-step guidance to its application. Being careful not to confuse illiteracy with incapability, one must be aware of the formidable obstacles that impede the dissemination of any new technology in tradition-dominated rural societies, geared as they often are to subsistence or semi-subsistence living and imperfectly linked to imperfectly organized markets and cultural centers. Information costs in this context may often loom larger than those associated more directly with development and application of the technology.

Except for a few approaches—e.g., small, homemade windmills and waterwheels, solar driers, hydraulic rams—with power capabilities near the bottom of the range considered, near-term (1- to 5-year) prospects for effective utilization of the energy sources examined in this report are not encouraging. While the necessary exercises have not been carried out here, an economic evaluation in the dimensions outlined above is likely to paint an even bleaker picture. In short, the developing countries, like the rest of the world, seem condemned to a continuing struggle with increasingly high relative prices for energy from conventional sources. The "energy crisis" will continue to be, in essence, a fossil-fuel crisis (specifically, an oil crisis). This conclusion should not be surprising; even in the industrialized nations, with their infinitely greater capacity for research and development, capital accumulation, skilled manpower, and ability to take risks, at least a decade or two must pass before any of what are now regarded as promising "unconventional" energy sources will play an important role in the energy-supply picture. Even where a particular energy technology holds promise, the prospect of short-term benefits to developing countries is doubtful; promising sources do not match well with the demands of energy users in developing nations. Again, this is to be expected, since the more promising prospects on the supply side are likely to reflect the pattern of energy demand in the industrialized countries where the research is taking place. That pattern is very different from energy usage or need in the developing countries. While generalizations about these diverse requirements should be made with caution, it is probably safe to say that the major demands for energy in developing countries relate to transportation and the variety of activities associated with food production.

As a recent report* stated, examination of the energy demands of agriculture reveals a much higher degree of net energy usage and a closer correlation of energy application to productivity than had been anticipated. Much of the phenomenal output of agriculture in the United States and other major food-exporting countries can be traced in large part to the massive application of power and fertilizer to land; labor inputs per unit of output diminished very rapidly as the pull of higher wages in manufacturing and services increased.

Since agricultural expansion in the LDCs is likely to take place in environments most unfriendly to agriculture–tropical forests, poorly or infrequently watered savannahs, or desert lands–energy inputs to achieve the increases of food production required to satisfy rural needs and an expanding rural population will be even more important than they were in the fertile, temperate plains.

More to the point, those inputs will continue to depend largely on petroleum and petroleum-based products. Although the world as a whole must find substitutes for petroleum-based fertilizers in the not-too-distant future, it will require these products in increasing quantities for some decades. Indeed, it is becoming increasingly apparent that the consumption of petroleum products in lower-valued uses (such as space heating) must be curtailed to ensure a continued supply of feed stocks for petrochemical

*Pimentel, David, *et al.* 1973. Food production and the energy crisis. *Science* 182: 443-449.

industries as a group. At present, the industrialized nations are experiencing difficulty in finding or converting to substitutes for petroleum to achieve this end. Since technological progress in the LDCs, with only minor exceptions, takes place because of a "trickling down" and adaptation of developments elsewhere, one should not expect the LDCs to experience less difficulty. The high cost of capital and the inherent inability, on both political and economic grounds, to take major development risks impose severe constraints on the freedom of developing nations, some balanced precariously on the edge of starvation, to experiment with new energy sources.

Transportation and communication, for example, are the keys to more effective market organization, to more efficient division of labor, and to improved access to natural resources. These are services a nation must buy with its investment in energy sources. But none of the energy sources discussed in this report, indeed none on any visible horizon (considering the constraints that in virtually all cases accompany the need), seems likely to make any significant contribution to transportation of the types and at the levels required to achieve transformation of an economic structure.

In short, the major contribution of energy to economic growth in the LDCs lies in areas that require conventional kinds of energy, and this will probably continue to be the case for at least the next few decades with no easy relief in sight on the side of supply. One cannot reasonably expect those developing nations with exportable supplies of fossil fuels to share them with their neighbors-in-need on any altruistic basis.

The more promising energy technologies discussed in this report involve conversion of the energy in wind and water to electricity, and the use of solar energy for heating and perhaps cooling. To the extent that these technologies relate to demands that are matters of comfort–light, radio, television, space heating, cooling–they will not contribute significantly or immediately to improvement in national economic productivity. Unless a development improves agricultural productivity, moves cash crops to market more efficiently, or brings in manufactured consumer goods at lower costs, it is not likely to have a significant effect on the overall growth potential of an economy emerging from the semi-subsistence straightjacket.

But there are applications of technologies that can indeed contribute to that overall economic growth. Improved communication, for instance, stemming from the availability of modest amounts of electricity, can make possible the dissemination of vital information on agricultural problems, public health, nutrition, and family planning. Solar crop driers can play a significant role in improvement of crop storage, increasing both the quantity and quality of the usable crop at the end of the storage period. The use of dung as fuel to replace the declining supply of firewood in many developing countries has seriously affected agricultural productivity by eliminating its use as fertilizer.* An approach to this problem—providing a fuel while preserving the fertilizer potential of dung—is the subject of a separate study.† An array of such small improvements, by relieving pressures on other energy sources that directly influence productivity, could have a long-term cumulative effect on a national economy and should not be discounted.

It needs saying, however, that there is a general high-capital intensity that is characteristic of economically significant energy supplies and that reflects the development pattern of the industrialized nations, which most LDCs appear to have been persuaded is the proper route to follow. This constitutes the chief barrier to unconventional approaches to the use of energy technologies in the LDCs. Thus, where capital and the skills to employ it profitably are scarce, governments are likely to confine major investments in energy to larger-scale projects that promise to contribute substantially to the economic productivity of a whole region or a nation. In its evaluation of the potential of a small-scale approach, given suitable technology, the panel was fully aware of this reality.

Electricity, for example, simply will not profoundly affect the economies of developing nations unless it is extremely cheap, and that means hydroelectric developments on a massive scale where topography and water supplies permit. Yet, relatively small projects using water flow as a direct source of energy or to generate electricity can be attractive in economic terms, even where there is no possibility of large-scale development, as is pointed out in the discussion of hydropower. Small dams can be constructed by local labor with local materials, and they combine long life with minimal maintenance over a considerable range of scales of construction.

Even so, it does not seem likely that significant changes in the use of hydropower will occur in the developing countries over the next 5 years. Most of the small-scale uses of hydropower have been known for centuries, and except for the benefit gained by introducing this ancient technology to areas where it may not be currently used, the most we really expect in the near-term future is a concerted effort to produce incremental gains in reliability and cost. Such gains are not trivial, but they do not suggest major opportunities to substitute local hydropower for imported fossil fuels.

Much the same comment can be made about the use of energy from wind. Wind is more readily available than falling water. For this reason, the LDCs

*Exacerbating the problem of fuel in rural areas is the growing shortage of firewood, a shortage that is reaching critical proportions in many countries. This problem is discussed in a recent publication: *The Energy Crisis: Firewood* by Erik P. Eckholm, Worldwatch Paper 1, Worldwatch Institute, Washington, D.C., September 1975.

†*Methane Generation from Human, Animal, and Agricultural Wastes.* In preparation. See p. 306.

may well find considerable overall benefit in the use of wind power as a prime source of mechanical power, as well as in the generation of electricity, by harking back to small-unit technologies that proved so useful in rural Europe and America not too long ago. The ability to build structures of proven characteristics at relatively low capital and operating costs, using local materials and manpower, still carries the same rural appeal. Wind power, used directly for large-scale generation of electricity, does not, however, appear any more economically attractive in the LDCs than in the industrialized countries, given the current state of the art. This is true not only because of the storage problem, but also because such a program would involve both capital costs and more elaborate power transmission and integration technology than LDCs can afford at projected costs for such systems.

The discussion of direct uses of solar energy suggests again that short-term gains appear limited to those obtainable from marginal improvements in processes already well known and long established throughout the world. Unfortunately, the fairly optimistic prospect for cost-competitive space heating (and, at higher cost, for air cooling) is of far less interest in the LDCs than in the industrialized nations. There are, after all, a good many cheap substitutes for space heating in the developing countries in temperate and northerly latitudes (e.g., traditional building and clothing design); and in the tropics the problem is obviously not very important. The use of direct solar energy in cooling might be attractive economically, as fossil-fuel costs continue to rise, in the major cities and in tourist-oriented facilities. But at prospective cost levels it certainly is not likely to affect the rural poor (who, in some parts of the developing world, constitute up to 90 percent of the population).

Despite the widespread occurrence of promising geothermal formations, the cost of exploration and development remains formidable, except in units too large to be suitable for the village economy. It does not seem likely that even the best of these fields can offer large amounts of electrical energy at costs comparable to those of prime hydroelectrical sites. That conclusion might well be altered, however, if the development of geothermal energy resources could be combined with the search for oil or the production of fresh water for municipal or irrigation use.

The major environmental obstacles to the development of geothermal energy in the industrialized countries—principally the problem of water residues that are very dirty indeed—may be less immediately pressing in some developing countries. However, even if the scale of the operation were large enough to reduce the unit-cost burden of heavy exploratory and developmental costs, geothermal energy could only be useful in supplying major cities and/or industrial enterprises requiring large amounts of cheap electrical energy. On such a scale, the management of residual waste waters could become a serious problem.

It is difficult to evaluate, in economic terms, the significance of the potential for energy produced by photosynthesis in the LDCs. The original aim of the panel's study in this regard was to concentrate on the production of combustible materials. Consequently, the panel did not include members with the expertise necessary for a discussion of engineering problems involved in technologies of exploitation, such as the destructive distillation of wood.

The economic issues involved in the utilization of plant materials as fuel are regrettably not dealt with in this report—with the exception of the microbiological production of fuels. Nevertheless, it should be pointed out that there are few places in the world where significant quantities of combustible or otherwise convertible materials could be grown that would not compete, directly or indirectly, with vitally needed food production. Direct competition with food production in the use of land is obvious. If, in addition, successful production of materials in sufficiently large quantities to make them desirable as a source of energy requires large amounts of fertilizer, the prospect will not be attractive, in view of the present situation of world food and fertilizer production.

The technology of utilization raises another question: Under what circumstances would it really be worthwhile to convert plant biomass into processed energy such as electricity or even as a primary heat source for manufacturing activities? Why not simply utilize it, as wood and vegetable fibers have been used since time immemorial, for local, small-scale, household fuel supplies, recognizing that a very high proportion of household fuel consumption in developing nations is for basic cooking, smoking, and drying processes? This question has been addressed in a recent report* that points out the dimensions of the problem of rapidly increasing scarcity of firewood in parts of Asia, Africa, and Central and South America—a scarcity of such proportion that it is intimately bound up with major deforestations and expansion of desert areas. This growing scarcity clearly indicates that expanded cultivation of trees and other combustible plant materials for direct use as fuel by households has not proved practical, and it seems unlikely that the "energy farm" approach would contribute as much to rural development as would expanded cultivation for direct use by households.

The section on photovoltaic production of electricity provides a review of the state of the art in this rapidly moving field, and a look into the future—but not the near-term future—of interest to the LDCs; for them the utilization of this technology is a distinctly remote prospect. Dramatic reductions in the capital costs of photovoltaic generation of electricity would admittedly bring the process within range of the rapidly rising fossil-fuel costs, without the problems that nuclear energy seems destined to generate

*Eckholm, *op. cit.*

for a number of years. Economies in manufacturing are certainly possible; nevertheless, the investment in photovoltaic research and development on the scale necessary to bring this technology within reach of the LDCs (to whom present fossil-fuel costs are still out of range) seems more easily justified if it is coupled to a mass market in the industrialized countries as well.

Science and technology have distinguished records of outrunning forecasts of progress. In the case of photovoltaics, however, it is certain that the major impetus will come from the developed countries, and the targets for application will be selected by them. The issue, then, becomes the relative rate of return from adapting to the needs of the LDCs a technology likely to be in a state of flux for some time, compared with the rates of return that might be earned in alternative energy investments. In this comparison, the near-term prospects for photovoltaic generation of electric power in developing countries do not appear favorable.

This study of the technological prospects for meeting the rural energy needs of the developing countries suggests that we are unlikely to see any really dramatic change in the definition and relative importance of "conventional" and "unconventional" sources in the next 10-15 years. In addition, technological changes that may occur toward the end of this period will not have significant effect upon the economies of developing countries for yet another decade or two unless a concerted effort is made to shorten the "trickling down" process (mentioned earlier). Therefore, we cannot escape the conclusion that there will continue to be an urgent need for more effective utilization of all energy sources for some time to come.

Conclusions and Proposals

This report provides information on the technologies necessary to exploit renewable energy resources. However, in the view of the panel, it serves a larger purpose. The report provides an examination of the technical and economic constraints that make the prospects for large-scale adoption of most of these technologies in the near future discouraging. The panel's work in connection with this study resulted in three fundamental and inescapable conclusions:

1. A variety of energy sources and technologies is indeed available as alternatives to conventional power systems.
2. With the exception of a few devices (e.g., homemade windmills, solar driers), there are no cheap alternative technologies of significance for either industrialized or developing nations, and there probably will not be any in the near future.
3. It is not enough that an energy source be available; the technology to put it to use must also be available. The benefits of any one of the suggested alternatives for producing energy could be multiplied many times if even a small amount of capital were invested (a) in developing the technology needed to use the energy and (b) in ensuring that the technology is properly integrated into the economy and the culture.

With this in mind, the panel concludes that certain actions might hasten the adoption of technologies that will make possible efficient and economic exploitation of at least some renewable energy resources in rural areas of the developing world. Since this study was intended to be primarily a technical assessment of certain energy technologies, the panel will not present firm recommendations on the subject, but rather proposals for further consideration.

The following activities are proposed as steps toward a solution of the problems of (a) interdependence of energy and economic development and (b) the consequences of pegging that development to conventional central-source power systems:

- Organization of workshops to evaluate the potential role of decentralized power systems for rural areas in developing countries;
- Organization of a pilot energy-oriented development program to assist rural areas in acquiring the needed energy technology and the means to exploit it usefully; and
- Establishment of regional institutes for research and development on technologies for exploitation of renewable energy resources.

WORKSHOPS

As a first step, a series of workshops should be held in Asia, Africa, and Latin America to discuss the potential economic and social benefits of developing and adapting small-scale decentralized power supplies that utilize renewable energy resources and that are aimed at rural communities. The workshops should give particular consideration to the two proposals that follow and should be structured to deal with specific country needs (giving full attention to cultural sensibilities). In addition to appropriate government officials and representatives from international and private funding agencies, the workshops should include people from the academic community—particularly from engineering faculties—and others who are responsible for project implementation in the field, including, when possible, village leaders known to be receptive to new ideas.

The panel members know that reports such as this, while perhaps interesting to those with related technical interests and useful as source material for other studies, generally are of no consequence to the people they are designed to help—certainly not in the short run. They are also painfully aware of the failures that have met previous attempts to introduce new energy technologies—such as solar cookers—into developing countries. If serious efforts to help LDCs acquire technologies to exploit their energy resources are to be attempted again, they should be undertaken with the participation at the outset of appropriate representatives of the proposed beneficiary communities.

In the panel's judgment, this proposal is the most important of the three and a necessary preliminary to the consideration of the two below.

PILOT ENERGY-ORIENTED VILLAGE DEVELOPMENT PROGRAM

A specific rural area (e.g., a village or group of villages) should be selected for study, as an economic unit, by a small team of technologists and rural-development economists. The purpose of this on-site study would be to identify methods—and to devise an appropriate scheme—consistent with national development objectives, to enable the chosen area to adopt appropriate alternative energy systems, including the importation of essential components not locally available. The team would accomplish this by providing information and training that would enable the community to make more efficient use of its own human and natural resources, or by providing capital equipment, as may be necessary.

This suggestion results from the panel's conviction that, when properly used, small—perhaps intermittent—amounts of power can be of critical value to a rural economy. The inability of developing countries to produce relatively small increments of energy is a major bottleneck to development progress in many places. Although provision of such amounts of energy is not a sufficient condition for improvement in economic and social well-being, it is a necessary condition.[1] Its efficacy, of course, is bound intimately to other economic, social, and political changes that must take place concomitantly. The panel believes the interdependence of these changes and energy supply to be so important that provision of energy technology presents a singularly opportune "point of entry" for a rural-development scheme.

The cumulative impact of the effective use of even small amounts of energy in a rural area can be considerable. Agricultural productivity, public health, communications, educational opportunities—all leading perhaps to a slowing down of the rural exodus—could benefit from the availability of even low-power devices.[2] For example, 100 watts available to run a small pump will, with reasonable efficiency, provide irrigation water at the rate of about 20 gallons per minute from a depth of 20 feet.* It could also provide safe drinking water from an aquifer hitherto impractical to tap, or permit treatment of an unsafe water supply. The communications equipment it could easily power might make emergency medical services available to an

*This is the equivalent of covering one hectare with one centimeter of water in about 24 hours. For a lift of one meter, however, a one-hectare rice paddy could be covered with water to a depth of two centimers in about 8 hours.

otherwise isolated community. By powering a village television receiver, as was done until recently in India (Project SITE—Satellite Instructional Television Experiment), it could provide vital information on agriculture, health, nutrition, and family planning. It certainly could provide adequate light for a household; a village of 200 families would require only 20 kilowatts, for example, to provide 100 watts of lighting per family.

None of the devices capable of providing modest amounts of power—with the possible exception of photovoltaics (solar cells)—nor the machines they could operate, need be considered high technology by today's standards. Nevertheless, they do represent a technological level usually beyond reach of a community that has little or no trade with the outside world and must make everything itself. Most of the alternative energy technologies discussed in this report (with the exception of homemade windmills, waterwheels, or solar crop driers, and perhaps the hydraulic ram) are probably inaccessible to most rural communities in developing countries. They are beyond reach, however, not because of lack of skill in the village, but because of the lack of capital to acquire the necessary materials and tools. Education may be wanting, but there is no dearth of capability. The lack of formal education or training has seldom been a barrier to the use of technology at this moderate level, if the community is shown how. Indeed, the village or farm dweller usually understands his problems better than others do, and if given the necessary financial and material resources, can adapt technology to those problems more effectively than most outsiders are willing to believe.

Rural mechanical genius is not an exclusive western characteristic; remarkable technical skills were manifest long before western science developed.[3] Good electric motors are being assembled by illiterate workers in surroundings that are relatively primitive by the standards of an industrialized country, but only because such components as copper wire, silicon-steel laminations, machined castings, bearings, and the like are available. Knowledge can be transmitted and skills transferred, but acquisition of a technology is almost impossible unless critical material bits and pieces can be obtained. There is no reason for self-sufficiency to be a goal in a situation where trade with outside sources of these materials exists or can be generated.

The panel is convinced that a developing country's rural economy will be helped by the creation or acquisition of sufficient capital to obtain the technology that, in turn, will allow the country to put its own human capabilities and natural resources to work. Its economic development will be further enhanced if the country can exploit its renewable energy sources in a manner appropriate to local requirements and consonant with local customs. The panel submits that if energy—and its use—is the one factor basic to economic development, then consideration of a development scheme focused on indigenous energy resources is more than warranted. In the absence of new means for providing energy, the situation for the energy-poor rural areas will not improve significantly; in the face of increasingly costly conventional energy sources, it will surely deteriorate.

Generalized economic studies and models of rural areas are no more useful in real situations than are purely technological studies. A village on the Mediterranean coast where there is wind, solar energy, and perhaps abundant rain, is radically different from a village in the rain forest that has water but little wind, or a village in India that at certain times of the year may have neither. Any attempt to find a universal model for all these situations will be so full of theoretical generalities as to be of little value to the specific microeconomies to which solutions must be applied. A microeconomy, however, can be effectively studied by specialists in rural economics and related disciplines working in tandem with technologists who know what can be done by capital infusion, by human labor, or by both. The proposal set forth here would test the hypothesis that if the means of adopting appropriate alternative energy systems are available, a rural economic unit can achieve a permanent improvement in its general standard of living and economic well-being. If the experiment is successful, the approach might be replicated elsewhere, in other rural economic units of varying characteristics, to permit a more conclusive assessment of the value of this approach in rural development strategy.

REGIONAL ENERGY RESEARCH AND DEVELOPMENT INSTITUTES

Regional energy research and development centers should be established to take the lead in development and adaptation of renewable energy-resource technologies appropriate to the region. We urge that these centers be staffed with indigenous personnel to the fullest extent possible and that the staffs include social scientists. Without such people, the panel fears that unfortunate experiences of the past, such as unsuccessful attempts to introduce solar energy devices (e.g., cookers), will be repeated and that the technologies discussed here will be rejected.

The final proposal stems from the panel's recognition of the need to adapt to local constraints both the technologies discussed and the devices available for their exploitation. In the course of the panel's discussions, the concept of international regional research institutes was advanced. This is not a new concept; it has been proposed in various forms in the past—first, perhaps, in 1951 by M. S. Thacker in a paper commissioned for UNESCO.[4] This proposal was for a World Power Commission whose function would be to oversee worldwide energy development. Concerned primarily with large-scale power, the commission was to be controlled by the industrialized nations. The idea was modified slightly in 1955 with the suggestion that the proposed international body include energy groups dedicated to assisting developing nations. In 1961, it was proposed that an international institute—the International Research Center of Solar and Aeolian Energy—to concentrate on nonconventional energy sources be established in Greece.[5]

In 1972, an NAS panel, in a report on solar energy in developing countries, revived the idea and recommended "the establishment of regional energy R and D centers" assigned the following functions:

"1. To help generate an information base on energy needs in the region;
2. To carry out the needed research and development to meet regional energy needs;
3. To field-test and provide feedback from users to the centers in order to improve energy processes and evolve useful solar (or other) processes;
4. To act as information, education, and training centers in cooperation with local universities; and
5. To maintain close contact with scientific and engineering developments and energy applications in other centers and in industrialized countries."[6]

In 1973, an International Working Party organized by UNESCO proposed the creation of an International Solar Energy Commission[7] that presupposed the creation of a network of regional international solar energy centers, and in 1974 the U.N.'s Advisory Committee on the Application of Science and Technology to Development proposed that the United Nations Environment Program (UNEP) "take the lead in initiating an energy institute which could . . . work with various governments, organizations, and individuals in the developing world. . . ."[8]

In the fall of 1975, the United States formally proposed the establishment of an International Energy Institute to help developing countries solve their energy problems.[9] The United States suggested that the institute might identify "current or new technologies most relevant [to the special needs of developing countries] and could help assess all countries' energy resources and requirements." Staffed by experts drawn from government, industry, and academic life in both industrialized and developing countries, it could provide training for local and regional technicians or specialists in energy problems. It could become a central point of contact where policymakers and experts could exchange ideas on plans and programs. Further, the institute could become a "first bridge between the massive effort the industrialized countries have now launched to develop alternative sources of energy, and the effort which the developing countries must now undertake."[10]

No significant move was made to implement any of the earlier recommendations, and the most recent proposal is still a matter of discussion within the U.S. Government. It is therefore timely, the panel believes, to emphasize the necessity for decentralizing not only the power supply systems, but also the proposed International Energy Institute. This decentralization is essential if a country is to take advantage of alternative technologies for exploiting renewable energy resources. An International Energy Institute would undoubtedly serve a valuable function as "a central point of contact" for coordination of research, energy-policy studies, analysis of energy systems, and training of developing country personnel in these and energy management skills. But development and local adaptation of alternative energy technologies are not likely to be accomplished in such a setting. Rather, the central institute should be associated with a series of regional institutes devoted to these ends. Successful local adaptation of new technologies is dependent on practical local input, which is likely to be obtained only under circumstances conducive to field testing and training in technical skills.

The work of these regional institutes should be directed toward lowering the cost of these technologies (so that they become financially accessible to developing countries) and using indigenous material and manufacturing capabilities whenever possible. The panel emphasizes once again that social scientists should be included on the staffs of such centers to give appropriate attention to economic and social factors in the cultural settings in which new technologies are being introduced.

The present cost of current technologies for exploitation of renewable energy resources is, in most cases, beyond the reach of rural communities in developing countries. The role played by energy in rural development is so important, however—particularly in improving the standard of living of rural populations—that the panel feels strongly that this situation should not be permitted to continue. A concerted attempt should be made to bring these technologies within reach of these populations, particularly in energy-poor countries, by increasing production, distribution, and appropriate adaptation and use. To this end, the panel offers these proposals in the hope that they may be usefully applied.

REFERENCES

1. Although awareness of this interdependence is only recently becoming widespread as a result of the "energy crisis," it is not a new concept. See, for example: Millikan, Robert A. 1939. Science and the world tommorrow. *Vital Speeches* 5:446-468. Also MacCurdy, George G. 1924. *Human origins: A manual of prehistory*. 2 vols. New York: Holt, reprinted by Johnson Reprint Corporation.
2. Merriam, M. F. 1974. *Engineering research for decentralized power: A plan for action*. Honolulu: East-West Center, Technology and Development Institute.

3. See, for example: Brown, N. L., and Pariser, E. R. 1975. Food science in developing countries. *Science* 188:589-593.
4. Thacker, M. S. 24 May 1951. *The role of energy in underdeveloped areas*. UNESCO Discussion Theme 1951 Energy in the Service of Man. Paris: United Nations Educational, Scientific, and Cultural Organization.

   See also Johnson, Reference 2 (Introduction) for a review of the history of this idea.
5. Golding, E. W. 1964. Methods of assessing the potentialities of wind power on different scales of utilization. In *Solar and aeolian energy*. Proceedings of the International Seminar on Solar and Aeolian Energy, 4-15 September 1961, Sounion, Greece. Athens: North Atlantic Treaty Organization. Distributed by Plenum Press, New York.
6. National Academy of Sciences/National Research Council. 1972. *Solar energy in developing countries: Perspectives and prospects*. Report of an *ad hoc* advisory panel of the Board on Science and Technology for International Development, Office of the Foreign Secretary. Washington, D.C.: National Academy of Sciences.
7. *Bulletin of the UNESCO Regional Office of Science and Technology for Africa*. 1974. IX(3) (July-September).
8. Billings, B. H. 1974. *A proposal to the United Nations Environment Programme for a programme in non-polluting energy*. Advisory Committee on the Application of Science and Technology to Development. U.N. Document No. E/AC.52/XX/CRP.7, 11 September. New York: United Nations.
9. Kissinger, Henry. 1975. Global consensus and economic development. Speech before the United Nations General Assembly, Seventh Special Session, September 1. Washington, D.C.: U.S. Department of State, Bureau of Public Affairs, Office of Media Services.
10. Kissinger, Henry. 1975. Energy, raw materials, and development: The search for common ground. Speech before the Conference on International Economic Cooperation, 16 December. Washington, D.C.: U.S. Department of State, Bureau of Public Affairs, Office of Media Services.

NONTECHNICAL SECTION

PART I

DIRECT USES OF SOLAR ENERGY

Heating, Cooling, Distillation, Crop Drying, and Power Generation

Current applications of solar energy that seem appropriate for use in rural areas, depending of course on local conditions, include:

- Heating of water (for domestic and commercial use);
- Heating of buildings;
- Cooling of buildings;
- Drying of agricultural and animal products; and
- Salt production by evaporation of seawater or inland brines.

From the technological standpoint, many other solar energy applications are possible. These include electricity generation, food refrigeration, cooking, production of pure water from salt water, and chemical manufacture. Their economic status is such, however, that near-term use cannot be expected; either the cost of conventional energy sources for these applications is well below solar costs, or there is very little need for such substitution in the economy of developing countries.

The applications listed are in use in various parts of the world today, but this is not to say that the technology does not need further development. It means simply that some equipment is commercially available, and applications in developing nations during the next 5 years can be expected to involve currently available equipment. Conversely, not every application for which commercial equipment is currently available can be expected to be adopted in rural communities in developing nations within the next 5 years without further significant research and development.

TECHNOLOGY CURRENTLY AVAILABLE

Within the next 5 years, only four of the applications listed can be reasonably expected to be adopted in rural areas of LDCs.

Agricultural uses of solar energy include the drying of various crops, employing both direct exposure to sunlight and the use of heated air passed through or over the material to be dried. These processes are available for immediate application, using locally available materials, labor, and expertise. Further, the reliability of the technology is proven, and recent research has resulted in design improvements. The costs of drying grains, fruits, lumber, and other products by solar-heated air have not yet been reliably appraised. Although they appear to be slightly more costly than conventional heating schemes, they are within economic reach.

Solar energy to produce hot water has long been used successfully for domestic hot-water supply. Since such systems are often economically competitive with others, the economic advantage of this system compared to those using fossil fuels will increase as the price of fuels rises. Where fuels such as wood or peat are locally available at little or no cost—though such supplies are diminishing—the capital cost of a solar-heated domestic hot-water supply cannot be justified.

Residential heating by solar energy often suffers economically wherever fossil fuel supplies are cheap and plentiful, a situation that appears to be vanishing rapidly. In the case of residential heating, hot-air systems have been used successfully instead of hot water. These systems are technologically unsophisticated, generally reliable, and can often be implemented with locally available materials, labor, and expertise.

The production of salt by evaporation of seawater or inland brines is an ancient practice sufficiently well known that it will not be treated extensively here. Suffice it to say that it is a current direct use of solar energy that is immediately available for use wherever salt-containing water occurs in conjunction with appropriate climate; the practical and economical use of this method does not depend significantly on technological improvements.

For the most part, these four applications appear to exhaust the available direct uses of solar energy that could be adopted by developing countries within the next 5 years. There is, perhaps, one other application that might be available toward the end of this period—the solar-driven cooling system employing the absorption cycle. About the only technology that will be available during this time, however, will be a redesigned lithium bromide water system. The reliability of the equipment is expected to become adequate within about 5 years, but the cost competitiveness is still open to question; depending upon local conditions, it may not be possible for LDCs in this time period to rely on local materials, industry, labor, and expertise. If the technology is found to be applicable, however, units of this type will probably make sense economically only if used in fairly large buildings where commercial activity provides the capital for defraying the high investment cost.

TECHNOLOGY SOON TO BE AVAILABLE

In the period from 5 to 10 years, one can expect the development of improved cooling systems both of the absorption type and the compression type. In the compression-type system, improvements and economies in the organic Rankine cycle and possibly in the Stirling cycle can be expected. With respect to absorption processes, research should include multistage systems and new working fluids. Some new modifications of absorption cycles will have to be proven, and there may also be success in developing desiccant systems suitable for regeneration by use of solar energy. Research and development in all these areas are needed.

It is possible that near the end of this period small solar-powered electrical generators may become practical. (This subject is more thoroughly discussed in the section on photovoltaics.) Especially for communities remote from electric networks, pumping of water, cooling, and perhaps lighting and communications may be solar operated. In areas of exceptionally high fuel price and abundant solar energy, this technology could be introduced if capital is available.

With respect to the large-scale generation of electrical power, there is little that can be accomplished in terms of in-place power in the next 10 years in the developing countries. This may not be the fault of the rate of technological progress so much as the fact that the very sophisticated industrial infrastructure and the considerable capital requirements may militate against successful implementation within this period. With respect to the technology for solar-driven electrical generating systems, there are numerous approaches in the research and development stage at present: Stirling, organic Rankine, and other cycles, as well as photovoltaic, thermionic, and thermoelectric systems. It is too early to single out one approach that is clearly superior, although some indications favor the "power-tower" approach. These approaches need further sorting out before a clear direction emerges. During the next 5-10 years, it is not expected that there will be sufficient agreement in the field to permit successful implementation of a single approach on a significant scale, even in the industrialized countries.

RESEARCH AND DEVELOPMENT

In the 0- to 5-year period, viable solar-energy applications in the LDCs can be implemented by the introduction of capital and the exploitation of local materials. Some materials will have to be imported, but the bulk of the material would be available in most countries. Capital availability is unquestionably the major factor in the near-term application of solar energy in the LDCs.

For the 5- to 10-year period, capital again is the limiting factor. But research and development on promising new applications and on improving and reducing the cost of solar systems already in use will also be needed. Support of these programs by the local governments, by intergovernmental organizations such as the United Nations and UNESCO, and by the international banking community can be effective. In addition to research on these systems in the technologically advanced countries, practical development projects in the LDCs should be of greatest benefit to the introduction of the technology. Monitoring of results by the sponsoring agencies and dissemination of these results through the existing technical and international reporting channels are necessary concomitants of any program of this kind.

Photovoltaic Devices

Direct conversion of the visible part of the solar spectrum—sunlight—to electricity is perhaps the neatest and most aesthetically pleasing of all schemes for the exploitation of solar energy. It is also, unfortunately, among the farthest from feasibility for widespread application in the developing nations.

Direct photovoltaic conversion can be achieved with basically simple devices that involve no moving parts, no additional sources of energy, and little if any maintenance. These photovoltaic devices, which have become known as solar cells, are based on the properties of certain crystalline solids that enable these materials to supply an electric current capable of performing useful work when the material is exposed to sunlight. Although the devices are indeed basically simple—consisting of a wafer of crystalline material to which two metal contacts are fastened—their fabrication requires the sophisticated technological capabilities of the industrialized nations. In fact, it is this very technology that is now developing methods that may (by reducing the cost of the product) make it economically feasible and justifiable for rural areas to make use of solar cells. Thus, with very few exceptions, developing countries can avail themselves of this technology only by acquiring its products through international trade. This will continue to be the case until more of the third-world nations develop the necessary technological and industrial infrastructure, and the industrial fabrication processes are sufficiently developed through large-scale production, to make transfer of the fabrication technologies themselves economically justified.

Solar cells were first used commercially in 1955. Their development since then has been primarily a by-product of space research. The very properties that make them attractive for use in developing countries—simplicity, low weight, efficiency, reliability, lack of moving parts—have made them indispensable in supplying power needs of spacecraft. Unfortunately, mass-production techniques (geared to high-volume demand and economic competition) have not yet been implemented in the manufacture of these devices. It is not surprising, therefore, that solar cells have remained so costly. The current concern over the availability and price of fossil fuels and the massive infusion of R and D funds into energy research may provide added stimulus to the recent trends toward cost-cutting, mass-production techniques for solar cells.

TECHNOLOGY CURRENTLY AVAILABLE

Photovoltaic systems with 10 percent solar-conversion efficiency and with peak power capacities from 1 watt to tens of kilowatts are available from manufacturers in Japan, the United States, England, France, and the Federal Republic of Germany (West Germany). Packaged systems for use in remote applications, such as navigational buoys and lighthouses, environmental monitoring stations, microwave relay stations, and forest ranger communications, are available as commercial products in the range of a few watts to a kilowatt; larger systems are available on a custom basis. Costs of a complete system are dominated by the photovoltaic array; prices of $20-$30 per peak watt, corresponding to $100,000-$150,000 per average kilowatt installed capacity, are now being quoted, though these costs are expected to decrease by as much as a factor of four within the next few years. These systems all incorporate silicon solar cells produced by modifications of the technology used to produce spacecraft solar cells and include batteries, voltage and current regulation, and other components such as DC/AC inverters as options.

During the coming 5 years, over $100 million will be spent by industry and government in the United States, Japan, West Germany, France, and England on the development of lower-cost terrestrial photovoltaic power systems. During this period the emphasis will be on the research and development of new processes for mass production of integrated solar modules and, by 1980, of integrated solar-cell arrays using cells made from mass-produced silicon ribbon. Arrays incorporating wide-aperture concentrators without diurnal tracking requirements will probably also be on the market. As a result of the intense research and development effort currently under way on these terrestrial systems, even greater cost reductions can be expected; solar cell modules with a cost of a few thousand dollars per average kilowatt (down by a factor of about 50), and having conversion efficiencies approaching 15 percent, could be available by 1980.

TECHNOLOGY SOON TO BE AVAILABLE

If the economic stimulus of fuel shortages and high prices continues, there is a good chance that commercial photovoltaic conversion arrays at costs of a few hundred dollars per average kilowatt will be available by 1985 or sooner. The integration of such arrays into a complete system incorporating energy storage, power conditioning, and transmission and distribution will be required before these are useful on any substantial scale for power generation greater than a few kilowatts per system. In this regard, availability of economically interesting storage systems (see chapters on storage) will be a crucial factor in determining the extent to which such technology is used.

In this price range there would be great interest in the potential use of such technology, perhaps in the establishment of local (village and community-size) "minigrids" with the eventual growth and interlinking of these into larger and more diverse electrical networks. However, if this technology is to be transformed into something that can meet the special technical, economic, and cultural constraints and needs of various developing countries, a deliberate and specific effort will be required; otherwise, the direct "transfer" to the remote village level of photovoltaic systems developed for integration into modern utility grids is unlikely.

PART II
INDIRECT USES OF SOLAR ENERGY

Wind Energy

The exploitation of "free" wind energy is as attractive in its way as photovoltaics are in theirs. It, too, is not free of problems, though for LDCs the problems may be more manageable. Wind was one of the earliest sources of power used to multiply the productive capacity of muscle. On the seas it has been used to propel ships, and on land it has served a variety of purposes on a village scale. These include:

- Pumping fresh water for domestic livestock and agricultural needs;
- Irrigating fields;
- Powering agricultural tasks, such as grinding corn, wheat, and sugarcane, and threshing, chaff cutting, and winnowing;
- Cutting wood;
- Pumping saline water in saltworks; and
- Generating electricity for a variety of purposes.

This chapter concentrates on small-scale wind machines producing up to 10-20 hp (7.5-15 kW).

TECHNOLOGY CURRENTLY AVAILABLE

A variety of windmills is currently available, either commercially or in the form of working prototypes that could be readily manufactured as a market developed. These include both vertical- and horizontal-axis rotational machines. Among the vertical machines are the Savonius rotor—particularly well suited to local construction—the Darrieus rotor, and the vertical sailmill. Horizontal-axis machines most common are 2-, 3-, and 4-bladed propellers usually associated with wind-powered electrical generators, the multiblade fanmill commonly used for pumping water, and the horizontal version of the sailmill characteristic of the well-known windmill of the Greek islands. In addition, simple windmills of the paddlewheel type, with either horizontal or vertical rotational axes, are used.

In terms of assessing the availability of this technology within the next 5 years, there are enough commercially built units currently available to permit immediate application, whether it be for pumping water, compressing air, or generating electricity.

The water-pumping units vary significantly in price from $4,000 to $8,000/hp ($5,000 to $10,000/kW) for fractional size units, to $1,000 to $2,000/hp ($1,300 to $2,600/kW) for units of 5-15 hp (7-20 kW). Few larger-scale pumping units are commercially available today.

Electrical generating units cost from $3,000 to $6,000/kW for fractional-kW generators or from $1,000 to $2,000/kW for units in the 5- to 15-kW range. One unit just going into production in the United States is expected to sell for about $500/kW. No larger-scale commercial units are currently being sold.

Most commercially available windmill units sold today are safe, dependable, require little maintenance, and are built with an expected life of 10-20 years (sometimes more), with proper maintenance.

In addition, a number of windmill designs are available that villagers themselves can construct, often utilizing locally available materials.* In small sizes, these units will produce mechanical shaft power at a cost of $1,000-$1,500/hp ($1,300-$2,000/kW). In general, however, they will be characterized by slightly higher maintenance and operational requirements—principally labor—than most commercially available units.

In developing areas, classical windmills will often cost much less than the figures quoted above, because a considerable amount can be done with locally available technologies.

*For example: "How To Construct a Cheap Wind Machine for Pumping Water." Leaflet No. L-5. Brace Research Institute, MacDonald College, McGill University, Ste. Anne de Bellevue 800, Quebec, Canada. ($1.00) Also: "Low Cost Windmill for Developing Nations" by Hartmut Bossel. Available from VITA (Volunteers in Technical Assistance), 3706 Rhode Island Avenue, Mt. Rainier, Maryland 20822, USA ($2.00)

RESEARCH AND DEVELOPMENT

Technological and economic improvements could be expected in 5-10 years if research and development can be encouraged. Such a program should include the following:

- Materials-engineering studies of existing commercial windmills, designed to improve production techniques and reduce costs;
- Development of windmill systems for the production of mechanical shaft power and the generation of electricity, utilizing as many indigenous or locally available materials and technologies as possible;
- Investigation and redesign, where necessary, of equipment to be powered by windmills to carry out the required unit operations—such as pumping, agricultural tasks (milling, threshing, grading), and electrical generation—with a view to their fabrication in local workshops;
- Development of decentralized power systems that are simple, reliable, and that may be serviced—if not fabricated—in village areas;
- Application of known technologies to adapt windmill systems to a variety of environmental conditions (such as winds, rainfall, snow, ice, temperature extremes), which may include development of new forms of rotors for small-scale applications (although significantly improved efficiency in this area should not be expected); and
- A program of public education in rural areas to show what windpower can do.

Such a program could be expected to result in the development and use of equipment capable of delivering 1-15 hp (750 W-11 kW) in a 10-mph (16-kmph) wind, from improvements in existing equipment.

The development of units in the 25- to 50-hp (20- to 40-kW) range would be more costly, but needs attention nonetheless. Considerable spinoff advantage can be expected from development programs currently under way in industrialized nations.

The research and development program outline would most efficiently be undertaken by a central organization, supported by multilateral funding and having an international staff. At present, there are several groups engaged in windpower research in the developed countries, all with different purposes and financing. Among them are academic groups, such as those at Oklahoma State University at Stillwater, Oklahoma, and at the Technische Hogeschool Eindhoven in the Netherlands; the nonprofit Brace Research Institute at McGill University, Montreal; the commercially supported group, Windworks, in Mukwonago, Wisconsin; and the Swedish Technical Board in Stockholm, a government body. In the developing countries, the Institute of Technology Bandung in Indonesia, for example, has done some development and field work on windpower devices.

The panel feels that the work of all these groups could be coordinated to the best advantage for windpower applications in developing countries only by the creation of a central research and development organization having this as a specific objective, working through regional energy research and development institutes; this is suggested in the section on the panel's conclusions. This organization would be in a position to develop coordinated collaborative design and testing programs with groups such as the Appropriate Technology Development Unit at the Ghandian Institute of Studies in Rajghat, Varanasi, India; the Solar Energy Laboratories in Karachi, Pakistan; the National Physical Laboratory in Jerusalem, Israel; and the Ministry of Electricity in Cairo, Egypt.

Such an organization would function with the collaboration of the Advisory Council for the Application of Science and Technology to Development (ACAST) of the United Nations, as well as with national, bilateral, and multilateral donor agencies, to facilitate the exchange of information on work in progress and ensure widest dissemination of results achieved.

Hydropower

THE HYDRAULIC TURBINE

The first power source extensively used (other than that available from the muscles of men and animals) was that derived from flowing water, either impinging directly upon paddles attached to a wheel or filling buckets attached to a wheel. These are the waterwheels, the precursors of the water turbine, whose origins can be traced to ancient Egypt, China, and Persia.

Waterwheels took over many laborious and monotonous tasks such as grinding grain for flour and animal feed, raising water for irrigation and water supply, textile manufacture, and metallurgical processing.

As the demands for power increased over the centuries, the simple waterwheel became larger. By the middle of the 19th century, the revolutionary development of the water turbine by Fourneyron in France began to displace the waterwheel as a power source; in fact, the steam engine had begun this displacement process a half century earlier.

The water turbine was vastly superior to the waterwheel from almost any practical point of view. It could generate much more power than the largest waterwheel in a much smaller volume or apparatus. It could perform adequately at high or low heads that could not be handled by waterwheels. It also could operate at a greater number of revolutions per minute than the waterwheel, principally by virtue of its smaller diameter.

In New England, during the last half of the 19th century, there were more than 50 manufacturers of water turbines supplying the needs of small, rural mill owners. It was not unusual, however, for a millwright to design and build his own water turbine; indeed there are a few "homemade" installations still in existence. By the end of the 19th century, the small water turbine in a mill was often belt-connected to an electric generator, principally for lighting purposes.

With the development of large-scale hydro- and thermal-electric central generating stations and the extension of electric power lines to rural areas, the manufacture of small water turbines began to decline rapidly. In recent years, however, no doubt because of the energy crisis, there is renewed international interest in the small-scale hydroelectric unit. Such units are currently available for use in developing nations.

By small-scale, low-head hydroelectric units, we refer to those capable of generating 5-15 kW at heads of 10-20 ft (3-6 m). While the discussion will be primarily in terms of generation of electricity, it must be recognized that the same apparatus can perform useful mechanical tasks directly; it may be connected via belts or gears to grain mills, pumps, wood- and metal-working machinery, and other machines of production.

Technology Currently Available

For use with low water heads, the fixed-propeller type of turbine or the more familiar Francis type* is more suitable than the more complicated types such as the Kaplan, which adjust themselves to the electrical load. Small hydroelectric generators are currently available from several manufacturers in the United States and Europe. Also, there appears to be vigorous manufacture and use of such turbines in the People's Republic of China, but this panel has no information regarding their availability for export.

Because of the limited production of such small hydroelectric units, their cost per kilowatt is quite high. Units capable of generating 10 kW cost about $1,000/kW. The cost per kilowatt rises for smaller capacity units. These costs do not include the cost of other necessary parts of the system, such as the dam, the piping connection (penstock) between the dam and the turbine, and the housing. With all costs considered, the cost of electricity at the generator would probably range from 8¢ to 10¢ per kilowatt-hour.

Research and Development

Significant cost reductions could probably be achieved in developing countries through the construction of propeller-type turbines from wood. (The more complicated Francis turbine does not lend itself easily to wood

*See p. 141 for description.

construction.) Reports from China, the Soviet Union, and Rumania indicate that this approach has been used successfully in these countries.

The propeller and the Francis turbines are a mature, time-tested technology; further advances can be expected to be limited to cheaper manufacturing methods and materials and to the incorporation of new technology (solid-state devices) in the electrical controls.

The study of hydropower developments of the past may also be helpful to LDCs. Actual small-scale mills using turbines for many rural needs may still be seen in the United States (New England), China, Switzerland, the Balkans, Norway, and Sweden.

THE HYDRAULIC RAM

The hydraulic ram is a simple, ingenious pump, operated solely by flowing water, that was developed in Europe at the end of the 18th century. Like the waterwheel, it experienced a period of popularity in the industrialized countries and then a decline coincident with the surge in production and distribution of economical electrical energy. There is still one hydraulic ram manufacturer in the United States who does business on an international scale.

The hydraulic ram can be described as a completely automated device that uses the energy in flowing water to pump part of that water to a height above that of the source. It can also be used to compress air for operating machinery. It is a simple device that contains only two moving parts.

Technology Currently Available

Manufactured hydraulic rams are currently available at costs ranging from $159 to about $2,700, depending on size.

Hydraulic rams can operate 24 hours a day for many months with no maintenance. No significant improvements are anticipated for the next 5 years.

Construction plans for making a hydraulic ram are available from USAID or from Volunteers in Technical Assistance.*

Additional research and development are suggested for the adaptation of the hydraulic ram to air compression for use in reciprocating engines and turbines.

*See p. vii for details.

THE HYDRAULIC AIR COMPRESSOR OR TROMPE

The trompe, or hydraulic air compressor, is a simpler device than the hydraulic ram. It has no moving parts, but utilizes hydraulic potential energy very effectively in the compression of air. This system, which would operate continuously, day and night, without attendance, could store energy in the form of compressed air that can be used to drive turbines and reciprocating engines that, in turn, can drive production machinery or electric generators.

It is suggested that large-scale studies of this simple, low-cost device be undertaken to verify the scant but impressive performance data available from the literature.

Photosynthesis as an Energy Source

The one renewable energy source on which mankind has relied since the discovery of fire is photosynthesis. Even though nature performs the initial function in the conversion of solar energy to an energy-rich biomass, the photosynthetic process is subject to technological advances as sophisticated in their own way as those that apply to photovoltaics.

Mankind has long found important and diverse uses for the by-products of the photosynthetic process—food, fiber, fuel, and shelter. In terms of the concerns of this panel, the exploitation of the by-products of the photosynthetic process is, in effect, a long-established technology for energy conversion and storage. The deliberate development of an agronomic energy resource, however, is not. The emphasis in this report is placed, therefore, on energy resources that can be developed by supplementing current agricultural practices in rural areas.

The space available to a village can be divided into arable and nonarable land; we are concerned here primarily, but not exclusively, with the more efficient use of the former. Although the panel concerned itself with culturing of plants as an energy feedstock, the terms of reference for this study did not permit considerations of the end use of the plant material produced. That final use clearly will depend upon the needs of the population and the methods available to convert the plant material to more useful energy forms, and will dictate many aspects of the specific research, development, and exploitation schemes pursued. (One conversion technology, however, is discussed in the section on microbial conversion.)

When agriculture is viewed from the point of view of human utilization of sunlight as an energy source, the criteria for choice and timing of crops differ from those conventionally used; factors commonly ignored now become very important. One of the first that should be considered is the energy content of the plant material. This is particularly important where there is a choice among alternative food or fiber crops having differing amounts of residue whose energy content can be exploited.

A second factor that should be considered is that maximum use of available sunlight requires that farmland be kept covered with green plants throughout as much of the year as possible. This involves, for example, multiple cropping, intercropping, nurse crops (i.e., later-maturing crops planted with early-maturing crops so that the former will have a substantial start when the latter are harvested), and the use of cool-, warm-, wet-, and dry-season plants in those seasons when traditional crops may not be planted. Of course, water supply or some other environmental factor may prevent year-round coverage, but bare ground must be covered with green plants—where there are no green plants, no photosynthesis occurs.

A third concept has recently emerged on a sound experimental basis. Most agricultural schemes convert into plant material less than 1 percent of the solar energy available during a growing season. However, some plants, because of an efficient photosynthetic system, can convert 2-3 percent of the incident solar energy. These plants, known as "C_4" plants,* require warm growing seasons and high levels of solar radiation. They are already widely grown in many developing countries, but in appropriate climates where they are not grown, they should be given serious consideration as crops, especially during periods of high insolation. These species have the added feature of being able to use water more efficiently and thus are more tolerant of drought than many common crop species.

Finally, there exists another group of plants known as CAM plants* that, by virtue of a genetically controlled metabolic characteristic, are among the plant world's most efficient water users and therefore do well in semiarid regions. They are slow-growing plants, but in dry areas, where plants could be harvested on long-term rotation (2-6 years), they could make the land more productive and utilize the available water more efficiently.

The choice of food and fiber plants to be grown is usually governed by tradition, which in turn is usually based on generations of experience with local environmental conditions. However, a careful reassessment of these factors, including such fundamental aspects as water supply, temperature, light intensity, and nutrient supply, should be made if maximum capture and storage of solar energy by plants is to be included among the goals of

*Examples are given in Part II of the Technical Section, chapter on photosynthesis.

agriculture. The ideas that have been briefly described are based on current practice and investigations of plants that are available now. These plants would need to be tested in given areas over several growing seasons, but well-known procedures for introducing new plants could be initiated now. In each area, the procedures to maximize energy capture would have to be worked out over a 2- to 5-year period, with a variety of plants and management techniques.

Microbiological Conversion of Plant Materials to Liquid Fuels

Another process in which biological phenomena are responsible for useful energy production involves the action of microorganisms on plant material substrates and the gathering of the fuel products that result.

Consideration, for example, of the ancient and honored art of fermentation of a variety of plant products to produce alcoholic beverages naturally leads to the idea that the alcohol could be oxidized outside, rather than inside, the body—thereby providing a useful source of sensible heat. Indeed, the single major obstacle to utilizing this idea in rural areas of developing countries is the technological difficulty in separating the alcohol—or other combustible liquids produced by fermentation—in sufficiently pure form to be used as a liquid fuel.

Production of ethyl alcohol by fermentation, as a chemical process, is quite similar to the microbial production of acetone, butyl alcohol, and isopropyl alcohol—all of which are flammable liquids. The processes all involve the growth of microbial populations under anaerobic conditions—i.e., not in contact with air—utilizing the nutrients found in various plant sources, and converting the carbohydrates (sugars and starches) of those plant sources to product systems such as: 1) ethyl alcohol; 2) acetone, butyl alcohol, hydrogen; 3) butyl alcohol, isopropyl alcohol, hydrogen; and 4) acetone, ethyl alcohol, hydrogen.

In three of the cases shown, a gaseous fuel is also produced as a by-product, although not in very large quantities. A similar anaerobic fermentation process does produce large quantities of methane, another gaseous fuel, but this process is the subject of a separate study and is not discussed in this report.*

All of the materials listed above have been produced by microbial systems on a commercial scale, particularly during the critical years of materials needs of the 1940s and, to some extent, in the early 1950s. However, with the possible exception of ethyl alcohol, they are now produced industrially, by nonmicrobial means, from petrochemical sources, and even the major part of the world production of ethyl alcohol is manufactured in this way.

TECHNOLOGY CURRENTLY AVAILABLE

While the chemicals listed can indeed be used as liquid fuels, that use depends on their availability in a sufficiently concentrated form; otherwise the water present will interfere with the combustion process (e.g., within an internal combustion engine) or simply dilute the energy density—the heat available from a given weight—to such an extent as to diminish the economic value of the material. Since the microbial processes produce these chemicals in water solution, mixed with a variety of other suspended and dissolved materials, recovery in a sufficiently concentrated form becomes the critical issue in evaluating the usefulness of the process. Because the technology required to effect the necessary separation is not likely to be available in rural areas, the panel does not feel that microbial processes can be considered for such use at the village level in developing countries within the next 5 years.

TECHNOLOGY SOON TO BE AVAILABLE

For fuel production utilizing microbiological processes to become realistic in rural areas of developing nations, research and development are needed on materials, equipment, and processing. To reduce capital investment and the need for sophisticated technical skills currently required, it will be necessary to explore the availability of local raw materials, to investigate the microbiology and physical chemistry of the fermentation systems, and to examine recovery operations and equipment.

**Methane Generation from Human, Animal, and Agricultural Wastes.* In preparation. See p. 306.

PART III
OTHER TECHNOLOGIES

Geothermal Energy

Geothermal energy, the natural heat contained within the earth, is generally too deeply buried to be of use. However, in areas of the world that have experienced recent volcanic activity, geothermal resources may exist that can be exploited economically. These are usually found in regions of hot springs or areas where the deep hot rock is known to be fractured or to have pore spaces that permit water or steam to be circulated to carry heat to the surface.

Geothermal energy is distributed widely over the world, but localized geographically; it is a resource that can be used only near natural occurrences. When it is present, however, it provides a source of energy for electricity generation, space heat, drying, and refrigeration that is competitive with other technologies.

Present technology for electricity generation depends on production and separation of steam from any associated hot water and utilization of the steam to drive turbines or for other direct uses. Hot water may also be utilized directly for space heat.

TECHNOLOGY CURRENTLY AVAILABLE

Technology for exploiting geothermal sources is in use in many countries.* Geothermal turbine systems appear to have lifetimes of about 30 years or more. The lifetime of wells appears to be at least 10-15 years—longer in the case of the hot-water wells used in conjunction with space-heating systems.

The power-plant costs range from $150 to $500 per kW capacity and from about $70 to $100 per kW for the wells. A minimum exploration program may cost about $100,000, but costs can easily reach $1 million if much exploratory drilling is required.

Geothermal power projects have proved to be reliable. Hot-water systems and dry-steam systems are in operation in Japan, New Zealand, Mexico, and the United States. Experience has shown that geothermal energy functions reliably 80-95 percent of the time.

Local fabrication of geothermal turbines and generators will depend on the existence of a moderately sophisticated manufacturing capability.

Drilling technology is relatively simple for shallow hot-water wells, but is much more complex for hot wells capable of producing boiling water and steam. Since oil- and gas-well-drilling equipment is universally available and is directly adaptable for geothermal drilling, this means that geothermal drilling is equally available. Some shallow geothermal drilling for hot water can be done with relatively simple cable-tool drilling rigs that are likely to be in use in many developing nations.

TECHNOLOGY SOON TO BE AVAILABLE

New geothermal technology is being developed mainly in the United States. A new device for hot-water or steam systems is the screw expander. This device employs two counter-rotating screws to serve as a positive displacement

*Reviews of the technology are available from several sources, including two major international symposia. A series of geothermal review papers has been issued as part of the proceedings of the United Nations Symposium on the Development and Utilization of Geothermal Resources, held in Pisa in 1970 and published as Special Issue 2, Volume 1, of the journal *Geothermics*. (Special Issue 2, Volume 2, Parts 1 and 2 of *Geothermics*, contains all the technical papers presented at the symposium.) The Second Symposium on the Development and Use of Geothermal Resources was held in San Francisco in May 1975, and abstracts of the 358 papers presented have been published by the Lawrence Berkeley Laboratory of the University of California, with the full proceedings to be published early in 1976. In addition, the UNESCO publication *Geothermal Energy* (1973) contains a useful collection of geothermal review papers.

device within which hot geothermal water boils; the steam expands, thereby rotating the screws, which may provide a source of mechanical or, by use of a generator, electrical power. A test program on a 1.25-megawatt device, sponsored by the U.S. National Science Foundation, is under way at the Jet Propulsion Laboratory of the California Institute of Technology.

Energy Storage

None of the intermittent energy sources discussed so far—certainly not those dependent on the vagaries of wind and sunshine—is well considered without the inclusion of storage technologies to hold energy, for instance, when the wind or the sun is down. Storage in those instances is vital; techniques of storage vary.

Energy can be stored in mechanical, electrical, chemical, and thermal forms, the last named being a special form of mechanical energy. We have excluded nuclear forms of energy storage because our concern is confined to applications in rural areas in developing countries.

There are few effective energy storage systems for near-term application in rural communities of developing nations. Stored solar heat for direct use as thermal energy is possible now and is, in fact, being used in water-heating and related systems (see the chapter on heating, cooling, etc.). In favorable geological situations, small hydroelectric pump-back storage is a possibility. Electrolysis to provide a simple fuel for small tractors, small short-range vehicles, refrigeration, etc., might provide the next step up if appropriate development is stimulated and if electricity is available from wind or other sources. Alternatively, batteries—probably lead-acid at least for the foreseeable future—can provide essential electricity for lights, radios, etc. A fundamental point to note is that electricity is often used for convenience in industrialized nations, whereas in rural communities of developing nations, for economic reasons, some other form of energy might be more useful. Compressed-air storage may be useful, but more exploration is needed to be sure. More specifically, mechanical energy can be stored in the form of kinetic energy (flywheels) or potential energy (hydraulic systems, compressed air, etc.). Hydraulic storage generally takes the form of hydroelectric pump-back systems, where water is pumped to a higher level during periods of excess generating capacity and, in turn, generates electricity in its downward flow during periods of short supply. It is feasible where 1) construction labor is cheap and 2) land is plentiful enough to be effectively used this way. Energy stored as compressed air usually involves the use of ordinary compressors but may be achieved with isothermal hydraulic compression systems. The latter tend to be much more efficient and are probably more applicable in developing nations because of their simplicity and lack of machinery with moving parts.

Electrical energy can be stored only in the form of electric or magnetic fields (batteries are chemical storage), but the technological requirements are extremely high.

Besides batteries, which convert to electricity the energy in chemical reactions, energy may be stored chemically 1) by electrolysis (or biochemical means discussed elsewhere), which produces hydrogen for direct combustion or for use indirectly in electricity-producing fuel cells, and 2) in the form of fuels produced by techniques such as bacterial processes or high-pressure, high-temperature, hydrocarbon synthesis processes.

TECHNOLOGY CURRENTLY AVAILABLE

Where economically feasible, hydroelectric storage systems are technically practical now. Although the generating components are available commercially, they are rather expensive for small systems. Making the systems less expensive is a combined function of increasing the volume of production of standard generating units and of using sites for storage dams where land and labor costs make reservoir construction feasible. Additional research and development are not likely to reduce costs significantly. (These generators are discussed more thoroughly in the chapter on hydropower.)

Compressed air systems for energy storage might well be practical for developing countries, and isothermal compression systems provide interesting possibilities. Isothermal hydraulic compression for energy storage died out around 1900, though there has been recent renewal of interest. Air turbines are available and the theory of isothermal compression is adequately known, so that working systems can be engineered now. They would be considerably more costly than diesel electric generators (perhaps two or three times as expensive), but of course would have no fuel costs.

Their application awaits the careful examination of needs for energy storage and the recognition, by those concerned with energy storage, that such systems can actually be built. These systems should have extremely long life when properly maintained—perhaps 50 years or more—and should be easily and inexpensively operated. Little research and development is needed on the technology, but some systems should be built to prove economic feasibility.

Whenever surplus electricity is available, and whenever hydrogen as a direct fuel can be used, high-pressure electrolysis should be considered. Electrolysis at pressures up to 300 atmospheres and at temperatures up to 250°C has been performed at extremely high efficiencies. The required pressure can be supplied by the gas generated, with no need for mechanical pumps; the elevated temperature can generally be achieved by using the heat developed within the system. In rural areas of developing nations, wind electric systems can be used to provide the electricity needed to make the hydrogen. Where fuel oil is not available, or where it costs a dollar or more per U.S. gallon delivered to the site where it is used, a combination of wind generators and electrolysis units should be seriously considered. Systems are technically feasible now to generate the hydrogen and could be available very shortly if the market appears promising.

Using the resulting hydrogen is another matter. Much experimental work has been done on operating internal-combustion engines with hydrogen fuel. Engines would be available if the market existed. Even now, short-term feasibility (within 5 years) could be a significant possibility. Such an industry could be stimulated into existence quickly by another more potential worldwide oil shortage of even moderate proportions. This total technology is "waiting in the wings." While it could be available to developing nations within 5 years, the developmental effort will have to take place in technically advanced countries. Considerable research and development—particularly development—are still needed. Heaters, grain dryers, and the like, should be engineered for this fuel in the near future. "Mini" tractors and "mini" trucks should also be developed to a prototype stage fairly soon.

Finally, energy stored in thermal form is, in fact, being used in developing countries now, generally for heating water or homes. This subject is discussed in the section dealing with heating, cooling, distillation, crop drying, and power generation with solar energy.

TECHNOLOGY SOON TO BE AVAILABLE

Apart from the prospective availability of technologies already mentioned, other energy storage sytems show little prospect of becoming available within

5 years or so, for use in developing countries, as a result of current or suggested research and development.

Storage of energy in electrical form (that is, in the form of electric or magnetic fields) is probably not practical now or in the foreseeable future for either developing or industrialized countries. Even if perfect dielectrics and perfect superconductors were assumed available, energy storage densities are simply too low. Economic projections are so discouraging as to preclude any suggestion here that the matter should be considered further.

In recent years, millions of dollars have been spent in the development of special types of batteries. Although most of this effort has been directed toward application in electric vehicles, if an economical new battery were developed it would have general applicability as an energy storage device. No such economical battery has yet grown out of this research, and it is difficult to see applicability in developing countries, within 10 years, of anything now on the horizon.

It is also possible to synthesize hydrocarbons through the technique of hydrogenating organic materials naturally available. This can be done at high temperatures (around 500°C) and high pressures (around 600 atmospheres), but such technology probably does not make sense in the context considered here, and applicability is probably more than 10 years away.

The use of flywheels for energy storage of more than a few seconds' duration awaits development of some of the super-flywheels that have appeared in the literature as technical proposals. Until someone builds one, we really will not know if they are feasible or not. At any rate, applicability probably is more than 5 years away, especially in developing countries.

RESEARCH AND DEVELOPMENT

Suggestions of areas where research and development would be expected to have a significant effect on availability have been mentioned in discussions of the specific technologies. Some general comments can be made here, however.

As is pointed out in the chapter on hydropower, materials research and development can be expected to cut costs of small hydroelectric generators to some extent. Although technically feasible now, production of hydrogen by high-pressure/high-temperature electrolysis on a small scale can likely be made more attractive in developing countries only by developments directed toward simplifying the technology involved in manufacturing these generators. The use of hydrogen in internal-combustion engines certainly depends on continued research and development in that field, and some engineering development is needed in heaters and grain dryers to adapt them conveniently to efficient use of this fuel.

Finally, although significant research and development are needed before flywheel storage, hydrocarbon synthesis, batteries, or electrical storage can be put to practical use, none of these technologies is expected to be available for application in rural areas in developing countries within 10 years.

TECHNICAL SECTION

PART I

DIRECT USES OF SOLAR ENERGY

Heating, Cooling, Distillation, Crop Drying, and Power Generation

Solar energy represents a vast energy source that is most available in many areas where population density is often low and where conventional energy resources may be expensive. Other than in agriculture, and in a limited amount of salt production by brine evaporation, there is no significant direct use in the developing countries. This is also generally true in industrialized countries. The additional applications in countries with advanced technological capabilities are for domestic water heating and for electricity generation in very small units, mainly through photovoltaic devices on space vehicles.

In contrast to the few useful applications of solar devices, there are many experimental and developmental solar energy utilization systems. These are being pursued in several of the industrialized countries, particularly the United States, the Soviet Union, Australia, Japan, France, Italy, Israel, India, and Canada. Experiments in these countries cover a broad range of potential uses of solar energy, including electric-power generation on a small scale, heating and cooling of buildings, water desalting by evaporation, cooking, food refrigeration, high-temperature materials processing, specialized solar drying and certain combinations of uses.

Of the two current commercial uses of solar energy in the industrialized countries, i.e., in domestic and institutional water heating and in miniature power systems employing photovoltaic cells, only the water heaters can be considered commercially feasible at this time. The extremely high cost of photovoltaic converters precludes their use for generation of electricity in terrestrial applications requiring more than a very few watts of power.

However, the possibility of substantial cost reduction of these systems may have a significant effect on their utilization in the near future. (This is discussed in the chapter on photovoltaic devices.)

Although not yet in general use, solar heating of buildings–particularly residences–in the United States is rapidly approaching commercial status. Solar heating systems are currently being installed, and several manufacturers are test marketing solar heating components and complete systems. This application can be expected to increase rapidly in favorable climates and should be considered a practical alternative to fuel heat in developing countries requiring space heating. Because of the current and imminent commercial availability of a variety of solar heating devices, manufacturers are listed in Appendix 2 to provide the interested reader with commercial sources of information. While the list is certainly not complete (solar energy is a field that seems to invite new commercial ventures of all sizes), it represents the variety of manufacturers and developers who have come to the panel's attention. It must be pointed out, however, that standard methods of evaluating performance have not been established–for instance, the U.S. National Bureau of Standards has not yet completed its work on standard methods for use by U.S. industry; consequently, specification details are not listed.

Because electric-power generation by photovoltaic devices is a subject that has accumulated extensive literature and experience, it is discussed in a separate chapter of this report.

SOLAR WATER HEATERS

Solar water heaters constitute a fairly recent development based on a common natural phenomenon: cold water in a container exposed to the sun undergoes a rise in temperature. In its modern form, the solar water heater is basically a flat-plate collector and an insulated storage tank. The collector is commonly a blackened metal plate with attached metal tubing and is usually provided with a glass cover and a layer of insulation beneath the plate. The collector tubing is connected by piping to a tank that stores hot water for use during non-sunny periods. When mounted on a roof or other suitable support, the collector absorbs solar radiation; by transfer of the resulting heat to water circulating through the tubing, hot water is supplied to the storage tank. In the most common designs, the storage tank is located above the top of the collector. The elevated position of the tank results in natural convection: water circulates from the collector to the tank and no pump is required. A typical arrangement is illustrated in Figure 1. Figure 2 shows how the system is modified when natural convection circulation ("thermosiphon") is not

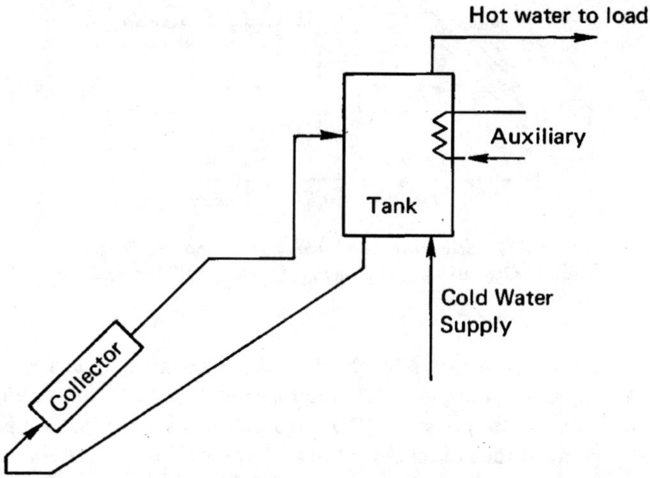

FIGURE 1 Schematic of a natural circulation solar water heater, with auxiliary energy added to the storage tank. Ref. 2 [Courtesy John Wiley & Sons, Inc.]

available. A solar collector area of 30-40 ft^2 (3-4 m^2) in combination with an insulated tank of 50-100 gal (200-400 l) capacity can provide 50-80 gal (200-300 l) of hot water at about 140°F (60°C) per average sunny day in a favorable climate. (This is normally sufficient for a family of four in the United States.) A useful summary of this technology has been provided from Australia.[1]

Past and Current Uses

Several countries have used solar water heaters for many years. They were first used extensively in the United States; thousands were made and put to use in southern California between 1900 and 1920 and in southern Florida between 1920 and 1940. New installations declined when the availability of natural gas provided a means of heating domestic water at lower cost. With the change in the price structure of fossil fuels, however, solar water heaters are again being manufactured in the United States on a small scale.

After 1950, water heater designs were greatly improved in Israel, Australia, and Japan. The general type described above was improved by incorporating better thermal contact between the black metal plate and the tubing through which water circulates, as well as by the development of black surfaces that provided greater net energy recovery from the sun. The basic collector and

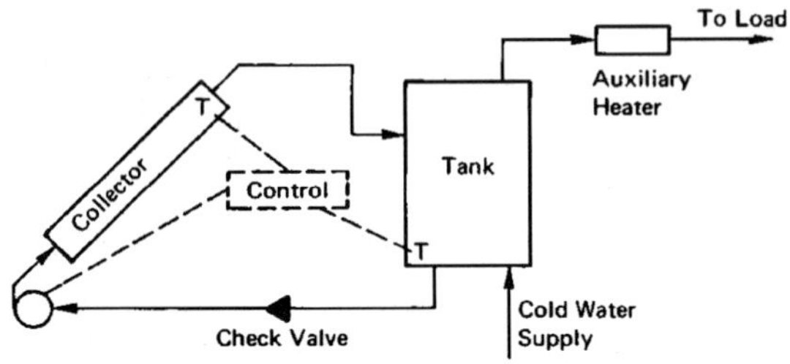

FIGURE 2 Schematic of a forced circulation solar water heater, with auxiliary on line to load. Ref. 2 [Courtesy John Wiley & Sons, Inc.]

system technology is reviewed by Duffie.[2] Commercial firms in Israel began manufacturing and marketing solar water heaters for local and foreign use; sales have been in the range of 10,000-15,000 units per year. At present, about 20 percent of the households in Israel have solar water heaters.

In Australia, solar water heaters were further improved (see below) by development studies in the 1950s and early 1960s, and several manufacturers went into production. They now have an established industry there and produce about 1.5 million units per year. In some districts, particularly in northern Australia, domestic water is heated almost exclusively by solar energy.

Current Technology (Available within 5 Years)

In Australian practice, the absorbers (collectors) for household-scale use typically measure 8.6-17.2 ft^2 (0.8-1.6 m^2), with an 80-gal (300-l) water storage tank, optionally fitted with a thermostatically controlled 1-kW electric booster, producing about 50 gal (180 l) of hot water on a good day. Domestic installations employ thermosiphon circulation, which allows the water to recycle through the absorber without pumps, but large commercial and industrial plants delivering some thousands of liters of hot water per day use forced circulation with thermostatic control. One group of residential structures in Darwin has nine separate banks of solar water heaters with a combined hot-water output of about 3,400 gal (13,000 l) per day. Solar water heaters are proving to be the most economical way of providing hot water in many parts of Australia, particularly in the Northern Territory where it is government policy to install solar water heaters in all government houses.

In Japan, hundreds of thousands of solar water heaters have been manufactured and sold, many of a design that combines the collector and storage functions to produce hot water for use at the day's end. By use of large-diameter plastic tubes–4 in. (10 cm), for example–laid side by side under a glass cover, water is heated and stored in the same containers.[3] In normal use, 20-50 gal (75-200 l) of cold water supplied to the unit in the morning are heated by the sun during the day so that by late afternoon the water in the solar heater has reached temperatures above 100°F (40°C). The hot water is normally fully used in the early evening for bathing and other purposes. These units generally operate without water pressure from the pipeline, whereas the type involving metal collectors and separate storage tanks may be operated at line pressure if so designed.

Experiments with solar water heaters have been undertaken in many countries, but to the best of the panel's knowledge, use in countries other than the United States, Israel, Australia, and Japan has, to date, been quite limited. Production of solar water heaters has commenced in Niger (Figure 3), and it is possible that small-scale manufacture is being undertaken in several other countries.

Cost

The cost of solar water heaters varies considerably with the quality of construction, but in general the better metal/glass collector units with insulated storage tanks involve a price for the collector of about $10-$12 per

FIGURE 3 Solar water heater, Niamey, Niger.

ft^2 (\$100 per m^2) in the United States. For a typical, family-sized installation of 30 ft^2 (3 m^2) and 50-gal (200-l) storage, a total installed cost of about \$300 can be assumed. The type of solar water heater in common use in Japan, providing hot water only in the latter part of the day and combining collector and storage, commands a price of \$100-\$150.

Reliability (Lifetime and Maintenance)

In its simplest form, the solar water heater is free of moving parts and requires a minimum of maintenance. It is essentially a trouble-free plumbing system with an expected life of 10 years or more; maintenance depends largely on the rate of corrosion or mineral deposits caused by materials dissolved in the water.

Research and Development

Water-heater technology is well established, and the needed development is principally to adapt the technology to the materials and manufacturing capabilities of the country in question. Hot water for hospitals, schools, industries, and other institutions, as well as for families, could become much more widely available. The nature of the equipment is such that it can be manufactured in developing countries and adapted to their conditions without difficulty.

HEATING AND COOLING OF BUILDINGS—SOLAR SPACE HEATING

Although residential heating is not a primary problem in most developing countries, comfort heating in some areas would certainly improve living conditions. Much traditional architecture has evolved, in part, to control or minimize temperature fluctuations in the structure. This has been done by designing openings to control admission of sunlight and designing the heat capacity of the structures to minimize undesirable (low or high) temperatures. Traditional building methods, however, have not always been followed in times of rapid population growth. Space heating in developing countries, particularly in buildings of nontraditional architecture, represents a limited problem, but one that can be solved by applying well-established principles. Modern scientific developments can be applied to building design to control all temperature extremes. These include reflective surfaces, controlled overhangs, use of insulating materials, and roof design.

Past and Current Uses

The technology of solar space heating where water is the medium is essentially an extension of the technology employed in solar water heating, except that energy has to be recovered from the tank through a heat-exchange surface (Figure 4). Reviews of the technology and its applications are available.[2,4] Although commercial solar heating has barely started, a substantial number of successful experimental installations have been made and operated for several years.[5] Collectors and storage units much larger than those employed in solar water heaters are necessary to provide a substantial portion of the heat requirement in a typical residence in a sunny climate.

In addition to the solar heating system, conventional energy systems are necessary for a high degree of reliability. In climates where low temperatures may be encountered, the conventional source should have the capacity to meet the residual energy need. A primary economic issue, then, is balancing the fuel savings made possible by the solar energy system against the capital costs. (Note that in developing countries another important economic factor may be the availability of capital.)

Systems employing air as the heat-transfer medium between the collector and a storage bin containing small rocks (Figure 5) have been used successfully. Solar heat is stored as sensible heat in the rocks and recovered, when needed, by passage of air over the rocks and thence to the rooms.

Several dozen experimental solar houses have been built and operated with the heating system comprising the collector, a heat-storage unit, an auxiliary heater, and appropriate heat-distribution-and-control systems. Many of these have been carefully engineered and their performance measured and reported; others have been put together by well-meaning enthusiasts. There has been notable diversity in concept and design.[6-10] Extensive economic studies have also been made that are applicable to industrial economies[11,12] and that indicate that solar heating should be able to provide energy at costs

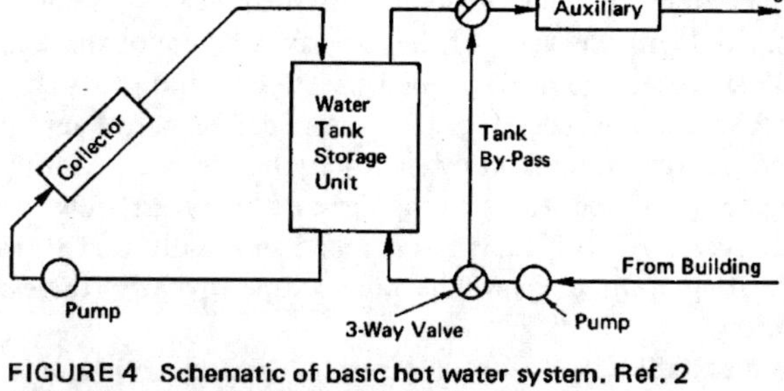

FIGURE 4 Schematic of basic hot water system. Ref. 2 [Courtesy John Wiley & Sons, Inc.]

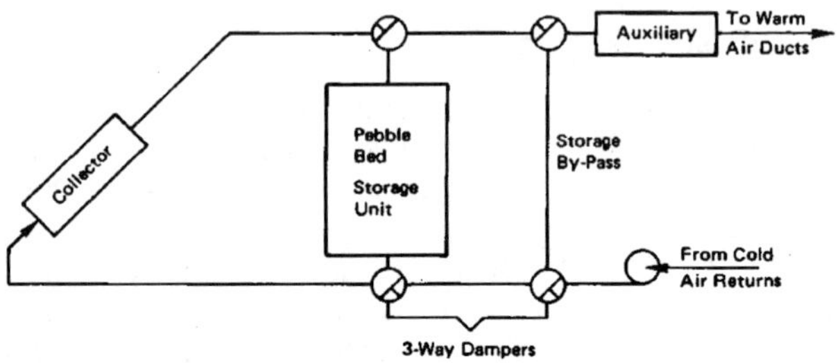

FIGURE 5 Schematic of basic hot air system. Ref. 2 [Courtesy John Wiley & Sons, Inc.]

competitive with other heating methods in many areas. A feasibility study of solar heating in Iran,[13] for instance, indicates that it may be economically practical in that country.

Current Technology (Available within 5 Years)

Several conclusions have been reached relating to application of solar energy in the United States that are useful in evaluating potential application in developing countries. First, solar heating systems have been built and routinely operated over many years, helping to meet the heating loads of houses. Second, the design principles of several systems are well established, and their performance can be predicted with a certainty limited more by variability of weather than anything else. Third, as fuel costs rise, solar heating becomes competitive with heating by fuels or electric heating in many areas of the United States. Indeed, hundreds of such residential systems have already been installed.

Cost

The main deterrent to widespread use of solar heating is the cost of the system compared with the value of the fuel saved by use of the solar facilities. In the industrialized countries, solar heating could not compete with conventional methods of heating with fuels such as petroleum, natural gas, and liquified petroleum gases. Only electricity has been sufficiently costly for solar energy to be competitive. During the past two years, however, the price of liquid and gaseous fossil fuels has climbed so rapidly that it now appears that solar heating of buildings will become competitive with these fuels in the near future.

Most studies indicate that in temperate climates it is not practical to provide 100 percent of the space-heating requirement by solar energy. Systems designed to provide large fractions of annual heating needs in locations with variable weather would be overdesigned much of the time, and interest and carrying charges on the overdesigned equipment would make the systems too costly. The primary reason solar water heaters have been competitive and solar space heaters have not, is the difference in load factor on the two systems. The solar water heater is used all year, whenever solar energy is available, whereas the space-heating system is used only during the winter months. The net useful output of the solar water heater, therefore, is 2-3 times greater for the water heater than for the space heater, per unit of collector surface—that is, per unit of capital cost. Because of this, as well as the several-thousand-dollar cost of a complete solar space-heating system, there has been very little application in industrialized countries and even less in developing countries. The recent severe increases in fuel prices have altered this situation in the United States, Japan, and elsewhere, where solar heating is now under rapid commercial development.

The fact that it is impractical to design a solar heating system as the sole heat source of a building may further limit its usefulness for space heating in the developing cold-climate countries. Whereas solar water heaters can be used on an "as available" basis—no hot water available at times—space heating is required virtually without interruption during cold seasons. Since it is possible that the greatest demand for heat will occur during periods of no sunshine, any solar heating system must include a conventional heating system of adequate capacity to keep the structure habitable. Since fuel of some sort must therefore be available wherever solar space heating is applied, its use in the developing countries may be limited.

It is clear that a major design problem is to achieve the most economical combination of solar energy and conventional fuels; studies indicate that optimum solar heating systems will provide perhaps two-thirds or three-fourths of annual heating at current fuel prices. Further escalation of fuel costs will push these fractions upward, possibly as high as 90 percent under some conditions.

Research and Development

Research and development efforts in solar heating have been aimed almost entirely at applications in the temperate climates of industrialized countries. Because of increasing fuel costs, rapidly expanding research and development activities are under way in the United States and other industrialized countries. Most attention has been given to applying solar heating to single-family residences; attention is now turning to problems of solar heating of multifamily dwellings and institutional buildings. A notable development of the past year or two is the emergence of industrial interest in manufacture and sale of equipment for solar heating. This is a prerequisite to widespread application; until local heating contractors can buy and install the equipment, the concept remains largely academic.

Solar heating in developing areas presents a set of problems that has received relatively little attention. Of several possible approaches, the first involves design of the structure to utilize such elements as fenestration, shading, insulation, building thermal mass, and selective paints to minimize heat needs. A second approach is to devise heating systems that can operate with no mechanical energy for fluid circulation. Systems have been proposed and studied (e.g., the Altenkirch solar house[14] and the dwellings at Odeillo, France[15]) using only natural circulation; there may be other possibilities worth considering for use in small residences (one such experimental house is shown in Figure 6). A third approach is to adapt solar heating systems (with mechanical circulation) that have been developed for use in industrialized nations, perhaps in larger buildings in urban areas where there are significant annual heating loads to be met.

The extent of space-heating needs in developing countries is virtually unknown. To gauge the potential role of solar energy, therefore, a region-by-region assessment of these needs, including studies of climate, fuel use, typical building designs, and other factors, should be undertaken.

FIGURE 6 Solar-heated house, Odeillo, France.

HEATING AND COOLING OF BUILDINGS—SOLAR AIR CONDITIONING

Conventional air conditioning involves compression of a vapor, its cooling and condensation to a liquid, and subsequent vaporization by expansion into a region of lower pressure. Heat is withdrawn from the air to be cooled by drawing from it the heat needed for vaporization of the liquid; the cycle is repeated by a return of the vapor to a pressure at which it can be condensed, rejecting the heat of vaporization at a temperature higher than that of the cooled region. The heat of vaporization is rejected to either the outside air or a stream of water.

Solar air conditioning can be accomplished in the same manner by conversion of solar energy to the mechanical form needed to operate a compressor. The need for mechanical power and its relatively high cost from solar energy can be avoided by use of the absorption refrigeration cycle.[2,16] In this process, the pressure of the refrigerant vapor is raised by heating instead of by mechanical compression; the vaporized refrigerant is recovered for recycling by absorption in a solution of the refrigerant and a salt. The low pressure of the expansion/vaporization region is maintained by the reduced vapor pressure of the refrigerant above the absorbent (solution), and the vapor is regenerated by allowing the solution of refrigerant and absorbent to flow into the generator where solar heat is applied.

This process has received some attention for application in temperate climates; other methods, such as air drying with desiccants regenerated by solar heat, are also possible. Technological feasibility seems assured, and economic feasibility appears promising. As with space heating, there are opportunities to use building design to increase comfort and reduce energy requirements.

Solar cooling has not been developed to the same extent as solar heating. Only two or three full-scale experiments have been carried out (e.g., the Colorado State University House[17]) and although they have been generally successful, their results indicate comparatively high costs as against conventional compression cooling. These studies, aimed primarily at applications in the United States, Australia, and Japan, are still at an early stage.

Most development efforts have been associated with supply of a solar-heated fluid to an absorption refrigeration cycle of the type described, generally using the lithium bromide/water system as the absorbent-refrigerant mixture. Temperatures of operation are compatible with efficient flat-plate solar collectors, so a commercial air conditioner can be adapted to a solar heat supply by substituting the solar-heated fluid for the conventional heat source.

In areas where capital is difficult to obtain—and then at high interest—solar cooling appears to have limited prospects except in commercial and public buildings where money may be more available. There is reason to expect that costs of solar cooling systems can be substantially reduced when further information and experience are at hand.

It is particularly important to note that solar air conditioning uses the same collector and storage system for cooling as for winter heating. Thus, the system in combination with solar heating, assuming climate conditions that require both heating and cooling during alternate times of the year, makes possible a considerably higher load factor on the expensive solar equipment.[18] This makes the economic picture more nearly like that of the solar water heater situation. However, the cost of absorption cooling is substantially higher than the cost of compression cooling.

Application in Developing Countries

Designing air-conditioning systems for use in developing nations presents problems of design and potential utility that may be difficult to address in terms used in industrialized nations. It has been suggested that air conditioning factories and offices would significantly improve the productivity of personnel. Air conditioning could be utilized by residences and community buildings as well. There are also technological problems of system design for operation without mechanical power. In addition to solar operation of mechanical or closed-cycle absorption coolers as noted above, there are other possible cooling methods, including open-cycle humidification-dehumidification,[19] cycles based on solar concentration of LiCl solutions by open evaporation,[20] intermittent cycles that can be adapted to air conditioning,[21] and cooling by nocturnal radiation.[8,9,22,23]

Building designs to minimize needs for cooling may be the most productive approach, particularly for small residential buildings.

The best methods of obtaining cooling with solar energy in developing countries are far from clear at this time, and the immediacy and extent of needs for air conditioning are not known. The importance of obtaining such information is evident.

SOLAR REFRIGERATION

Closely related to air conditioning, solar refrigeration is generally intended for food preservation or for storage of biological and medical materials. There have been experiments in several countries including the United States,[24,25] Sri Lanka,[21] France,[26] and the Soviet Union[27] on solar-operated coolers, using absorption cooling cycles. Most of these are aimed at household-scale food coolers or small-scale ice manufacture. A single experiment in Mexico with a solar-operated, household-scale cooler was unsuccessful. Although the machine was simple in design, it was too complicated to operate and therefore was not usable by the people for whom it was intended. In the Soviet Union, a household refrigerator employing an ammonia-absorption cycle and a pressurized solar collector is under development.

Research and Development

There are many possible refrigeration cycles and systems, such as intermittent absorption cycles, that can be considered for solar-powered refrigeration. Nevertheless, even so basic a question as the best scale on which to operate solar refrigeration for developing countries has yet to be established; community-scale systems may offer advantages of larger-scale, reduced-educational requirements for users, and (if appropriate) the possibility of distributing ice to smaller users (i.e., households). However, they require a substantial investment in capital and organization of community training and maintenance programs.

There are a number of open questions regarding solar refrigeration. Nevertheless, its application has the attractive possibility of increased utilization of available foodstuffs if refrigeration could be successfully provided. Research and development are needed because of the lack of a practical solar refrigerator at any price and of any understanding of the cost of systems that could be developed. An objective and critical analysis of the need for, and the prospect of, practical solar refrigeration should be the first step in any development effort.

SOLAR STILLS AND DESALINATION

The use of solar energy for desalting seawater and brackish well water has been demonstrated in several moderate-sized pilot plants in the United States, Greece, Australia, the Soviet Union, and several other countries. This century-old process—the basin-type still—has been modified and adapted to modern materials. It consists of a shallow pool of brine from which a slow evaporation of water takes place. In its modern form it is covered by sloping sheets of glass; water condensing on the underside of the cooler glass covers runs to troughs at the lower edges and to storage. Excess brine that has not evaporated is run to waste as salt water is supplied to the basin. The idea was first applied in 1892 at Las Salinas, Chile, in a plant supplying drinking water for animals working in nitrate mining and transport. The Las Salinas plant reportedly operated for 30 years. Modern developments in solar distillation have been directed to the use of materials and designs for economical and durable construction, to reduce product-water cost. A complete review of solar distillation is available,[28] and the United Nations has published a less detailed, but useful, state-of-the-art report.[29]

Current Technology (Available within 5 Years)

Solar distillation is used on a small commercial scale to supply small towns and motels in isolated areas in Australia (Figure 7) and small communities in the Mediterranean basin (Figure 8) and the Caribbean. These are areas where either an adequate source of fresh water is unavailable or the cost of fresh water brought in by transport is too high. Although small community-scale solar stills are nearly competitive with other desalination systems, the process must be regarded as experimental. Still under study are problems of construction materials, energy and material flow, and physical design to maximize output and minimize cost.

Some designs have been rather well standardized, and considerable operating experience is available upon which to judge their utility and costs.

FIGURE 7 Glass-covered solar still, Coober Pedy, Australia. (Constructed in 1966; total area 38,000 ft^2 [3,530m^2])

FIGURE 8 Solar still on the island of Patmos, Greece. (Constructed in 1967; total area 95,000 ft^2 [8,826m^2])

Cost

Solar distillation requires relatively large capital investment per unit of capacity and, in properly designed and constructed systems, a minimum of operating and maintenance costs. Product-water costs, therefore, depend primarily upon still productivity, service life, capital cost of the installation, and amortization and interest rates.

Since productivity of a solar still is dependent on the intensity and duration of the insolation it receives, it is subject to both diurnal and seasonal variations, which must be taken into account in any analysis of needs and costs. Solar-still productivity is conveniently referred to, in round numbers, as being typically 0.1 gal/ft^2/day (4 l/m^2/day) for a clear summer day, with seasonal fluctuations expected to range down to half that on clear days in mid-latitudes (15°-45°). Thus, experience shows that a still will yield about 25 gal/ft^2 (1,000 l/m^2) annually, with some variations dependent on climate and design. A typical lifetime for a still constructed of concrete, glass, and other durable materials is 20 years or more, although other still designs involve less durable materials that must be renewed periodically.

The lack of more general use of solar stills is almost entirely the consequence of the high capital investment required and the resulting high cost of water produced. In recent years, durable solar stills have been built for a unit cost of \$1.50-\$3/ft^2 (\$16-\$32/m^2). If a durable solar still can be constructed for $1/ft^2 ($10.75/m^2), including materials and labor, typical yields would produce water at a cost of about $3-$4 per 1,000 gal (80¢-$1.05 per 1,000 l).

These figures may be compared with typical freshwater costs of 5¢-25¢ per 1,000 gal (1¢-5¢ per 1,000 l), and costs of desalted water from large fuel-fired evaporation plants in the vicinity of $1-$1.50 per 1,000 gal (25¢ per 1,000 l). There seems to be little prospect, therefore, that large solar distillers will be competitive with large desalting plants supplied with conventional energy sources unless fuel prices escalate greatly.

Nevertheless, a significant advantage of the solar still is the flexibility it offers in choice of size (capacity). In situations where a community or an industry requires small quantities of water—say, less than 50,000 gal (200,000 l) per day—the solar still may be more economical than the conventional desalting plant. The latter, operating at this low rate rather than at millions of gallons per day, requires a heavy investment per unit of output, with the result that the cost of water produced from it can be several dollars per 1,000 gal. Thus, since cost of the solar desalting plant appears to be competitive in small installations, there is the opportunity to use such plants a) in small communities where potable water is unobtainable except at very high cost; b) in certain industrial and commercial applications where materials must be processed in a region where all available water is brackish; or c) for watering of livestock in areas where grazing is possible if water is supplied. It is for these moderate-volume requirements that solar stills have been built.[30]

Reliability (Lifetime and Maintenance)

Distiller productivity can be predicted[29] for those designs in which problems of mechanical failure or corrosion are minimal. Proven designs are those that use standard, durable construction materials such as glass, concrete, asphalt, and corrosion-resistant metals,[26] although designs that rely in part on new plastic materials are also being investigated.[31] Construction, maintenance, and operation do not require high levels of skill in working with complex machinery, and little day-to-day attention is needed. In short, designs are now available for serviceable solar stills that can be used with reasonable confidence.

With most designs it is possible, to some degree, to use locally available materials and labor in construction.[32]

Research and Development

The basin-type solar still combines the solar energy collection function and the distillation function in a single unit. Separation of these functions would allow regenerative, or multistage processes, to be solar operated. This development is dependent upon significant progress in solar-collector technology (i.e., a "breakthrough"). Notable improvement in performance and cost production may result from refinements in basin-still design. In this area particularly, the use of new materials with unique properties, locally available materials, and new designs is potentially important.

Combined energy-source systems, in which solar energy to the still is augmented by waste heat from, for example, intermittently operated diesel or gasoline engines, may reduce the cost of product water from the still.[33] Multiple-function plants, in which water production is integrated with water use, are also being considered, for example, in an integrated system for energy, water, and food production.[34]

Finally, multipurpose systems, producing some combination of water, salt, and possibly power, can be conceived. Within the limits set by relative markets for these commodities, such systems could be significant, but only after considerable research and development.

Evaluating Solar Distillation

The following checklist is presented as an aid in assessing the relative value of solar distillation as a method of meeting a particular water need.

Climate. If the solar radiation climate is good, i.e., skies are generally clear, a solar distillation plant may be feasible. Productivity of stills is a direct function of solar radiation received on a horizontal surface.

Scale of need. Solar distillation now appears suited to water requirements on a relatively small scale, i.e., less than about 50,000 gal (200 m^3) per day. For larger demands, or demands that are anticipated to increase in the near future, other desalination methods are now more economical.

Site. The site for the still installation, besides having unobscured solar radiation, should be near the water users. Brackish surface water, groundwater, or seawater must be available nearby. Competing uses for sites should also be considered.

Estimate of still size. A well-designed, well-constructed still in a good climate should produce about 25 gal per ft^2 of still (1,000 l per m^2) per year. This yield, compared with annual water needs, will give a first approximation of the required solar-still area.

Design. One or more designs can be selected or developed consistent with local conditions, available materials, and skills of the local work force.

Estimate of monthly yield. Based on the selected design, monthly yields of distilled water can be estimated either from experimental data on similar stills in other locations or through the use of estimation methods such as those outlined in the *Manual of Solar Distillation*.[28] These yields should be based on monthly mean radiation levels and temperatures (the most widely available form of the pertinent meteorological data) such as those available in *World Distribution of Solar Radiation.*[35]

Rainfall contribution. If local conditions and regulations permit, rainfall runoff from the glass covers of the still can be recovered. This can be estimated from average monthly rainfall data (if rainfall is not highly variable) or minimum monthly rainfall (if it is variable), applying a reasonable recovery factor. Useful monthly rainfall collected can then be added to still yield to provide an estimate of month-to-month production of useful water from the still.

Monthly needs versus production. A comparison of month-to-month distribution of water production and water needs will indicate water-storage requirements, modifications in still size, or possible uses of supplementary sources.

Estimate of cost. With the foregoing information, the appropriate data on still designs, first costs, service life, maintenance and operating requirements, and local information on interest rates and other economic factors, the cost of delivering water from the solar stills can be estimated and compared with alternatives.

Other considerations. Considerations not directly quantifiable in terms of cost may also be important. These could include, for example, sociological and personal factors, resistance to change, distance of users from the potable-water supply, protection of the still from vandalism, sales methods or other costs of distributing the water, subsidies, and/or fuel savings.

In general, additional research and development would involve further adaptation of existing technology to the specific needs of developing countries. Adaptations would include design modifications that allow use of locally available materials and locally manufactured components. Studies of this type could improve the economics of solar distillation, widen the areas in which it might be useful, and thereby contribute to the solution of water-supply problems, at least for small communities in appropriate climates.

SALT PRODUCTION

Solar evaporation of seawater or brines has been a traditional method of obtaining salt. It remains important today—on both a small and large scale—in many nations. The basic concept is simple: in areas where evaporation exceeds rainfall, a shallow pond of brine is exposed, which results in evaporation of water and ultimately in crystallization of salt. Solar evaporation is used in many developing countries, for instance, India,

Pakistan, Mexico, Colombia, and Chile. It is also an important industrial process in the United States, Israel, and elsewhere.[36] Modern developments have been concerned with improved pond construction and salt-harvesting techniques.

There appears to be little further research on traditional solar evaporation processes that cannot be done by the industries using the process. The suggestion has been made, however, that further studies of "solar ponds" might lead to improved salt production and by-products of power or distilled water.[37] Solar ponds are large-area brine ponds approximately 1 m deep, in which vertical gradients of salt concentrations are maintained so that the most concentrated and most dense solutions are at the bottom of the pond. Heat is generated at the bottom of the pond by absorption of solar radiation–transmitted by the water–by the pond's black bottom. In spite of the temperature rise of the brine at the bottom, its specific gravity remains higher than that at the top if a sufficiently large concentration gradient is maintained. Under these conditions, convection mixing is minimized and the bottom layers of brine, which may reach 80°-90°C (175°-195°F) while the top remains at 25°C, may then be withdrawn from the pond and used for power generation, salt production by multiple-effect evaporation of brine, water distillation, and so on. To maintain the concentration gradient against the slow upward diffusion of salt, the surface must be slightly "washed" with fresh water, and concentrated brine supplied at the bottom.

It must be recognized that a "solar pond" multipurpose facility would involve high capital cost and that the development of the process and its potential application are probably less attractive in the developing countries than in industrialized nations.

SOLAR CROP-DRYING

Of all the direct uses of solar energy, sun drying of crops is perhaps the most ancient and widespread. The customary technique involves spreading the material to be dried in a thin layer on the ground to expose it to sun and wind. Copra, grain, hay, and fruits and vegetables are still dried in this manner in many parts of the world, including the industrialized countries. In recent years, innovations have been adopted, particularly for fruit drying, in which fruit is placed in carefully designed racks to provide controlled exposure to solar radiation and wind and to improve material handling. Improved process control and product quality have resulted.

In recent years, the term "solar drying" also has come to mean the process whereby agricultural materials are dried not by direct exposure to the sun–and wind, rain, insects, vermin, and birds–but by means of solar-heated air in more protected surroundings. The process is of special interest in the case of the soft fruits; these are particularly vulnerable to attack by insects, as the sugar concentration increases during drying.

A large portion of the world's supply of dried fruits and vegetables continues to be prepared by sun drying. Doubtless this process is the cheapest and simplest way to dry crops in regions having abundant sunshine and where the post-harvest season is characterized by low relative humidity and little or no rainfall. In the case of green lumber—where direct exposure to the sun tends to produce curling and warping in many woods—and in the humid tropics, kiln-drying can be accomplished with solar-heated air. The cost of this process is not yet well established.

To remain stable in storage, agricultural crops should be dried to a moisture content of 12-15 percent by weight. The relative humidity of the air that will be in equilibrium with crops of that moisture level varies from crop to crop and ranges from about 48-60 percent for fruits,[38] grain, and hay.[39] Marked decreases in relative humidity (RH) can be achieved with rather minimal increases in air temperature, and the temperature change needed to produce any desired change in RH can be ascertained from standard psychrometric charts. Adequate drying of crops can therefore be achieved in humid climates by raising the temperature of the air circulating among the items to be dried.

Current Technology (Available within 5 Years)

Although there is no significant commercial manufacture of solar crop dryers, experimentation over the past 20 years has produced a number of designs. These range from the use of solar-heated air in more or less conventional air dryers to a combination of direct drying and air drying by placing the materials to be dried in flat-plate collector-dryers. Among the former are various designs developed in the United States,[39] Turkey,[40,41,42] Canada,[43,44,45] Brazil,[46] and Australia.[47] Combination collector-dryers have been designed and used successfully in India[48] and Trinidad.[49,50]

In the developing countries, solar dryers should be usable by farmers with limited technical skill and small capital resources. The equipment should be as simple and cheap as possible. Locally available materials and relatively unskilled labor may be adequate for constructing "do-it-yourself" dryers of some types. Because of seasonal, short-term use, crop dryers may have to be designed so that a variety of crops, maturing at different times, can be dried in sequence by use of the same equipment. If made and sold commercially, solar dryers will have to be made available at prices well below those of typical solar collectors so that the cost can be recouped in relatively short periods of use.

Because of the wide variety of circumstances encountered, the potential economic advantages of solar crop-drying must be assessed on an individual basis. This will depend principally on the size of the crop to be dried and the prevailing weather conditions. It has been shown that the addition of a solar air heater to an unheated air-drying system can shorten drying time by 50-75 percent. Where the choice is between these two systems, the cost of the solar dryer will be largely offset by the saving in cost of the circulating fan as a consequence of the lower power requirement. In the case of lumber drying in Australia, economic analysis based on current designs and materials has shown solar heating to be slightly more costly than conventional heating systems.

Research and Development

Development of solar drying can conceivably benefit from further development of collector-dryer combinations and flat-plate air heaters and energy-storage systems to supply hot air to dryers. Research in design and control of these processes, for the particular crops or other materials to be dried, could lead, in developing countries, to other practical applications that could result in improved utilization of food supplies.

POWER GENERATION

Solar energy for power generation, either in the form of electricity or mechanical work, has been the subject of extensive research in the Soviet Union, United States, France, Italy, and Israel.[51] In addition to the photovoltaic approach to be discussed in the following chapter, efforts to generate power by use of heat engines have been the most numerous. All of these efforts have been more or less dependent on means of collecting and/or concentrating incident solar energy. A large assortment of solar collectors has been operated in numerous experimental assemblies to produce steam or other vapor, which then has been employed in engines of various types to generate electricity or mechanical power. The largest such unit was a plant built in Egypt in 1913, which developed about 50 hp (37 kW).[52]

Most of the work has been directed toward the use of reflecting surfaces to concentrate the solar energy onto a small receiver/boiler, which permits the development of much higher temperatures than is possible with a flat-plate collector. High-pressure steam for electric-power generation or, with precise equipment, extremely high-temperature heat for chemical and metallurgical processing can thus be produced. Concentrators in the form of paraboloids (dishes), parabolic cylinders (troughs), and other shapes have been investigated.[53] A design first studied in the Soviet Union and now being investigated in the United States, involves a large array of flat mirrors mounted on the ground and continually oriented to reflect the sun onto a high-pressure steam boiler. The boiler is located at the top of a high tower having a turbogenerator at its base.[54]

Research and development specifically aimed at solar power generation in the less industrialized regions has been carried on in the Soviet Union, Africa, and, to a limited extent, in Israel. Notable among these efforts are several small solar pumping units of French manufacture in West Africa, based on a design by Girardier,[55] and a similar installation in Mexico. An example of one of these installations in shown in Figure 9. Flat-plate water heaters supply hot water to a boiler in which propane is vaporized and from which this vapor is supplied to a reciprocating engine. Several installations of 1- to 3-kW capacity are in service. Present overall efficiencies of less than 1 percent could be substantially increased by design improvements. Although costs of power produced by this system are high, comparison with diesel units does not show a large difference when diesel engine life is assumed shortened by the lack of maintenance and repair personnel customary in remote areas.

The Soviet program has provided numerous types of experimental solar concentrators up to 5 m (16.5 ft) in diameter, and electric-power units of 1- to 3-kW peak output.[56] Rankine engines operated by steam and a pressurized

FIGURE 9 Solar pumping station in Africa. (Flat-plate collector, organic-fluid Rankine engine; one of several built in the early 1970s) [Girardier, Ref. 55]

Stirling engine have been used experimentally in conjunction with these concentrators. Comparatively high technical skill would normally be required in the installation and maintenance of these systems.

Costs of solar power units are a major deterrent to their use; economic application would require major design and material savings or usage in situations where the only alternatives are high-priced and short-lived small generators requiring costly fuel.

Other developments in this field will be treated in Appendix 4.

REFERENCES

1. Commonwealth Scientific and Industrial Research Organization. 1964. *Solar water heaters, principles of design, construction and installation*. Division of Mechanical Engineering, Circular No. 2. Victoria, Australia: Commonwealth Scientific and Industrial Research Organization.
2. Duffie, J A., and Beckman, W. A. 1974. *Solar energy thermal processes*. New York: John Wiley and Sons.
3. Tanishita, I. 1970. Present situation of commercial solar water heaters in Japan. Paper presented at International Solar Energy Society Conference, 2-6 March 1970, Melbourne, Australia.
4. Beckman, W. A., and Duffie, J. A. 1974. Solar heating and cooling of buildings. In *The energy crisis and energy from the sun*. Symposium on Solar Energy Utilization, 20th Annual Meeting, Institute of Environmental Sciences, 30 April 1974, Washington, D.C. Mt. Prospect, Illinois: Institute of Environmental Sciences.
5. Löf, G. O. G. 1964. Use of solar energy for heating purposes: Space-heating. In *Proceedings of the United Nations Conference on New Sources of Energy: Solar Energy, Wind Power and Geothermal Energy, 21-31 August 1961, Rome, Italy*. Vol. 5: *Solar energy II*, pp. 114-139. U.N. Document No. E/Conf.35/6. New York: United Nations.
6. Engebretson, C. D. 1964. The use of solar energy for space heating: MIT solar house IV. In *Proceedings of the United Nations Conference on New Sources of Energy: Solar Energy, Wind Power and Geothermal Energy, 21-31 August 1961, Rome, Italy*. Vol. 5: *Solar energy II*, pp. 159-177. U.N. Document No. E/Conf.35/6. New York: United Nations.
7. Löf, G. O. G.; El Wakil, M. M.; and Chiou, J. P. 1964. Design and performance of domestic heating system employing solar heated air—the Colorado solar house. In *Proceedings of the United Nations Conference on New Sources of Energy: Solar Energy, Wind Power and Geothermal Energy, 21-31 August 1961, Rome, Italy*. Vol. 5: *Solar energy II*, pp. 185-197. U.N. Document No. E/Conf.35/6. New York: United Nations.
8. Yanagimachi, M. 1964. Report on two and a half years' experimental living in Yanagimachi solar house II. In *Proceedings of the United Nations Conference on New Sources of Energy: Solar Energy, Wind Power and Geothermal Energy, 21-31 August 1961, Rome, Italy*. Vol. 5: *Solar energy II*, pp. 233-247. U.N. Document No. E/Conf.35/6. New York: United Nations.
9. Bliss, R. W. 1964. The performance of an experimental system using solar energy for heating, and night radiation for cooling a building. In *Proceedings of the United*

Nations Conference on New Sources of Energy: Solar Energy, Wind Power and Geothermal Energy, 21-31 August 1961, Rome, Italy. Vol. 5: *Solar energy II*, pp. 148-158. U.N. Document No. E/Conf.35/6. New York: United Nations.

10. Hay, H. R. 1973. Energy technology and solarchitecture. *Mechanical Engineering* 95(11):18-22.
11. Tybout, R. A., and Löf, G. O. G. 1970. Solar house heating. *Natural Resources Journal* 10(2):268.
12. Löf, G. O. G., and Tybout, R. A. 1973. Cost of house heating with solar energy. *Solar Energy* 14:253-278.
13. Bahadori, Mehdi N. 1973. A feasibility study of solar heating in Iran. *Solar Energy* 15:3-26.
14. Adler, S.; Levite, G.; and Tabor, H. 1964. The Altenkirch solar-cooled house. In *Proceedings of the United Nations Conference on New Sources of Energy: Solar Energy, Wind Power and Geothermal Energy, 21-31 August 1961, Rome, Italy*. Vol. 6: *Solar energy III*, pp. 60-65. U.N. Document No. E/Conf.35/7. New York: United Nations.
15. Trombe, F.; Robert, J. F.; Cabanet, M.; and Sesolis, B. 1975. Some performance characteristics of the CNRS solar houses. Paper presented at the International Solar Energy Society Conference, July 28-August 1, Los Angeles, California.
16. Chung, R.; Duffie, J. A.; and Löf, G. O. G. 1963. A study of a solar air conditioner. *Mechanical Engineering* 85:31.
17. Ward, D. S., and Löf, G. O. G. 1975. Design and construction of a residential solar heating and cooling system. *Solar Energy* 17:13.
18. Löf, G. O. G., and Tybout, R. A. 1974. The design and cost of optimized systems for residential heating and cooling by solar energy. *Solar Energy* 16:9-18.
19. Dunkle, R. V. 1965. A method of solar air-conditioning. *Mechanical and Chemical Engineering Transactions*, Institute of Engineers, Sydney, Australia MC1(1):73-78.
20. Baum, V. A.; Kakabaev, A.; Khandurdiev, A.; Klychiaeva, O.; and Rakhmanov, A. 1973. Utilisation de l'énergie solaire dans les conditions particulières des régions à climat torride et aride pour la climatisation en été. In *Proceedings of the International Congress. The Sun in the Service of Mankind, 2-6 July 1973*, UNESCO House, Paris, pp. EH-96:1-13. Washington, D.C.: Research Applications Directorate, U.S. National Science Foundation.
21. Chinnappa, J. C. V. 1962. Performance of an intermittent refrigerator operated by a flat-plate collector. *Solar Energy* 6(4):143.
22. Hay, H. R., and Yellott, J. I. 1969. Natural air conditioning with roof ponds and movable insulation. *American Society of Heating, Refrigerating and Air-Conditioning Engineers Transactions* 75 (Part I):165-177.
23. ______. 1970. A naturally air conditioned building. *Mechanical Engineering* 92(1):19-25.
24. Williams, D. A.; Chung, R.; Löf, G. O. G.; Fester, D. A.; and Duffie, J. A. 1958. Cooling systems based on solar refrigeration. *Refrigeration Engineering* 66:33.
25. Sargent, S. L., and Beckman, W. A. 1968. Theoretical performance of an ammonia-sodium thiocyanate intermittent absorption refrigeration cycle. *Solar Energy* 12:137.
26. Trombe, F., and Foex, M. 1961. Production de glace à l'aide de l'énergie solaire. In *Applications thermiques de l'énergie solaire dans le domaine de la recherche et de l'industrie*. Les Comptes-rendus des Colloques Internationaux du Centre National de la Recherche Scientifique, Montlouis, 23-28 June 1958. Paris: Centre National de la Recherche Scientifique.

27. Kakabaev, A.; Klychiaeva, O.; and Khandurdiev, A. 1972. Refrigeration capacity of an absorption solar refrigerator plant with flat glazed solution regenerator. *Applied Solar Energy (English translation of "Geliotekhnika")* 8(1-2):90.
28. Talbert, S. G.; Eibling, J. A.; and Löf, G. O. G. 1970. *Manual on solar distillation of saline water*. Office of Saline Water, Research and Development Progress Report No. 546. Washington, D.C.: U.S. Department of the Interior.
29. United Nations Economic and Social Affairs. 1970. *Solar distillation*. U.N. Document No. ST/ECA/121. New York: United Nations.
30. Morse, R. N., and Read, W. R. W. 1968. A rational basis for the engineering development of a solar still. *Solar Energy* 12:5.
31. Hay, H. R. 1973. Plastic solar stills: Past, present, and future. *Solar Energy* 14:393-404.
32. Lawand, T. A. 1968. *Engineering and economic evaluation of solar distillation for small communities.* Technical Report No. MT-6. Ste. Anne de Bellevue, Quebec, Canada: Brace Research Institute, McGill University.
33. Proctor, D. 1973. The use of waste heat in a solar still. *Solar Energy* 14(4):433-449.
34. Jensen, Merle H., and Eisa, Hamdy M. 1972. Controlled-environment vegetable production; results of trials in Puerto Peñasco, Mexico, 1968-1970. Tucson: University of Arizona, Environmental Research Laboratory.
35. Löf, G. O. G.; Duffie, J. A.; and Smith, C. O. 1966. *World distribution of solar radiation*. Engineering Experiment Station, Report No. 21. Madison: University of Wisconsin.
36. Spiegler, K. S. 1955. Solar evaporation of salt brines in open pans. In *Solar energy research*, F. Daniels and J. A. Duffie, eds., p. 119. Madison: University of Wisconsin Press.
37. Tabor, H., and Matz, R. 1965. Solar pond status report. *Solar Energy* 9:177.
38. Szulmayer, W. 1973. Thermodynamics of sun-drying. In *Proceedings of the International Congress. The Sun in the Service of Mankind, 2-6 July 1973, UNESCO House, Paris,* pp. V-24:1-10. Washington, D.C.: Research Applications Directorate, U.S. National Science Foundation.
39. Buelow, Frederick H. 1964. Drying crops with solar heated air. In *Proceedings of the United Nations Conference on New Sources of Energy: Solar Energy, Wind Power and Geothermal Energy, 21-31 August 1961, Rome, Italy*. Vol. 5: *Solar energy II*, pp. 267-272. U.N. Document No. E/Conf.35/6. New York: United Nations.
40. Akyurt, M., and Selçuk, M. K. 1973. A solar drier supplemented with auxiliary heating systems for continuous operation. *Solar Energy* 14:313-320.
41. Selçuk, M. K. 1967. The final report of the solar fruit drier. *COMPLES Bulletin* (Cooperation Mediterranéenne pour l'Energie Solaire, Marseilles) (12):91-101.
42. Selçuk, M. K.; Ersay, Ö.; and Akyurt, Mehmet. 1973. Development, theoretical analysis and performance evaluation of shelf type solar driers. Paper presented at the Winter Annual Meeting of the American Society of Mechanical Engineers, 11-15 November 1973, Detroit, Michigan.
43. Lawand, T. A. 1965. A description of two simple solar agricultural dryers. *COMPLES Bulletin* (Cooperation Mediterranéenne pour l'Energie Solaire, Marseilles) (9):51-56.
44. ——. 1966. A solar-cabinet dryer. *Solar Energy* 10(4):158-164.
45. ——. 1966. *How to make a solar cabinet dryer for agricultural produce*. Leaflet L.6. Ste. Anne de Bellevue, Quebec, Canada: Brace Research Institute, McGill University.

    Also available in Arabic (Leaflet L.7) and Spanish (Leaflet L.8) translations.

46. Ghosh, B. N. 1973. A new glass-roof dryer for cocoa beans and other crops. In *Proceedings of the International Congress. The Sun in the Service of Mankind, 2-6 July 1973, UNESCO House, Paris*, pp. V-30:1-17. Washington, D.C.: Research Applications Directorate, U.S. National Science Foundation.
47. Read, W. R. W.; Choda, A.; and Cooper, P. I. 1973. A solar timber kiln. In *Proceedings of the International Congress. The Sun in the Service of Mankind, 2-6 July 1973, UNESCO House, Paris*, pp. V-28:1-10. Washington, D.C.: Research Applications Directorate, U.S. National Science Foundation.
48. Kapoor, S. G., and Agrawal, H. C. 1973. Solar dryers for Indian conditions. In *Proceedings of the International Congress. The Sun in the Service of Mankind, 2-6 July 1973, UNESCO House, Paris*, pp. V-29:1-11. Washington, D.C.: Research Applications Directorate, U.S. National Science Foundation.
49. Headley, O. St. C., and Springer, G. F. B. 1973. A natural convection solar crop drier. In *Proceedings of the International Congress. The Sun in the Service of Mankind, 2-6 July 1973, UNESCO House, Paris*, pp. V-26:1-10. Washington, D.C.: Research Applications Directorate, U.S. National Science Foundation.
50. Satcunanathan, Suppramaniam. 1973. A crop drier utilizing a two pass solar air heater. In *Proceedings of the International Congress. The Sun in the Service of Mankind, 2-6 July 1973, UNESCO House, Paris*, pp. V-27:1-17. Washington, D.C.: Research Applications Directorate, U.S. National Science Foundation.
51. Tabor, H., and Bronicki, L. 1954. Small turbines for solar energy power package. In *Proceedings of the United Nations Conference on New Sources of Energy: Solar Energy, Wind Power and Geothermal Energy, 21-31 August 1961, Rome, Italy.* Vol. 4:68-81. U.N. Document No. E/Conf.35/5. New York: United Nations.
52. Jordan, R. C., and Ibele, W. C. 1955. Mechanical energy from solar energy. In *Proceedings of the World Symposium on Applied Solar Energy, 2-5 November 1955, Phoenix, Arizona*, pp. 81-101. Menlo Park, California: Stanford Research Institute.
53. Zarem, A., and Erway, D. 1963. *Introduction to the utilization of solar energy*. New York: McGraw Hill Book Co.
54. Baum, V. A.; Aparisi, R. R.; and Garf, B. A. 1956. High-power solar installations. *Teploenergetika* 3(6):31-39. (In Russian)
55. Girardier, Jean-Pierre; Alexandroff, M. G.; and Alexandroff, J. M. 1973. Les moteurs solaires et l'habitat pour les zones arides: Realisations actuelles et perspectives. In *Proceedings of the International Congress. The Sun in the Service of Mankind, 2-6 July 1973, UNESCO House, Paris*, pp. E-82:1-10. Washington, D.C.: Research Applications Directorate, U.S. National Science Foundation.
56. Starodubtsev, S. V.; Umarov, G. Ya.; and Kordub, N. V. 1965. A 2.7 meter diameter vacuum film solar concentrator. *Applied Solar Energy (English translation of "Geliotekhnika")* 1(1):16.

Photovoltaic Devices

Solar cells, usually in the form of thin films or wafers, are semi-conductor devices that convert from 3 percent to somewhat under 30 percent of incident solar energy into DC electricity, with efficiencies depending on illumination-spectrum intensity, solar-cell design and materials, and temperature. A solar cell behaves very much like a low-voltage (≃0.5-volt) battery whose charge is continuously replenished at a rate proportional to incident illumination. Connection of such cells into series-parallel configurations permits the design of solar "panels" with high currents and voltages as high as several kilovolts. Combined with energy-storage and power-conditioning equipment, these cells can be used as an integral part of a complete solar electric conversion system. Following their invention as practical devices in 1955, they have been used primarily for providing electrical power to spacecraft, though there have been some specialized terrestrial applications.

The extraordinary simplicity of a solar-photovoltaic system (Figures 10 and 11) would make it appear a highly desirable energy system for terrestrial purposes, both in industrialized and developing nations. The attractive attributes of photovoltaic arrays include the absence of moving parts, very slow degradation of properly sealed cells, possibility for modular systems at sizes from a few watts to megawatts, and extreme simplicity of use. The exceedingly high costs of development and fabrication of spacecraft solar arrays, however, have discouraged any serious thought of widespread terrestrial use of such a technology at the present time, in spite of the potentially attractive characteristics of such systems. A complete spacecraft solar-cell array costs anywhere from $500,000/kWe (average) for the Skylab

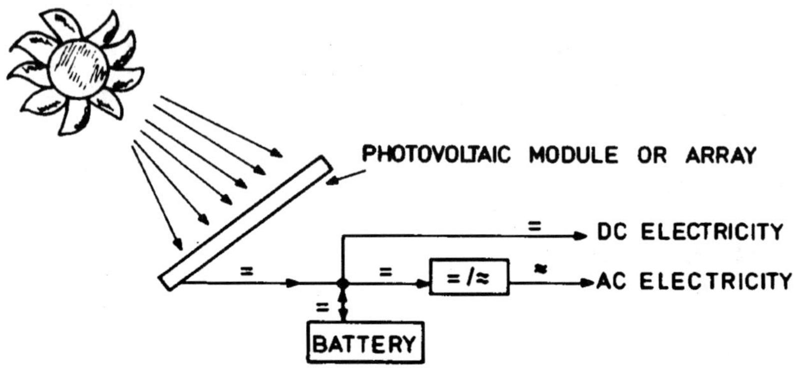

FIGURE 10 Simplified photovoltaic system.

arrays (10 kWe) to several million dollars per average kilowatt for early Mariner spacecraft arrays. An example of such an array is illustrated in Figure 12, where a Mariner IX spacecraft is shown with four large panels designed to deliver 400 W (total) of DC electrical power with an incident solar illumination of 690 W/m^2 (Mars orbit).

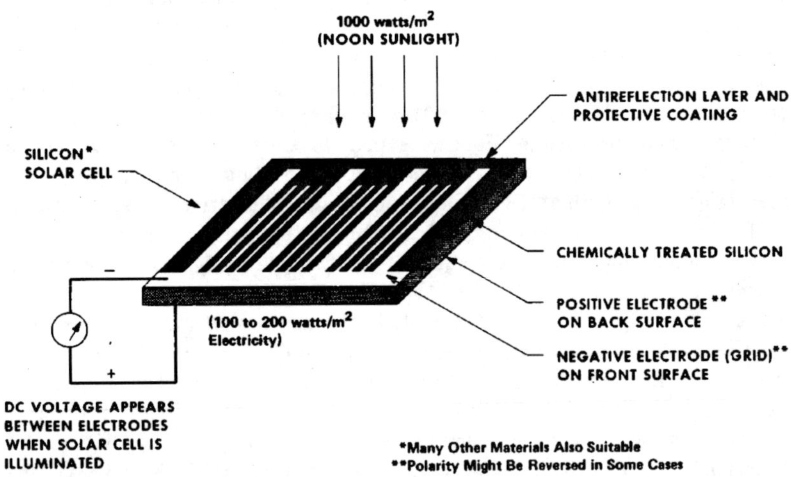

FIGURE 11 Direct conversion of sunlight into electricity—simplified representation of solar-cell operation.

There is now good evidence that, with appropriate technological developments and mass-production techniques, the cost of such solar arrays can be lowered (perhaps as early as the mid-1980s) to the point where a complete system–solar conversion, storage, power conditioning and transmission/distribution–can compete on a life-cycle cost basis with other large-scale energy-system alternatives, and perhaps be useful even in small-scale applications in remote rural areas.

Substantial programs for the development of commercially interesting photovoltaic systems in the United States, West Germany, Japan, and elsewhere have recently been initiated and, coupled with important developments in the past few years, now provide some concrete basis for such a prognosis. Important recent events include the development of continuous production of ribbon silicon suitable for solar cells, improvements in efficiency and stability of CdS solar cells, and the development of inexpensive, wide-aperture concentrators (Winston collector).[1]

FIGURE 12 Mariner IX spacecraft, showing four silicon solar-cell panels.

PRINCIPLE OF OPERATION

The operation of the solar cell is based on the photovoltaic effect—the creation of charge carriers within a material by the absorption of energy from incident ionizing radiation. The materials within which this occurs most usefully are the semiconductors—those materials whose properties are somewhere between those of conductors and insulators. They are characterized by the fact that their valence electrons are not free to move about and conduct current as is the case in metals, but are normally confined to electron-pair bonds between the atoms in the crystal. In semiconductors, however, the band of energy levels normally occupied by valence electrons is sufficiently close to the band of energy levels available to conduction electrons that, by absorption of the energy of a photon, an electron can jump the gap between the valence band and the conduction band and become a carrier of electric current. The resultant vacancy ("hole") created in the residual electron-pair bond can be filled by an electron from a neighboring bond, which has the effect of a positive charge or "hole" migrating in a direction opposite to that of the electron flow.

In order for the conduction of electricity made possible by the creation of hole-electron pairs by ionizing radiation to provide useful work, three conditions must be met. First, the number of pairs of charge carriers created must exceed the number normally present at a given temperature. Second, the material must contain an internal inhomogeneity that keeps opposite charges separate until they are permitted to recombine by flowing through an external circuit. Third, the mean diffusion distance (lifetime) for holes and electrons before they recombine must be greater than the distance to the "collection point," i.e., the inhomogeneity.

The inhomogeneity is a region within the device separating materials that differ as to the type of charge carrier that predominates in the conduction of electricity: n-type semiconductors if the predominant carrier is negative (electrons) and p-type if it is positive (holes). The p or n characteristic can be imparted to a semiconductor by incorporating into the crystal lattice minute quantities of an atom with one less or one more electron than each matrix atom contributes to the electron-pair bonds with its neighbors.* In the case of silicon, for example, four valence electrons participate in these bonds. If it is "doped" with atoms having three valence electrons, such as boron or gallium, at any given lattice site the "dopant" becomes an acceptor of an electron in order to complete the bonds to its neighbors, thus providing a hole as a charge carrier and creating a p-type region. On the other hand,

*This is generally accomplished by diffusion from the surface at an elevated temperature or by ion implantation at low temperatures.

phosphorus, arsenic, or antimony, for example, with five valence electrons, will act as electron donors in silicon and form an n-type region.

The three types of p-n junction encountered in semiconductor devices are the metal-semiconductor barrier, the homojunction, and the heterojunction. An example of the former is the galena-"cat's whisker" arrangement used in early crystal-set radio receivers; the homojunction is a p-n junction within a semiconductor, and the heterojunction is a p-n junction between two different semiconductors.

Although they are rather easily fabricated, metal-semiconductor junction ("Schottky barrier") devices (Figure 13) have not, in the past, been particularly interesting as solar cells because their efficiencies are low, they do not utilize as much of the solar spectrum as other types, and they suffer from the inherent problem that any radiation impinging on the metal serves only to raise its temperature. Improved efficiency (10-15 percent), as the result of recent experimental work with these devices, has rekindled interest in them, however, because of their low cost.

Most solar-cell development has taken place with homojunction and heterojunction devices. Figure 14 illustrates some typical characteristics of solar cells.

Candidate Materials and Configurations

Literally dozens of materials, alone or in combination, possess the semiconductor properties required for high-efficiency (>.10) conversion of solar radiation to electricity. A number of these have been investigated as possible commercial solar-cell materials, and three of these—silicon, cadmium sulfide,

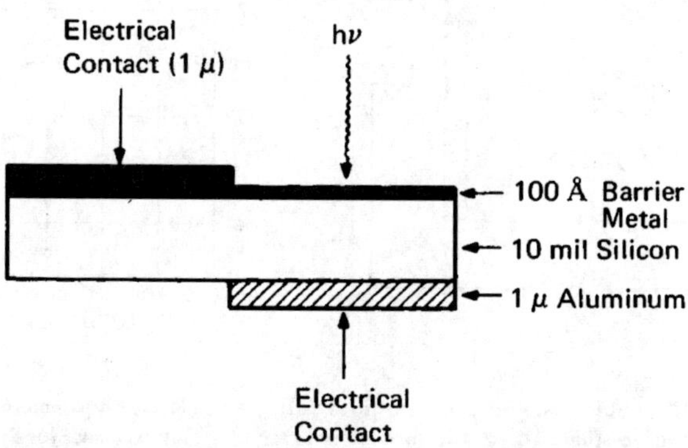

FIGURE 13 Structure of a Schottky barrier solar cell (not drawn to scale).

and gallium arsenide–have all been successfully used in spacecraft applications. Others are in experimental states of investigation. Still others, though known theoretically to be potentially interesting candidates, have yet to be thoroughly studied for these applications. Table 1 includes a brief partial summary of these and their status.

In addition to the possible materials and combinations, there are many possible configurations and processes for achieving them. Configurations

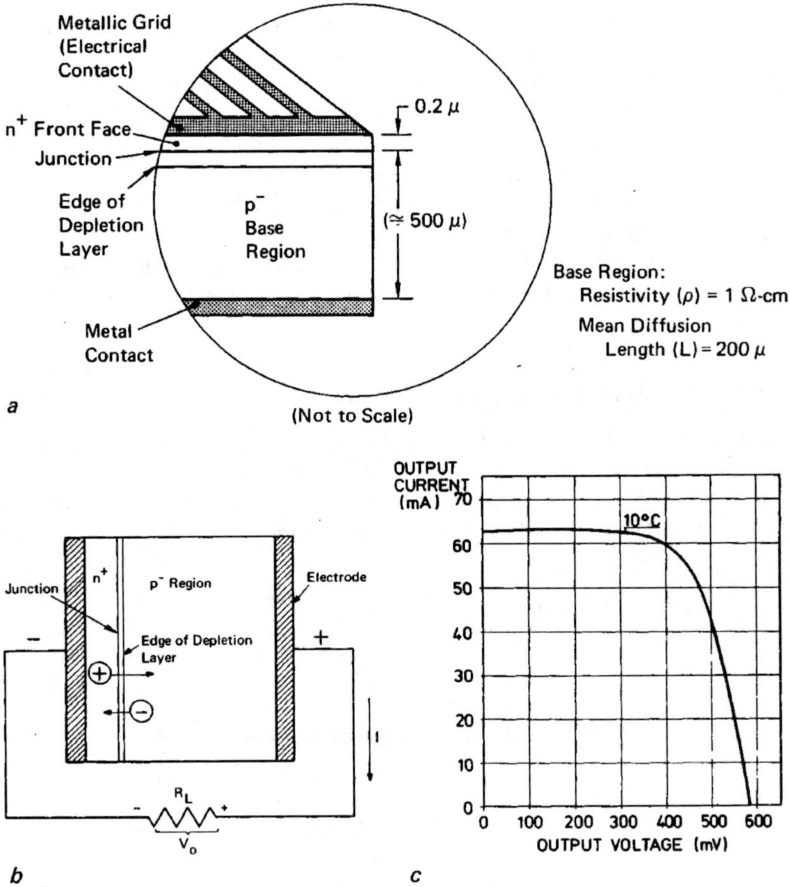

FIGURE 14 Typical characteristics of junction solar cells. *a*, Physical characteristics of a typical n^+/p solar cell. *b*, Illuminated solar cell with an external load. *c*, Typical solar cell I-V characteristics.

include the use of elemental material in thick and thin films (silicon and selenium), variations in junction design including the possibility of "vertical" junction cells to permit high-voltage operation, ternary compound materials such as GaAlAs for increased efficiency, and the use of graded bandgap materials also to increase the potential efficiency above that possible with constant bandgap materials. The various possibilities are discussed in detail in the current literature.[2]

Processes for forming the semiconductor junction include diffusion at high temperatures, evaporation to form a Schottky barrier layer on the surface of a semiconductor (such as silicon), and chemical epitaxial growth of layers (silicon, GaAs, or GaA1As), as well as ion implantation. Base materials can be formed through single-crystal growth by various methods including dendritic web growth, Czochralski growth, and a ribbon generating method called EFG (edge-defined film growth). (See "Current Technology" below.) Thin films can be formed by sputtering, evaporation, vapor deposition, and other techniques. Electrodes can be attached by evaporation, silk screening, and application of metal "lace."

These examples merely illustrate the many combinations of materials, cell designs, and fabrication processes possible. Although theoretical investigation indicates that over a dozen possible material combinations can yield high conversion efficiencies, and that certain fabrication processes (such as EFG and thin-film formation) can lead to economically interesting cells in principle, the search for a practical near-optimum combination of these will probably take the better part of a decade and perhaps $100 million or more in funding; indeed, development costs may be as much as $1 billion. However, if an adequate level of funding is sustained, and outstanding people from industry, universities, and other centers continue to be involved, the achievement of an economically interesting terrestrial solar-cell system seems inevitable.

PAST AND CURRENT USES OF TERRESTRIAL PHOTOVOLTAIC SYSTEMS

Since the invention of the silicon solar cell in 1955, there have been roughly a hundred different terrestrial applications of photovoltaic solar energy conversion systems, ranging from scientific experiments to operational applications by industry and government. Installations from a few watts to over a kilowatt peak power have been made in Africa, South America, Mexico, the United States, Canada, Europe, Japan, Southeast Asia, and the Middle East. These have provided power for lighthouse navigational and warning lights; radio, microwave, and television relay stations; aids to

TABLE 1. Some Candidate Materials for Terrestrial Solar Cell Fabrication

Material	Efficiency[(a)]	Status[(b)]	Type[(c)]	Commercial Availability[(d)]	
				0-5 years	5-10 years
AlSb	(e)	experimental	III-V	no	perhaps
InP	.05	experimental	III-V	no	perhaps
GaP	.03	experimental	III-V	no	perhaps
GaAs (Al)	.16	experimental	III-V	perhaps	perhaps
GaAs ($Ga_xAl_{1-x}As$)	≃0.25	experimental	III-V	no	perhaps
$CdS(Cu_2S)$	.05-.08	advanced[(f)] development	II-VI	perhaps	perhaps
CdTe	.05-.06	advanced development	II-VI	no	perhaps
SiC	.03	experimental	IV-IV	no	unlikely
Si	.15-.18	commercial	elemental	yes	
ZnSe	(e)	theoretical possibility	II-VI	no	perhaps
$CuInS_2$	(e)	theoretical possibility	$II\text{-}III\text{-}VI_2$	no	unlikely
$CuInSe_2$	(e)	theoretical possibility	$II\text{-}III\text{-}VI_2$	no	unlikely
$AlInS_2$	(e)	theoretical possibility	$III\text{-}III\text{-}VI_2$	no	unlikely
Zn_3P_2	(e)	theoretical possibility	$II_3\text{-}V_2$	no	perhaps
Cu_2O	(e)	theoretical possibility	$II_2\text{-}VI$	no	perhaps

[(a)]Efficiency of devices as measured under Air Mass Zero (AM0—see footnote, page 99) conditions of 1,400 W/m^2 incident solar radiation (i.e., with spectral distribution unmodified by atmospheric absorption or scattering—the characteristics of a space environment) and cell temperature approximately 23°C.

[(b)]"Experimental" refers to cells that have been fabricated in very small numbers under research laboratory conditions. "Advanced development" refers to devices that have been fabricated on a larger scale (many thousands of devices) under conditions that more or less simulate industrial production. "Commercial" refers to devices that are commercially available and technically suitable for use in rural communities.

[(c)]Roman numerals refer to the group in the periodic table to which each of the elements in the material belongs.

[(d)]The estimates of commercial availability reflect the opinions of a number of experts as reported in the recent open literature. As a result of the recent acceleration in funding for commercial development on such devices (mainly supported in the United States by the Energy Research and Development Administration [ERDA]), these estimates may prove to be conservative.

[(e)]Theoretical efficiencies of these devices are in the range of 10-20 percent under AM0 navigation on offshore oil platforms; weather-monitoring stations; remote educational-television sets; highway-emergency call boxes; aircraft-warning lights at airports; and remote communications stations for forest management. (Some examples are shown in Figures 15 and 16.) The present annual commercial market for photovoltaic arrays is perhaps 100 kWe (peak), and is divided among Japanese, American, and European (French, British, and West German) manufacturers (Appendix 3).

Experimental and Prototype Systems

The categories of application of photovoltaic systems for terrestrial use include scientific tests and demonstrations, quasi-commercial or prototype commercial applications, and fully commercial applications.

Experimental or demonstration uses of solar cells began in 1955 when Bell Laboratories and the Bell Telephone Company installed a solar-powered, rural-telephone carrier system in Americus, Georgia.[4] The system was operated for about 6 months—as a technical demonstration and publicity effort. In 1973, combined photovoltaic (CdS) and thermal collectors were integrated into a laboratory/house at the University of Delaware[5] to explore the nature of residential solar electric/thermal systems connected to a local electric-utility grid. At the California Institute of Technology, scientists from the geology department are using surplus spacecraft solar panels (from Ranger and Mariner spacecraft), suitably modified for protection against weather, to power remote scientific geological stations in California and Mexico.[6] And the Mitre Corporation (McLean, Virginia) is developing a 1-kW solar electric/hydrogen system to demonstrate the use of solar-generated electricity combined with electrolytic hydrogen as a secondary energy carrier.[7] All of these applications have been largely scientific in nature; they have not involved exploration of near-term markets for photovoltaic applications, although the work at the University of Delaware will eventually lead to an evaluation of combined photovoltaic/thermal solar collectors for building applications.

Other experimental systems have been installed in the Chilean desert as a joint University of Chile/RTC (la RTC Radiotechnique-Compelec, France) project, in Iran (at Pahlavi University in Shiraz), as well as in France, Africa, the Soviet Union, India, Japan, England, and West Germany.

---

conditions at 23°C. These materials have been suggested by Loferski[3] as the prime candidates for further research for high-efficiency photovoltaic conversion.

(f) Although cadmium sulfide cells are actually commercially available, their operation is so unreliable and so poorly understood that in terms of rural use they should be considered to be in a state of advanced development.

FIGURE 15 Water pump powered by silicon solar-cell array. Illustration shows one of four panels required to power the ¼-horsepower motor that operates the pump. The array size is 36 X 39 in. (91.4 X 99 cm) and it produces 4 amperes at 12 volts DC. [Courtesy Spectrolab, Inc.]

Quasi-commercial or commercial prototype systems are those in which the initial installation was made in order to determine the operating economics of the system and to make a comparison with other available energy systems. Such applications have generally been in situations where there has been a need for remote power in the 1- to 100-W range and where the cost of replacement of batteries, transportation of fuel, or remote power lines was prohibitive. Such applications include, for example, remote radio beacons; radio, television, and microwave booster and repeater stations; and warning and navigational lighting on offshore oil platforms. A number of installations made on a prototype basis have led to commercial installations, following successful operation of the prototype.

Examples include the first remote solar-cell application in Japan, which provides power for a 150-MHz VHF repeater station on Mt. Shinobu,[8] and the installations of Motorola solar-cell-powered telecommunications equipment by the California Department of Forestry in the late 1960s.[6] In both cases, commercial installation followed the economic and technical success of the initial installations.

CURRENT TECHNOLOGY (AVAILABLE WITHIN 5 YEARS)

For purposes of this report, commercial systems are those that are produced as a regular product line by a company, and commercial applications are those in which such products have been purchased by some organization because the solar option was the most economical on an annual cost basis. The current commercial market for terrestrial photovoltaic systems could be characterized as one in which some combination of high reliability, low or no maintenance, zero fuel requirements, and noiseless operation, at power levels below 1 kWe (peak), justify, on an economic basis, the use of photovoltaic systems.

These characteristics of photovoltaic systems are not necessarily advantages over conventional power systems if transportation on foot, or by horseback, Jeep, helicopter, etc., for fueling and maintenance purposes to remote locations is easily available, if batteries can be purchased and installed

FIGURE 16 Silicon solar cells used for communications. (Photovoltaic panel used by the U.S. Bureau of Land Management to power a VHF radio repeater atop Jack's Peak, Nevada. The array size is 36 × 24 in. (91.4 × 61 cm) and it produces 2.5 amperes at 12 volts DC.) [Courtesy Spectrolab, Inc.]

each year, power lines can be laid, or noise insulation installed where required. Each of these has some specific cost for a given application and a geographic location. Hence, there are situations where the characteristics of photovoltaic systems translate directly into economic advantages. (An economist would say that in these cases the market is operating normally.)

In 1961, Kobayashi[8] reported that it was more economical, on a life-cycle cost basis, to provide remote power at levels up to 50 W by a solar cell/storage battery combination costing $130/peak watt than to run a power line 1 km. Today, Spectrolab, Inc., has sold over 100 systems for remote power for navigational and warning lights on offshore oil rigs in the United States.[9] Several states in the United States, including California, Nevada, and Oregon, are purchasing photovoltaic power systems for remote radio repeater stations and other similar applications; sailboat owners are purchasing solar arrays to keep batteries charged during long voyages; and the French are providing solar television sets in Niger (see below) for educational purposes. Small solar-powered radios were marketed by Motorola in the early 1960s and, at the extreme end of the luxury market, a West German company has recently introduced a solar-power (trickle-charged battery) electronic cigarette lighter for several hundred dollars! As mentioned earlier, the total world market for diverse commercial and spacecraft applications is roughly 100 kWe (peak) per year and, at present prices for photovoltaic arrays, it is expected to grow to perhaps 3 times that within 3 years.[10]

Of the various types of photovoltaic devices that have been manufactured, only silicon solar cells have become an established commercial product (Table 1). CdS cells have been used in space applications by the United States and France; GaAs cells were used by the Soviet Union in near-sun deep space probes. Nevertheless, silicon cells are the only devices that can be made reliably in quantity, are stable with time, and whose operation is sufficiently well understood to enable properties to be predicted and electrical characteristics to be designed in.

Commercial systems discussed in this section are available from manufacturers in the United States, Japan, West Germany, France, and England. Manufacturers will provide either individual silicon solar-cell modules appropriately encapsulated in rugged supports or complete systems including batteries and power-conditioning equipment.

Most of the devices and systems currently available have been developed and used for specific applications where conventional energy sources are not available, but the need for modest amounts of power is critical in terms of commercial or military operations. Thus, they are too costly for applications in rural areas of developing countries except where a specific and critical need might occur. A summary of some current manufacturers of photovoltaic arrays and an example of typical product data available are listed in Appendix

3 for the benefit of those concerned with use of such systems. Only a few typical examples will be mentioned here; the manufacturers themselves are probably the best source of information regarding practical applications in various locales.

Solar-powered signaling from remote stations in the United States. In 1973, the Tidelands Signal Corporation of Houston, Texas, fabricated a complete aid-to-navigation, warning-light system, including silicon solar arrays fabricated by Solar Power Corporation (Massachusetts), and installed the system on an offshore oil platform on the Texas Gulf coast. The lighting system consists of 1 2-mi fog signal and 4 5-mi lamps. Energy consumption is about 25 amp-hr/day x 12 volts = 300 Whe/day. Previously, this lighting system was powered by 40 1.2-volt, 3,300 amp-hr primary batteries. The total weight was 2,000 lb and these were replaced annually. The solar generator system incorporates 80 photovoltaic modules (1.5 W peak under AM 1* illumination, 25°C) in a 4 x 5-ft (1.22 x 1.52-m) array. The information available implies a retail cost of roughly $20/peak watt and indicates that at such prices for a terrestrial array (sealed, ready to install), such arrays begin to compete with primary and secondary batteries in markets traditionally served by this hardware.[11] The report on this system claims that "solar cell/secondary battery systems clearly compete on a life-cycle cost basis, although the specific numerical details are not discussed."

The National Aeronautics and Space Administration (NASA) Lewis Research Center and the National Oceanic and Atmospheric Administration (NOAA) are cooperating on a project to design, fabricate, and install a number of solar power systems for remote atmospheric monitoring stations.[12,13] Two installations, one in Virginia (Sterling) and one in California (Mammoth Mountain), were built in 1973 and further installations are expected. As a precommercial application, solar arrays have been fabricated by NASA/Lewis using commercial solar cells in arrays made of modules containing 48 (6 x 8) circular silicon solar cells each and supplying 3 W (AM 1) per module. The arrays at each installation contain 20 modules, for a peak power of 60 W, and are encapsulated in fluorinated ethylene propylene (FEP) sheets.

Solar-powered educational television sets in Niger. This example is one of the very few documented applications of photovoltaics in a developing country.[13] Although similar programs of educational television for use in outlying rural schools have been or are being instituted in other developing

*The air mass ratio is the ratio of the distance light travels through the atmosphere to the length of the atmospheric path if the source were at the zenith. Thus, in this case, Air Mass One (AMI) implies that the sun is directly overhead.

countries (e.g., Ivory Coast, Brazil, India), none of these seems to have involved the use of solar cells as a power source.

Télévision Scolaire du Niger (TVSN) was created in 1966 as part of a program to upgrade the level of primary education in the country. This educational television system is intended principally for primary schools located in regions without electricity. Reception is assured through the use of television receivers especially designed to operate in climates of extreme heat or cold. These sets are transistorized, will receive at a wavelength of 61 cm (492 MHz), and are designed to operate on batteries supplying 35 W at 34 volts DC (±15 percent). With average battery life of 2,000 hours, it was calculated that one hour of television reception cost 1.38 francs per receiver.

In order to develop a more economical source of energy, the technical services of TVSN and the Office de l'Energie Solaire (Niamey) in 1968 installed an experimental solar panel to power the television set at a school near Niamey. The experiment demonstrated the practicality of providing solar-powered television reception in Niamey during the entire school year (October to June). After an applications study by the ORTF (Office National de la Radio-Télévision Française), six installations followed in 1972, and by 1973 some 800 students in 22 classes were receiving instruction via solar-powered television reception of broadcasts from the production center in Niamey.

As a consequence of the encouraging results of this experiment, the Government of Niger decided on the progressive establishment of a network of solar-powered television sets, with plans to reach 80 percent of the population with educational programs by 1985.[13]

Reliability (Lifetime and Maintenance)

Photovoltaic arrays have no moving parts, and the basic physical mechanism that accounts for the photovoltaic property has a lifetime measured in thousands of years for silicon. (That is, it is basically related to the rate at which impurity atoms, which form the p-n junction, diffuse through the crystal lattice, degrading the junction.) Techniques have been developed to encapsulate silicon solar cells in a clear silicone material that provides excellent shock insulation and protection from environmental effects. The closer such encapsulation approaches true hermetic sealing, the closer the lifetime of the device will approach that of the base material.

Such reliability may be critical if such systems, when they appear economically attractive, are to be rapidly and widely adopted. People are slow to put total reliance on innovations until a long period of testing and experience has gone by.[14]

In the case of suitably designed solar-cell arrays, the level of maintenance required to provide very high reliability (on the order of one failure per 10 years of operation) is probably both low and inexpensive. It principally involves protection of the transparent surfaces from extreme abrasion and periodic cleaning of the surface and perhaps the electrical connections. Modules could conceivably be developed for which maintenance would consist of only occasional cleaning. A number of photovoltaic systems have operated for close to a decade with no cleaning and with no observable degradation, despite such relatively dirty industrial atmospheres as the Cleveland Airport, for example.

Little labor is required for upkeep because of the low maintenance requirements of solar cells. Apart from skills and materials normally needed for battery maintenance, only a periodic cleaning of the solar array's surface with water—and perhaps soap and a cloth—is needed to maintain a solar-cell installation. Thus, there is no need for a supportive infrastructure for maintenance and repair, specialized training of maintenance personnel, or special tools, etc. (Compare this with the minimum tools, training, and access to spare parts required for the simplest internal-combustion engine/generator combination.)

Cost

The economics of the large-scale energy systems currently used in the industrialized nations (and to a considerably lesser extent in developing countries) differ substantially from the economics of small-scale energy systems that might be used in developing countries. In both sets of circumstances, however, the basic capital costs of various system alternatives must be established before any procedure to calculate final costs of energy delivered to the ultimate user can be employed. The costs of energy from a solar-energy conversion system include many factors beyond the capital cost of the solar conversion module. Capital or initial costs include, of course, the costs of the array modules, with support and orientation structures, plumbing (if forced cooling is used), and other elements such as batteries, inverters and other power-conditioning equipment, and hardware for local distribution of electricity. Additional capital investment costs include provisions for replacement parts, tools, materials for cleaning surfaces and inhibiting corrosion (where necessary), and possibly backup systems such as inexpensive internal-combustion engines plus generators, and occasional use of fuel. Other costs will of course include packaging and transporting the system elements to site, fees and tariffs for importation, and labor costs for assembly and operation of the system. Still additional costs include the development of a local infrastructure to handle replacements, training personnel to use the equipment, development and printing of instruction manuals, and possible additional costs associated with local institutional factors such as the need to monitor how much electricity each member of a settlement is drawing from the system. Other social costs might include payment to, or redirection of, people who make their living delivering kerosene or other fuels that are totally or partially displaced by the solar systems. A partial list of such cost elements is shown in Table 2.

Finally, as with any other proposed investment, the cost of capital and fixed charges will be an important factor in determining the cost of energy. In a photovoltaic system where the costs of the system operation depend primarily on the total capital investment in the delivered system, the interest rates applied to the loans will be particularly important, since the amortized cost of electricity will be almost linearly proportional to the interest rate.

The cost of a cell can be unambiguously expressed in terms of the cost per unit area of the finished device. The actual cost of energy produced in a working environment will depend on such factors as the efficiency of the cell as a function of temperature and of illumination intensity and wavelength, insolation patterns, and other environmental factors. Since the realistic applications of such cells will be in integrated modules, the final costs must be determined in terms of the performance of these modules and not of the cells alone. The current price for individual silicon solar cells is approximately $10,000 per peak kWe ($40,000-$60,000 per average kWe) and the cost of a completed array (with or without batteries and power conditioning, since these are relatively cheap) is $30,000-$70,000 per peak kWe ($120,000 and up for average power). Production of solar arrays or modules at prices more interesting in terms of rural use in developing countries (under $1,000 per average kWe) will depend on major cost reductions in the production of the semiconductor "blank" and its conversion to the finished cell, and in techniques for combining the cell and all of the necessary supports, contacts, connections, covers, etc., into the final module or array.

Of the various techniques currently under development, only one seems likely soon to result in cost reductions, in the cell blanks and finished cells, of sufficient magnitude to bring total costs below $1,000/kWe within 5 years. This is a technique, currently undergoing commercial development by Mobil Tyco Solar Energy Corporation (Waltham, Massachusetts), for the production of continuous silicon ribbon of high-enough quality (i.e., purity and freedom from lattice imperfections) to produce solar cells with conversion efficiencies in excess of 10 percent under standard conditions. The process is known as the EFG or Edge-defined Film-Growth technique. In this technique, a "seed" crystal of silicon is dipped in a bath of molten silicon and a film is pulled through a capillary die (Figure 17) to produce a ribbon. (A prototype array of cells fabricated from this material is shown in Figure 18.) Ribbons of 1-in. (2.54-cm) width with thicknesses down to .008 in. (0.2 mm) have been

TABLE 2. Cost Components for a Photovoltaic System

Capital Cost	Continuing Costs
Equipment	*Equipment*
Solar conversion modules including mechanical supports, heat transfer (active or passive), orientation mechanisms, concentrators, etc. Batteries	Replacement components for damaged system elements Replacement of batteries (3 to 5 years) and other elements due to corrosion and other forms of degradation, engines after 3 years
Power conditioning (inverters, voltage regulation, current stabilization, transformers, etc.)	Tools, manuals, etc., which are needed continuously and which break or wear out (or are stolen, sold, or otherwise made unavailable)
Local transmission and distribution components, including cables, plugs, and connections, switches and relays, etc.	
Transportation	*Maintenance and Operation*
Packaging for shipment	Labor for maintaining equipment, possible costs for night-time protection
Transport from sources to LDCs; (for those components not produced locally)	Labor for operating systems, including handling billings or other techniques for dividing up local support of the system
Internal transport	
Fees	*Capital Costs*
Import duties Taxes Hidden costs	Interest on capital borrowed to purchase systems
Local Support	*Local Taxes and Other Fees*
Spare parts Tools Manuals Training	Possibility of taxes or fees of various kinds imposed locally
Array Deployment	*Fuel*
Cost of land Labor and materials for deployment On-site structures for housing storage batteries, power-conditioning equipment, etc.	Fuel costs for backup system(s) that may be required to reduce risk of solar-system outages to acceptable levels

continuously pulled at rates of 1-1½ in. (2.5-3.7 cm) per minute, and with continuous lengths of up to 80 ft (24 m). It has been estimated that finished solar cells could be produced for the cost of $165/kWe (peak, AM1, 10 percent efficiency, .004 in. or 0.1 mm thick) or between $500 and $825 per kW average.[15] These estimates are based on anticipated reduction in loss of silicon during fabrication by a factor of 2, a further reduction—by a factor of 3—in the amount of silicon used (as a consequence of the reduced thickness of the wafer), and a factor of 3 reduction in the cost of the silicon itself as a consequence of the processing technique. In addition, the possibility exists of further significant reductions in cost by the use of solar collectors that would concentrate the light of a larger area onto the cell, thus effectively increasing the area of the solar cell. Figure 18 shows a prototype array of silicon-ribbon solar cells of the type being developed for commercial production.

It should be noted that at the current stage of development, because of the concentration of lattice dislocations in the ribbon, it is not yet possible to obtain these hoped-for efficiencies (10-20 percent) under standard conditions in solar cells made, by direct diffusion, with silicon ribbon. Such efficiencies

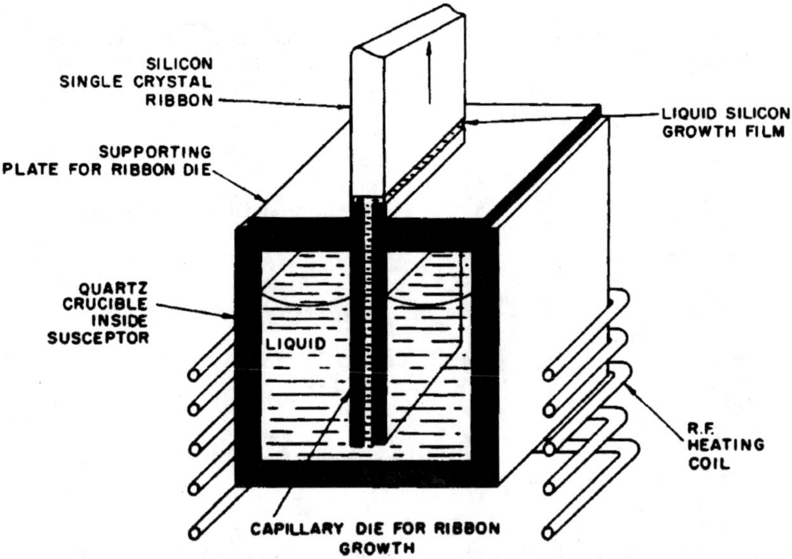

FIGURE 17 Schematic of solar-cell silicon ribbon growth (EFG—"Edge-defined Film Growth") from capillary die. [Courtesy Mobil Tyco Solar Energy Corporation]

FIGURE 18 Terrestrial photovoltaic array (prototype) fabricated from EFG ribbon silicon. [Courtesy Mobil Tyco Solar Energy Corporation]

have been obtained, however, when an epitaxial layer of silicon is grown on EFG ribbon and the junction is then formed in this layer.*

Other techniques for the development of low-cost manufacturing processes for production of silicon solar-cell arrays have been explored in considerable detail by Wolf.[17] Among those being examined, for example, is generation of polycrystalline silicon sheet by rolling or extrusion followed by sheet recrystallization by floating zone regrowth.

RESEARCH AND DEVELOPMENT

Although the development of silicon-cell solar arrays at prices under $1,000/kWe average was discussed in the previous section, it is really difficult

*In work recently reported, for example, a conversion efficiency of 10 percent was obtained with an EFG-ribbon cell on which a 30-μ layer of silicon had been epitaxially grown, while a conversion efficiency of only 6.2 percent was obtained with a cell made from ribbon alone.[16]

to predict whether such a price breakthrough will occur within the next 5 or 10 years. It seems likely that by 1985 such arrays will be available for about \$500/kWe average, but the likelihood of their availability any sooner depends on worldwide interest in, and commercial demand for, such systems.

Integration of solar cells and optical concentrators has been explored in the past. The potential advantage resulting from the reduced cell area needed for a given electrical output might mean a reduction in cell needs by a factor of 4-10; it seems, therefore, that work in this field might usefully be encouraged to compound the savings from new fabrication techniques. This would be the case especially in situations where cooling water is available to keep the silicon temperature from rising beyond the point where the cell efficiency is seriously impaired. (At 200°C the theoretical maximum efficiency is reduced to about 5 percent from a theoretical limit of about 22 percent at 25°C.)

In the case of materials other than silicon, there are some interesting candidates from which highly efficient solar cells might be commercially manufactured within 10 years (Table 1) but, again, whether a sufficient price breakthrough will occur to make them reasonable power sources for developing countries is hard to say. Gallium arsenide is particularly interesting in that, because of the nature of its bandgap, incident sunlight is absorbed at a depth of about 1-2 μ (compared to 100-200 μ for silicon), which makes it possible to use much less material per device than with silicon, thus perhaps compensating for its higher cost. Furthermore, because of its electronic properties, GaAs has a higher theoretical efficiency than silicon, its bandgap (1.35 eV) is closer to the optimum for sunlight (average photon energy $\simeq$1.5 eV) than is silicon's (1.1 eV), and it can operate efficiently at higher temperatures, thereby making the potential use of concentrators more promising.* Thus, the combination of factors—thinner wafers, smaller areas, and higher efficiencies (as high as 19.1 percent with a concentration factor of 1735)[18]—makes gallium arsenide a material that could compete with silicon as a source of efficient solar cells. However, its accessibility in terms of cost depends strongly on how much further research and development are stimulated by worldwide demand.

PHOTOVOLTAIC SYSTEMS IN DEVELOPING COUNTRIES

The attractiveness of a photovoltaic system to a developing country will depend on the economic significance of that application to those who have to

*Although there are many other direct bandgap materials, besides GaAs, that could be considered as candidates for similar reasons, more industrial effort has already been devoted to investigations of gallium arsenide than to any of the others.

pay for and maintain it. In some cases, this may be some agency of government, a donor agency such as the U.S. Agency for International Development (AID), or an international agency such as the United Nations Development Program (UNDP) or the International Bank for Reconstruction and Development (World Bank). In others, it will be the local inhabitants themselves. A detailed analysis of the value of various energy-related or energy-derived (specifically electrical energy) services in various cultural and geographic environments is required before a useful assessment of the potenial market for solar power systems can be made (unless the cost of these systems drops to the point where it is the cheapest alternative available for large-scale power generation). The nature and size of various markets in developing countries will depend on the delivered cost of the photovoltaic systems as well as on the value of electricity-derived services. Part of the required analysis would be an economic assement of the value associated with the following features of photovoltaic systems:

- High reliability;
- Low maintenance requirements;
- Zero fuel requirements;
- Intermittent output without storage, continuous output with storage; and
- Modularity (when one piece of the system goes out, the rest can continue to function, which is not the case with generators).

Reliability and maintenance have already been discussed. Photovoltaic systems share total independence of fuel requirements with all other alternative-energy technologies discussed in this report; the intermittent nature of their output—without storage—is another characteristic shared with many of these technologies. The modularity of photovoltaic systems, however, is unusual and worth special attention.

The modular nature of photovoltaic systems permits the users to gain experience with a relatively small investment. This is a crucial aspect of rapid diffusion of an innovation.[14] When large investments in innovations are required, they may never be adopted due to the lack of opportunity to test them at an acceptable level of financial risk. Systems can grow as the affluence of the local community grows, and system elements could be designed to permit the development of local "grids" as neighboring systems grow and eventually become contiguous. Loads can grow with supply, meaning essentially full amortization of the investment. Finally, a modular power system means that one or a few photovoltaic elements can fail and the system can continue to operate. Replacements can be obtained at the most convenient and least expensive time.

REFERENCES

1. Winston, Roland. 1974. Principles of solar concentrators of a novel design. *Solar Energy* 16:89-95.
2. See, for example:
   a. *Photovoltaics specialists conference: Conference records*. 1972-1975 eds. New York: Institute of Electrical and Electronic Engineers.
   b. *The Photovoltaic power and its applications in space and on earth*. 1973. Special volume of the *Proceedings. International Congress of The Sun in the Service of Mankind, 2-6 July 1973, UNESCO House, Paris*. Bretigny-Sur-Orge, France: Centre National d'Etudes Spatiales.
   c. *Proceedings of the Symposium on the Material Science Aspects of Thin Film Systems for Solar Energy Conversion, 20-22 May 1974, Tucson, Arizona*. U.S. National Science Foundation, RANN–Research Applied to National Needs, Document No. NSF-RA-N-74-124. Tucson: University of Arizona, Optical Science Center.
   d. *Workshop proceedings. Photovoltaic Conversion of Solar Energy for Terrestrial Application, 23-25 October 1973, Cherry Hill, New Jersey*. 2 vols. U.S. National Science Foundation, RANN–Research Applied to National Needs, Document Nos. Vol. I: NSF-RA-N-74-013A, Vol. II: NSF-RA-N-74-013B. Pasadena: California Institute of Technology, Jet Propulsion Laboratory.
3. Loferski, J. 1974. Some problems associated with semiconductor materials for large scale solar energy conversion. In *Proceedings of the Symposium on the Material Science Aspects of Thin Film Systems for Solar Energy Conversion, 20-22 May 1974, Tucson, Arizona*, p. 237. U.S. National Science Foundation, RANN–Research Applied to National Needs, Document No. NSF-RA-N-74-124. Tucson: University of Arizona, Optical Science Center.
4. Pearson, G. 1964. Applications of photovoltaic cells in communications. In *Proceedings of the United Nations Conference on New Sources of Energy: Solar Energy, Wind Power and Geothermal Energy, 21-31 August 1961, Rome, Italy*, Vol. 4:236-241. U.N. Document No. E/Conf.35/5. New York: United Nations.
5. Böer, K. W. 1973. The solar house and its portent. *Chemical Technology* (July): 394-400.
6. Costoque, E., and Vivian, H. 1969. *Solar energy rechargeable power systems*. Jet Propulsion Laboratory Document No. 650-681. Pasadena: California Institute of Technology, Jet Propulsion Laboratory.
7. Johnson, Augustus. 1975. Personal communication.
8. Kobayashi, M. 1964. Utilization of silicon solar batteries. In *Proceedings of the United Nations Conference on New Sources of Energy: Solar Energy, Wind Power and Geothermal Energy, 21-31 August 1961, Rome, Italy*, Vol. 4:229-235. U.N. Document No. E/Conf.35/5. New York: United Nations.
9. Rosenblatt, A. I. 1974. Energy crisis spurs development of photovoltaic power sources. *Electronics* 47(7):99-111.
10. Ralph, E., and Wolf, M. 1975. Personal communications.
11. Kelly, B. P.; Eckert, J. A.; and Berman, E. 1973. Investigation of photovoltaic applications. In *The photovoltaic power and its applications in space and on earth*, pp. 511-518. Special volume of the *Proceedings. International Congress of the Sun in the Services of Mankind, 2-6 July 1973, UNESCO House, Paris*. Bretigny-Sur-Orge, France: Centre National d'Etudes Spatiales.
12. Forestieri, A. F. 1973. Photovoltaic terrestrial applications. *Workshop Proceedings.*

Photovoltaic Conversion of Solar Energy for Terrestrial Application, 23-25 October 1973, Cherry Hill, New Jersey, Vol. II: 150-152. 2 vols. U.S. National Science Foundation, RANN–Research Applied to National Needs, Document No. NSF-RA-N-74-013 B(Vol. II). Pasadena: California Institute of Technology, Jet Propulsion Laboratory.

13. Polgar, S. 1973. Alimentation de téléviseurs par photopiles pour la télévision scolaire du Niger. Paper presented at the International Solar Energy Society Conference, 2-6 July 1973, Paris.
14. Rogers, Everett M., and Shoemaker, F. Floyd. 1971. *Communication of innovations–a cross cultural approach*. 2d ed. New York: Free Press.
15. Mlavsky, A. I. 1974. Ribbon growth of silicon. In *Proceedings of the Symposium on the Material Science Aspects of Thin Film Systems for Solar Energy Conversion, 20-22 May 1974, Tucson, Arizona*, pp. 71-83. U.S. National Science Foundation, RANN-Research Applied to National Needs, Document No. NSF-RA-N-74-124. Tucson: University of Arizona, Optical Science Center.
16. D'Aiello, R. V.; Robinson, P. H.; and Kressel, H. 1975. Epitaxial silicon solar cells. *Technical digest*. Abstract of papers presented at the Annual Meeting of International Electron Devices, 1-3 December 1975, Washington, D.C., pp. 335-338. New York: Institute of Electrical and Electronic Engineers.
17. Wolf, M. 1974. Methods for low cost manufacture of silicon solar arrays. Paper 74-WA/Ener-4. New York: American Society of Mechanical Engineers.

    ______. 1975. Cost goals for silicon solar arrays for large scale terrestrial applications–update 1974. *Energy Conversion*, Oxford, England 14:49-60.

    ______. n.d. Progress in new low cost processing methods. Philadelphia: University of Pennsylvania.
18. James, Lawrence W. 1975. III-V compound heterojunction solar cells. *Technical digest*. Abstract of papers presented at the Annual Meeting of International Electron Devices, 1-3 December 1975, Washington, D.C., pp. 87-93. New York: Institute of Electrical and Electronic Engineers.

PART II
INDIRECT USES OF SOLAR ENERGY

Wind Energy

Wind has been described as "the greatest terrestrial medium for harvesting, harnessing, and conserving solar energy."[1] The source of wind is, of course, in the atmospheric temperature differences generated by the sun, which, in turn, give rise to pressure differentials. The wind is a mechanism for dissipating, as kinetic energy, the potential energy accumulated in those pressure differences.

Wind energy has been used for thousands of years to propel boats and ships and to provide rotary windmill power to reduce the physical burdens of man. The windmill is an invention that dates from the earliest times of recorded history.[2] There is evidence that the ancient Egyptians used windmills as early as 3600 B.C. to pump water to irrigate their arid fields and to grind grain. The early Persians ground grain with a vertical-axis sailmill. Europeans imported the technology from the East and were probably the first to introduce the horizontal-axis mill around the 12th century. Almost all subsequent development has been with this latter type, primarily because of the low rotational speeds and greater efficiency of the horizontal-shaft mill. Up to the 19th century, the millwright's craft in Europe was esteemed, and mills were built that would work without fail for centuries, although their sails had to be replaced every 15-20 years.

Traditionally, rural populations used windmills principally for the following applications:

- To pump water for domestic use such as watering livestock and irrigating crops;

- To perform agricultural tasks such as grinding corn, crushing sugarcane, threshing, and wood cutting;
- For specialty applications such as moving saline water in salt works; and
- For the generation of electricity (in more recent times).

At the village level in many cultures, classical, traditional forms of windmills are a familiar and established technology. In recent years, there have been attempts by organizations in both industrialized and developing regions to meld wind energy and more advanced technologies into forms that can be used by rural communities. The work of several organizations in the United States and Canada is elaborated upon in this report, but the list is by no means exhaustive; these examples are cited only to illustrate the type of work that has been carried on in these fields.

The Asian experience with windmills dates to ancient times; only a few characteristic examples will be mentioned. The horizontal windmill seems to have been introduced into China during the Sung period (960-1280).[3] Eventually it evolved into something quite different from the Persian type, using the principle of the luffing sail adapted from the junk for its vanes. In the People's Republic of China, windmills are currently used for irrigation on small holdings, using windmill-driven, scoop-bucket systems. The windmills are always constructed of locally available materials—bamboo and wood—and are generally built by the user, although perhaps at one time they were produced by craftsmen for distribution to others. With minor modifications, the same device is used to grind beans and rice and for shelling crops, as well as for pumping.

In Thailand, simple windmills constructed locally, using wooden or bamboo poles with cloth sails, have been used for many years to move water; more recently, a two-bladed, wooden-propeller type of windmill has been introduced. Here, the water is lifted by scoop-buckets attached to an endless belt, or by "water ladder." The buckets scoop water from shallow wells and raise it to the top of the rotating wheel. Each bucket, in turn, turns over on its downward journey, discharging its water into a wooden chute. The water is then lifted up the "water ladder" to a point well above ground level and flows to channels dug in the ground to water crops. The "ladder" consists of a series of wooden flights fitted to an endless chain driven by mechanical linkage to the windmill. The flights are closely fitted to an inclined wooden trough and are able to "scrape" water up the trough, elevating it about 1 m (Figure 19).

The windmill as we know it in the West was derived from watermills—the mechanism is basically that of a watermill turned upside down. Unfortunately, no early designs have survived, for in the earliest days parts were set out full size on the workshop floor and made to templates, much as is done today. The shaft horsepower of these windmills was around 55 (41 kW) at a wind speed of 17.5 mph (28 km/hr).[4] It is interesting to note that with the four-blade type of construction commonly used, the ratio of tip speed to wind velocity was usually very low.

FIGURE 19 Windmill-driven water ladder—Thailand. [Courtesy W. Heronemus] *a,* Drive shaft and gear of a Thai water ladder in use in the salt works near Samut Songkhram. *b,* Thai water ladder driven by a two-bladed windmill.

In Europe, the earliest machines made to utilize the wind, dating back to the 12th and 13th centuries, were made in Holland and England. Constructed of wood, iron, and stone, these machines were made by unknown millwrights who, for the most part, designed the mills they built empirically, yet did excellent jobs of design and execution. Subsequently, well-known engineers, including John Smeaton, Sir William Cubitt, and Sir William Fairbairn, became interested in the practical use of windmills. These early windmills in England were used for grinding corn, whereas the Dutch machines were used for pumping to reclaim land. Later, in England, windmills for pumping water in the Fens—Norfolk and Suffolk—came into general use. Farmers have used similar windmills in Russia for hundreds of years; in more recent years the Soviet Union has been devoting much effort to developing large windmills to produce electricity.

Simple sail windmills have been used for centuries on many Greek islands such as Crete and Rhodes, where the primary use is for pumping water. (A modern version of this device is shown in Figure 20.)

FIGURE 20 Sail windmill. [Photo courtesy Brace Research Institute. Windmill and tower designed by Windworks]

Small windmills traditionally have been used in Portugal to irrigate small holdings. The windmills themselves, about 15 ft (4.9 m) in diameter, may be made of metal and are often mounted on very small towers. In the rural areas, a scoop-bucket is used for irrigating. In the cities, windmills are frequently fixed to the roofs of houses to raise water from a sump at ground level to storage on the roofs, when there is insufficient pressure from the community supply to raise the water to a convenient height. The pumps used for this purpose are simple, compact, diaphragm pumps.

In the United States, homemade windmills were a common sight, particularly in the semiarid Midwest, at the turn of the century.[5] They were generally built by people with limited economic means and little or no access to the technological and industrial infrastructure that was already well on the way to development in the United States by that time. Because they represent a technology indigenous to rural areas, depending on inexpensive, locally available materials, and are typical of the ingenuity and inventiveness of farmers the world over, it may be useful to devote some space to their description; the technology remains appropriate.

Among the windmills constructed by farmers for their own use, the following types were the most common:

Jumbo windmill. The jumbo windmill was a kind of paddlewheel operated by the wind in much the same way that the overshot waterwheel is driven by water. (See "Hydropower," p. 137) Its operation and construction are apparent from the illustration in Figure 21. Its efficiency varied from 15 to 40 percent, depending on the design, and a typical design delivered up to 1 hp (750 W), depending on the wind speed. Variations in design included the number of blades ("paddles"), adaptation to use with existing buildings to eliminate construction of the shielding box (Figure 22), and use of canvas sails to form the "screw jumbo" (Figure 23).

Merry-go-round windmill. Representing an attempt to increase the size—and therefore the power capabilities—of windmills constructed of wood, the merry-go-round windmill was a vertical-axis paddlewheel. It relied on a series of movable shutters surrounding the wheel that could be positioned to shield the blades from the wind on the "return" half of the wheel's rotation. When the shutters were closed, the mill resembled a closed circular cylinder. Figure 24 illustrates a version with a shutter able to accomodate itself automatically to shifting wind direction by virtue of the attached "rudder." The larger merry-go-round windmills were generally mounted on the ground; smaller models were mounted on simple wooden towers (Figure 25).

Battle-ax windmill. This mill, mounted on a wooden tower, consisted of four, six, eight, or more arms, each with a fan-like blade, fastened to a horizontal axis. When viewed edge-on while rotating, the assembly resembled

FIGURE 21 Model of a jumbo windmill with wind guard or cut-off. [Source: Barbour, Ref. 5]

a series of battle-axes slashing in opposite directions—hence the name. Again, the construction was inexpensive and simple, as illustrated in Figure 26. Figure 27 illustrates a battle-ax windmill arranged to drive a crosscut saw and equipped with a hand-operated brake to control the speed. Battle-ax windmills are rugged mechanisms capable of performing useful work with a minimum of maintenance. For example, the mill illustrated in Figure 28, built by a farmer in Nebraska, was capable of pumping 930 gal (3,520 l) per hour from a well 42 ft (13 m) deep, when driven by a 13.5-mph (21.7-km/hr) wind. This is an output of 125 W (0.17 hp) of useful work.

Holland windmill. As is obvious from the name, this is a version of the familiar Dutch windmill that, in its traditional form, consisted of four slatted fans covered with cloth and fastened to a horizontal axis mounted in a rotatable structure that allowed the direction of the axis to change with the wind direction. In the modifications that were developed in the American Midwest, however, the number of fans was increased to six or eight, the cloth sails were replaced by wood, and the mills were generally mounted on existing farm buildings.

In addition to the types described, several types of turbines were built having greater sophistication of design—for instance, variable-pitch blades—and using many materials that had to be purchased.

While these homemade mills were indeed common, the ubiquitous, commercially manufactured fan-mill was very popular for pumping water and generating electricity, and rapidly became a familiar sight throughout rural America. Thousands of these fan mills are still being used for pumping water on farms, primarily for watering livestock. Wind-powered electric generators such as the Jacobs Wind Electric Plant were made and sold by the thousands. This equipment obviously fulfilled a need in rural communities and is currently experiencing a revival in the United States.

CURRENT TECHNOLOGY (AVAILABLE WITHIN 5 YEARS)

Windmills as a source of mechanical power continue to be an appropriate technology for rural applications. For their construction and installation, they use locally available materials and labor and utilize local energy sources. The village craftsmen who built them and the local inhabitants who used them had the process entirely under control. When use of the windmill to generate

FIGURE 22 Gang of jumbo windmills erected above corncribs or sheds. [Photograph of model. Source: Barbour, Ref. 5]

FIGURE 23 Model of screw jumbo windmill, with canvas sails and arms braced with twisted wire. [Source: Barbour, Ref. 5]

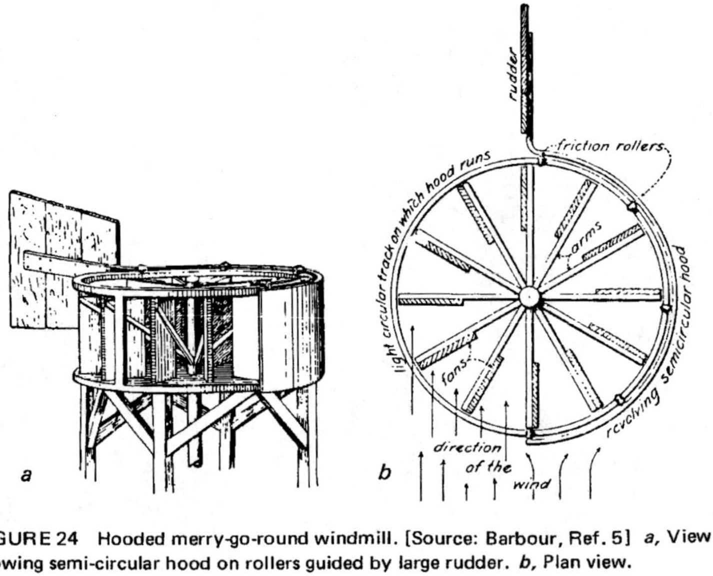

FIGURE 24 Hooded merry-go-round windmill. [Source: Barbour, Ref. 5] *a,* View showing semi-circular hood on rollers guided by large rudder. *b,* Plan view.

FIGURE 25 Small mounted merry-go-round windmill near Lincoln, Nebraska. [Source: Barbour, Ref. 5]

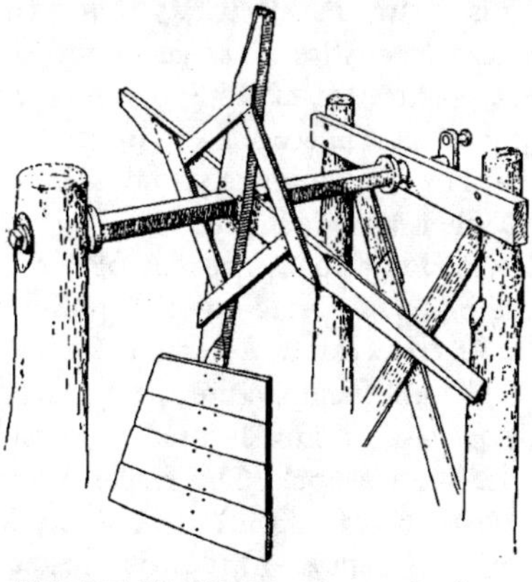

FIGURE 26 Working parts of a four-fan battle-ax windmill on the farm of Matthew Wilson, Overton, Nebraska. [Source: Barbour, Ref. 5]

FIGURE 27 Battle-ax windmill driving cross-cut saw, built by A. G. Tingley, Verdon, Nebraska. [Source: Barbour, Ref. 5]

FIGURE 28 Battle-ax windmill pumps. [Source: Barbour, Ref. 5]

electricity is contemplated, however, the technology required is somewhat more than can be assumed to be available in all rural areas. Not only is equipment required to generate, control, and distribute the electricity, but the necessary details of matching the operating characteristics of the generator to the properties of the windmill require some engineering expertise. Although these skills are certainly available in many localities, in some cases it may be useful to consider systems already available. For this reason, a survey of current technology in forms commercially available is included here.

Technology available in the United States. Prior to the advent of rural electrification, there were dozens of manufacturers of wind-powered pumps and electric generators in the United States. As far as can be determined, only two continue to make windmill pumps, and both manufacture the typical, multiblade, "American" fanwheel. These pumps are primarily used on open cattle ranges in the West to provide water for livestock. The Heller-Aller Company* makes systems using fanwheels from 6 ft (1.83 m) to 12 ft (3.66 m) in diameter, in increments of 2 ft (.61 m), and Dempster Industries makes 6- and 8-ft (1.83-m and 2.44-m) fanwheels (Figure 29). Both systems include

*Details on currently manufactured windmills are given in Appendix 5.

a reciprocating cylinder pump that will deliver up to several hundred gallons per hour, depending on the depth of the well, the diameter of the pump cylinder, and the wind velocity. The Heller-Aller machines range in price (in the United States) from about \$400 to almost \$1,200, exclusive of tower; Dempster's prices on a similiar basis for windmills of the same size range from about \$460 to \$2,000.

Until recently, only one company in the United States continued to make a wind-powered electrical generator. The "Wincharger," for many years a familiar sight in the countryside (Figure 30), is still manufactured by Dyna Technology, whose normal production is a 12-volt model designed for use with storage batteries. Typical cost for such a system, including the windmill, generator, governor assembly, electrical controls, handbrake, and 10-ft (3-m) tower, but without battery, is about \$450 in the United States. This system will produce 200 W at maximum wind velocity (23 mph) and thus represents an initial cost of about \$2,200/kW.

Recently, as a result of the increasing concern about fuel consumption, several other U.S. companies have begun offering windmill generator systems. Most of these companies seem to be distributing machines manufactured in Europe or Australia, though others are producing prototype machines, in relatively small quantities, based on variations in design of propellers and generators or alternators. Although in most cases prices of these systems make them questionable on economic grounds for village use in developing

a

b

FIGURE 29 "American" fanwheel windmills. *a*, "Baker" windmill. [Courtesy Heller-Aller Co.] *b*, Dempster windmill. [Courtesy Dempster Industries, Inc.]

countries—not to mention problems of replacement parts, maintenance, and repair—the possibility of cost reduction and manufacturing simplification that may result from increased demand makes listing them potentially useful. Appendix 6 lists as much information as we have been able to gather at this time.

One development likely to be available commercially within 5 years represents a departure in design of both the windmill and the electrical system. Moreover, the system is designed to permit construction of most components in LDCs. A multiple-blade fanwheel (Figure 31) has been developed, based on the principle of the bicycle wheel, in which the outer rim and the central hub are held in place by tension members, i.e., wire spokes. The blades, of airfoil design, are aluminum sheets wrapped around pairs of adjacent spokes, with their pitch determined by the placement of the ends of

FIGURE 30 "Wincharger"
[Courtesy Dyna Technology, Inc.]

FIGURE 31 Bicycle-wheel turbine. [Courtesy *Popular Science*, © 1974, Times-Mirror Magazines, Inc.]

the spokes on the rim. The result is a great saving in weight compared to conventional construction, and two companies are embarking on production of this fanwheel as part of electrical generating systems. The American Wind Turbine Co.* will be producing a mill coupled to a rim drive instead of a gear box; by this arrangement efficiencies greater than 40 percent have been obtained. This system can supply either DC or AC of arbitrary, constant frequency, independent of the wind velocity or the shaft speed of the generator. The result is achieved by modulating the field excitation of a small high-frequency (i.e., high-speed) alternator and converting the modulated

*Details on windmills under development are listed in Appendix 6.

output to DC or AC with simple solid-state circuitry. The operation of the system is analogous to "detection" of conventional audio-frequency amplitude modulation of a radio-frequency carrier. In this case, the carrier varies in frequency as the alternator shaft speed varies, but so long as the wind speed remains above a certain minimum, the filtering circuit will remove the high-frequency components and provide an output dependent only on modulation frequency.[6] Use of a high-speed alternator provides the added advantage of greater power output for a given size and becomes practical with the rim-drive arrangement that the new fanwheel design makes possible. Cost figures for this system are not yet available.

A more conventional use of this turbine is being made by American Energy Alternatives, Inc., in a prototype DC system with models capable of supplying 1,500-5,000 W. Costs per kilowatt (in the United States) are expected to vary from about $1,500 for the 1.5-kW model to about $600 for the largest (5-kW) system.

Another system in the prototype stage, planned for production by Independent Power Developers, has a three-blade propeller windmill with DC output at voltages from 24 to 240 volts and rated at 15-18 kW. Cost estimate of the complete system, less tower, is about $8,500, or between $500 and $600/kW.

Technology available in Europe. Windmill-operated generators and pumps

FIGURE 32 Aerowatt generator. [Courtesy Pennwalt/Automatic Power]

FIGURE 33 Elektro generator. [Courtesy Elektro GmbH]

are currently available in Europe from several manufacturers. The French firm, Aerowatt, produces five models of generators (Figure 32), ranging from 24 W to 4.1 kW in capacity, at a wind speed of 16 mph (26 km/hr). Their equipment, while capable of providing maximum output at reasonable wind speeds, is nonetheless quite expensive, ranging from $2,000 for the smallest to $20,000 for the largest (in the United States). The Italian firm of Domenico Sperandio is reported to be manufacturing seven models of wind-powered generators, ranging from 100 to 1,000 W, but no further details were available at the time of writing. Wind-driven alternators in single- and three-phase models are manufactured by Elektro GmbH in Switzerland, in sizes ranging from 50 W to 5 kW (Figure 33). At costs ranging from $1,020 to $3,840 (in the United States) for the basic system, they seem less expensive than the Aerowatt machines, but still represent a capital investment beyond the means of most villages. Finally, the West German firm Maschinenfabrik Ludwig Bening manufactures three models of propeller-type windmills designed to operate reciprocating pumps. Three-, four-, and five-bladed mills can be connected to a variety of pumps to supply water in amounts ranging from 686 gal (2,600 l) per day from 52.5 ft (16 m) at a wind speed of 6.7

mph (10.5 km/hr), to 3,800 gal (14,500 l) per day from 23 ft (7 m) at a wind speed of 17.9 mph (54.5 km/hr). (On a 24-hour basis, this is the equivalent power output range of about 5-12 W.) Prices average about $2,530 for the complete assembly, including pump and guyed tower. This firm also makes a wind-powered generator supplying 24-400 W, depending on the wind speed, at 24 volts DC, with costs ranging from about $200 to $3,300 (at German port of export) depending on the tower.

Technology available in Australia. Wind-driven generators manufactured by the Dunlite Electrical Co. are of two sizes, 1,000 W and 2,000 W (Figure 34). The smaller model uses a 12-volt DC generator and the larger a 115-volt DC generator; both models include an inverter to supply 115 volts AC (60 Hz). Without tower and batteries, prices are $1,729 and $2,805 (in the United States) respectively.

Technology available in Latin America. In Argentina, the Fabrica de Implementos Agricolas manufactures a fanwheel-powered reciprocating cylinder pump of a type formerly made in the United States (Figure 35). This company is, in fact, the licensee of the U.S. firm (Aermoter) that still distributes the finished package but no longer manufactures either the windmill or the pump. The fanwheels vary in size from 6 ft (1.83 m) to 16 ft (4.88 m) in diameter, at prices in the United States ranging from $465 to $4,235 (excluding the cost of the tower). Pumping capacities are similar to those of the Dempster windmills.

FIGURE 34 Dunlite generator. [Courtesy Dunlite Electrical Co. Pty. Ltd.]

FIGURE 35 Aermotor windmill pump. [Courtesy Aermotor/Braden Industries, Inc.]

Technology available in South Africa. P. Andrag and Sons., Ltd., are the South African licensee of Aermotor and manufacture the Aermotor 8- and 10-ft (2.5-m and 3-m) fanwheel windmill heads for use with water pumps. These are sold in South Africa for $620 and $950 respectively, for the windmill head, gear box, and 4-post stub tower.

Reliability and Cost

Technology provides only part of the answer to the question of how reliably power can be extracted from the wind, and at what cost. In the first place, not all of the power in a wind stream can be extracted by a windmill. The power in a stream of moving air is proportional to the product of the cross-sectional area of that stream and the cube of its velocity, and it can be shown that the maximum fraction of that power that can be extracted by a windmill is 59.3 percent. As a practical matter, horizontal-axis wind devices achieve less than 70 percent of this theoretical maximum; that is, they are able to extract somewhat less than 42 percent of the power in the wind, and vertical-axis machines are significantly less efficient.

Second, although the wind is a free and inexhaustible source of energy, it does not always blow, and when it does, the velocity may be too low to allow the windmill to begin producing power, or too high for the mechanical design that is practical for average wind speeds. In a review prepared for the Food and Agriculture Organization (FAO) Golding described the problem as follows: "At any specified point on the earth's surface the wind, at a given instant of time, can have a speed ranging from zero to as much as 125 miles per hour (56 m/sec) or more. Even the windiest places have their calm spells while the least windy are occasionally subjected to storms which bring high winds. One cannot, therefore, rely absolutely on power from the wind at a given moment however windy the chosen site may be and, on the other hand, one cannot safely reduce the mechanical strength of a windmill, destined for a relatively calm area, to any great degree. Nevertheless, the probability of there being power available at a particular time is much greater at the windy site [than] is the probability of a machine, erected there, having to withstand severe mechanical stresses fairly frequently."[7]

Finally, the cost of a windmill, whether operating a pump, a generator, or used to grind grain or perform other mechanical tasks, can be evaluated either in terms of the capital cost of the components or—usually more realistically—in terms of the amount of useful power produced. This evaluation depends strongly on the average wind speed over the year and, of course, on the minimum wind speed at which the device starts producing useful power. Generally, the minimum speed is about 6 mph (2.4-3 m/sec). Manufacturers usually list outputs at the "rated" wind speed, which is generally about 15 mph (6.5-7 m/sec) although some large generators may be rated at higher speeds. The amount of useful power produced also depends on the ability of the user to use it when it is produced. In most parts of the world the wind blows at random times, and if the energy cannot be used when the wind provides it and cannot be stored against the time that it is actually needed—as electrical energy in storage batteries or as water pumped to an elevated reservoir, for example—the usefulness of even an elegant wind device is severely compromised. Therefore, one of the essential first steps is to obtain information on average wind speeds at the site contemplated, including data on the length of time during the year that the wind blows with a speed greater than a given minimum. Velocity-duration curves drawn from these data are useful in calculating the amount of power actually in the wind at a given site and, if necessary, the annual energy output. Such wind data may have to be measured at the proposed site, but more often than not, sufficient information to judge the reliability of the wind may be obtained either from the national meteorological service or the World Meteorological Organization. Details on the calculations may be obtained from standard references such as Golding (References 4 and 7).

Mechanically, the windmills discussed in this chapter are far superior to fossil-fuel-powered engines in terms of reliability. Apart from their independence of fuel supply, they require minimum maintenance (usually an annual inspection and lubrication suffice) and seldom require repair. Some designs are self-regulating and even under the most adverse wind conditions require no care or attention.

The cost of a wind system is closely linked to the type and height of the tower on which the machine is mounted. A large part of that cost is the cost of the tower, which is dependent on its design and the materials of construction. Because tower requirements—and hence permissible designs and materials—are so much a function of local circumstances, the cost of such a system cannot be comprehensively discussed here; it must be evaluated for each situation, based on requirements specified by the manufacturers of the various wind machines.

LOCAL IMPLEMENTATION

The use of local technology in design, construction, and installation of windmills depends on the knowledge, skills, and materials available to the local community. Aside from the simpler—and less efficient—home-built machines described earlier, and locally devised improvements in assembly of purchased systems, other contributions to reducing the cost of wind systems could certainly be made.

Detailed technical information on a variety of home-built windmills is available from several sources in the United States and Canada. Some of these plans were specifically developed for local construction in developing countries. Others, although not designed with that specific purpose in mind, might be adapted to such a situation in certain cases.

Information in the first category is available principally from two sources: Volunteers in Technical Assistance (VITA) in the United States, and the Brace Research Institute in Canada.* VITA will supply plans for several types

*VITA, 3706 Rhode Island Avenue, Mt. Rainier, Maryland 20822, USA; Brace Research Institute, Macdonald College, McGill University, Ste. Anne de Bellevue, Quebec, Canada.

of windmills; for example, one built from scrap automobile parts including a rear axle, voltage regulator, and generator or alternator. The mill itself is a modified fanwheel, the blades of which feather automatically as wind speed increases, and is designed for local fabrication. With several options, the entire unit will supply from 500 W to 3 kW; the cost of materials will vary with the value and availability of scrap automobile parts on the local market. All labor for fabrication, assembly, and installation can be supplied locally.

Brace Research Institute has a number of publications available that are aimed at windmill construction in developing countries. They will supply details for a 10-hp (7.5-kW) system based on a 32-ft (9.8-m) diameter, three-blade rotor connected to a truck rear axle. The entire assembly, including the tower, is designed for local construction, using as many local resources as possible. Brace will also supply details for the construction of a 25-ft (7.6-m) diameter canvas sailmill, a 42-ft (12.8-m) tower built of timber, and plans for the Savonius rotor windmill, a vertical-axis device with an S-shaped cross section developed in the 1920s by S. J. Savonius, a Finnish

FIGURE 36 Savonius rotor windmill, Sri Lanka. [Courtesy Brace Research Institute]

engineer. The Savonius rotor has proved to be a popular design for local construction, since it can be built from an oil drum that is cut in half lengthwise* (Figure 36).

Information in the second category—that is, developed for home construction in the United States—is becoming available from a number of sources, primarily as a result of a growing concern about consumption of fossil fuels; detailed plans for some of these designs are available.** The panel has not examined these plans to evaluate them for their appropriateness to materials easily available in developing countries; they are mentioned because they may be a source of ideas for the interested technologist.

It should be emphasized that the materials and skills needed for tower construction would be available in most rural communities. In rural United States, for example, it was long assumed that even though a farmer might purchase a wind machine, he would probably build his own tower. Manufacturers' catalogs often included detailed plans for tower construction, examples of which are given in Appendix 7. It should be borne in mind, however, that production of wind machines requires access to suppliers of many parts—control equipment, alternators, generators, and a host of incidentals such as bearings and brushes—not manufactured in many developing countries, but needed to make a finished product.

RESEARCH AND DEVELOPMENT

With the renewed interest in this ancient technology, ideas that have not been exploited, such as the vertical-axis Darrieus rotor† (Figure 37), are being revived. In addition, many new ideas are being generated, along with prototypes of machines that represent attempts to put these ideas into practice.[8]

*An interesting use of the Savonius rotor is the centrifugal windmill pump designed by Professor John D. Isaacs (Scripps Institution of Oceanography), which uses as a pump a rotating vertical pipe with a horizontal arm.

**For example: (1) Plans for a 12- or 16-ft, 3-blade downwind windmill, coupled to an automotive alternator, are available from Helion, Box 4301, Sylmar, California 91342, USA; (2) plans for a three-bladed variable-pitch windmill coupled to an automobile generator, battery, and inverter are available from Sencenbaugh Wind Electric, P.O. Box 11174, Palo Alto, California 94306, USA.

†Invented by J.G.J.M. Darrieus and originally patented by him in 1927, the catenary rotor was subsequently patented in the United States in 1931 (US 1,935,018, Dec.8).

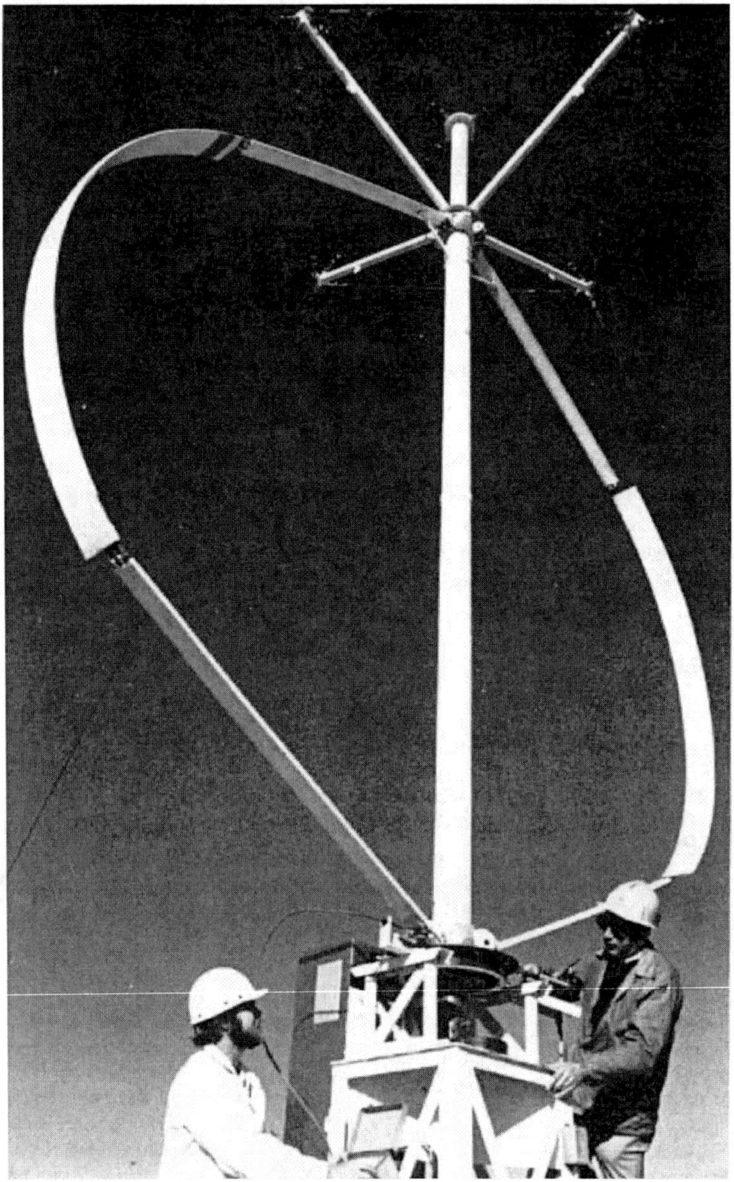

FIGURE 37 Vertical-axis wind turbine (Darrieus rotor). [Courtesy of Sandia Laboratories/ERDA]

Although a new organization has recently been formed* to coordinate some of these activities, it is primarily a trade organization concerned with manufacture in the United States. Because of the significant potential usefulness of this technology to rural areas in developing countries, a central research and development organization should be formed. With multilateral support, an international staff, and working through regional institutes, this organization would keep abreast of new theories and developments and test prototype machines developed anywhere in the world that show promise for use in developing countries. The staff of such an organization, supplemented by expert consultants, should be able to produce a coordinated and rational program of wind-power exploitation based not only on assessment of new developments, but also on their own original research and development. (At present, there is little coordination of most relevant material in the entire field of alternative energy sources.)

An excellent example of the kind of work such an international organization could do for developing countries is provided by a study of wind power in Thailand recently produced for the U.S. Agency for International Development.[9] The result of an on-the-spot survey by an expert in wind technology, this report details two proposed windmill designs that fit the needs of the Thai farmer and represent a significant increase in agricultural production (through improved irrigation). The proposed fanwheel systems, 6 and 12 m (20 and 40 ft) in diameter, are designed to fit existing indigenous manufacturing capabilities. They are also essentially independent of foreign imports for materials or parts—including the pump—and could be sold at a cost well within the reach of a typical Thai farm family.

Maintaining and encouraging development of the ancient art of the wind machine can result in some revolutionary ideas for supplying the energy needs of rural areas. Parallel with new design concepts for rotors, design of towers that can be built economically must form part of overall wind-system design; otherwise, on the basis of cost, rotor-design improvements may be nullified. A central organization with a competent, development-oriented staff working through regional research and development institutes, entrusted with the capacity to develop new equipment suited to local needs, seems a necessary first step.

REFERENCES

1. Fuller, R. Buckminster. 1974. Energy through wind power. *The New York Times* January 17.
2. Lanoy, H. 1944. *Les aeromoteurs modernes*. Paris: Girardot & Co.

*American Wind Energy Association, 21243 Grand River, Detroit, Michigan 48219, USA.

3. Needham, J. 1965. "Power-sources and their employment (III), wind force; the windmill in east and west." In *Science and Civilization in China*. Vol 4: *Physics and physical technology, part II: Mechanical Engineering*, pp. 55-568. London: Cambridge University Press.
4. Golding, E. W. 1956. *Generation of electricity by windpower*. New York: Philosophical Library.
5. Barbour, E.H. 1899. *Wells and windmills in Nebraska.* Water-supply paper, No. 29. Washington, D.C.: U.S. Geological Survey.

   An excellent review of U.S. windmill technology at the turn of the century—unfortunately out of print. Much of the material on farm windmills in the United States was drawn from this source. See also Davidson, J. B., and Chase, L. W. 1908. *Farm machinery and farm motors*. New York: Orange Judd Co.
6. Allison, H. J.; Ramakumar, R.; and Hughes, W. L. 1972. A field modulated frequency down conversion power system. Paper presented at the 1972 meeting, IEEE (Institute of Electrical and Electronic Engineers). *Transaction on Industry Applications* IA9(2):220-226.
7. Golding, E. W. n.d. *Windmills for water lifting and the generation of electricity on the farm.* Agricultural Engineering Branch, Land and Water Development Division, Informal Working Bulletin, No. 17. Rome: Food and Agriculture Organization.
8. One of the more interesting approaches currently under development is the vortex augmentor concept. In this approach a delta-type surface is interposed between the windmill and the wind, with an angle of attack between 10° and 15°. In the vortices created as a "tail wind" the total wind velocity can be increased by as much as a factor of two. This effect can be used advantageously in three ways: (a) to increase the power output of a given windmill at a given velocity (the power varies as the cube of the wind velocity); (b) to decrease the size of the rotor needed to provide a given power at a given wind velocity; or (c) to utilize wind at velocities lower than those at which standard wind-operated generators normally begin supplying energy. Sforza, Pasquale M. 1975. Vortex augmentor concepts for wind energy conversion. In *Proceedings of the Second Workshop on Wind Energy Conversion Systems, 9-11 June 1975, Washington, D.C.*, Frank R. Eldridge, ed., pp. 433-442. U.S. National Science Foundation, RANN-Research Applied to National Needs, Document No. NSF-RA-N-75-050. McLean, Virginia: The Mitre Corporation.
9. Heronemeus, W. E. 1974. *A survey of the possible uses of windpower in Thailand and the Philippines*. Prepared for the U.S. Agency for International Development, Office of Science and Technology. Amherst, Massachusetts: University of Massachusetts, Department of Civil Engineering.
10. Merriam, Marshall F. 1972. Windmills for less developed countries. *Technos* (April-June) 9-23.

    Also available as: *Is there a place for the windmill in the less developed countries*? Technology and Development Institute Working Paper Series, No. 20. Honolulu: East-West Center.

    This paper contains a good general discussion of windmills and a useful economic analysis.

Hydropower

Although hydropower can be used to perform mechanical work directly, or indirectly through hydroelectricity, the discussion in this section focuses primarily on hydroelectric applications–reflecting major use worldwide. Direct mechanical applications are included in the discussion where appropriate, and are treated separately.

Hydropower has as its energy source the potential energy of a supply of water at some superior elevation. This energy is converted within a turbine into mechanical energy that, in turn, is converted into electrical energy within a rotational electric generator.

The power theoretically available from a waterpower site is equal to the weight of water flowing in a given time multiplied by the drop in elevation (the "head") of the water. This may be considered as either horsepower or kilowatts, depending on the mathematical constants employed.

The actual power obtained is of course less than the theoretical power. It is calculated by multiplying the theoretical power by various factors that represent the efficiency of each element of the hydroelectric system. These elements include:

- The apparatus conveying water to the turbine (canal, penstock, fittings, and valves);
- The turbine and its discharge piping (draft tube);
- The electrical generator; and
- Any indirect connection between the turbine and generator (gears or a belt).

The overall effect of these efficiencies is to reduce the theoretical power by a factor of 0.6-0.8.

These energy-conversion processes and the means of carrying them out represent a mature technology with established paradigms within which no major surprises have occurred for over 50 years.

At the beginning of this century, there were several hundred manufacturers of small water turbines in the United States. Today, only one remains in business offering a full line of small turbines "off the shelf." This situation may be attributed to the fact that small-scale water power has been superseded by 1) the internal-combustion engine, 2) the steam engine, and 3) the extension of the distribution systems of large central power plants into rural areas.

In several European countries and China, however, the manufacture of low-head, low-power hydroelectric sets appears to be a vigorous, though limited, activity. Indeed, in the book *Water Power Development* by Mosonyi (1960) there is a chapter devoted to midget water power stations in which the power output is less than 100 kW.[1]

HYDRAULIC ROTATING PRIME MOVERS

In their historical or evolutionary sequence the rotational hydraulic prime movers are:

- Waterwheel (overshot, undershot, and breast);
- Tub wheel/flutter wheel/roue à cuve; and
- Turbines.

Waterwheels generally rotate in a vertical plane and are distinguishable from turbines by the fact that no changes in water pressure occur across the vanes of the undershot wheel or the buckets of the overshot and breast wheels.

The *overshot wheel* (Figure 38) has "buckets" at its periphery that are filled with water at the top of their rotation and empty as they approach their lowest position. Although efficiencies of 80 percent were obtained with these wheels, their speed was slow, they generated little power for their size, and they were incapable of operating with rising tailwater. Therefore, along with the less-efficient breast and undershot wheels, they became obsolete. There appears to be only one commercially operating overshot waterwheel in New England. It is used to grind grain, make cider, and generate electricity, and was manufactured by the now-defunct Fitz Water Wheel Company of Hanover, Pennsylvania. It is 13 ft (3.96 m) in diameter, 6 ft (1.83 m) wide, and is said to be capable of developing 30 hp (22.4 kW) at 8 rpm.

The *breast wheel* operates with "buckets" as does the overshot wheel, except that it receives water at about radial height. Its efficiency is around 60 percent.

The *undershot wheel* (Figure 39) is sometimes called a current wheel, since it receives its energy from the impact of a flowing stream of water on its flat or curved radial vanes. Its maximum efficiency is about 20 percent, but this level of efficiency requires that the wheels be built with sides or shrouds to minimize the effects of turbulence, or that the wheel receive water via a channel slightly wider than the wheel itself (a Roman improvement dating to about the third century A.D.).

The *tub wheel* or *roue à cuve* is a variant of the undershot wheel in which the wheel operates in a horizontal plane. An inclined open or closed channel conveys water to it from an elevation of up to about 10 ft (3 m) so as to impinge at right angles on inclined straight blades. It was widely used in mountainous regions of Europe (Scandinavian, Alpine, and Balkan areas) for the grinding of grain. In the Shetland Islands, tub-wheel-powered grist mills were a communal facility. They were often made entirely of wood; their efficiency was about 20 percent.

A tub wheel was in operation in Cilleyville, New Hampshire, from about 1910 to 1960 and was used to grind grain for chicken feed. Its wheel, 7 ft

FIGURE 38 Overshot wheel. [From: "Water Power on the Farm," Bulletin No. 60, Fitz Water Wheel Co., Hanover, Pennsylvania, 1923]

FIGURE 39 Undershot wheel. [From: "Water Power on the Farm," Bulletin No. 60, Fitz Water Wheel Co., Hanover, Pennsylvania, 1923]

(2.14 m) in diameter, rotates on a 5-in. (12.7-cm) steel shaft and receives water under a head of 6 ft (1.83 m) through a 2-ft (0.61-m) closed, square, wooden channel or penstock bound with iron straps. A modern version of the tub wheel is manufactured by the Ganz Company of Budapest.[1]

The *water turbine* differs from the waterwheel in that there is a contrived steady flow of water through passages created by the turbine blades or buckets. The blades are so designed that water enters the passages smoothly and leaves with minimum velocity and energy, having transmitted most of its available kinetic energy to the rotor.

There are three well-established classes of turbines generally classified according to the names of their developers and promoters. They are, in chronological order of their appearance:

1. Francis (reaction, full admission)
2. Pelton (impulse, partial admission)
3. Propeller (reaction, full admission)
   a. Nagler (fixed propeller blades)
   b. Kaplan (variable pitch propeller)

Each class dominates a range of heads and speeds, and each has a unique form of runner or rotor.

Essentially, the Francis, Nagler, and Kaplan turbines are linear descendants of the simple reaction turbine of Hero of Alexandria, dating from about 100 B.C. Their development represents the efforts of inventors to create more power from a given space.

The Pelton turbine may be seen as the direct descendant of the undershot waterwheel or the tub wheel (Figure 40).

The *Francis turbine* has water entering the rotor around its complete circumference (Figure 41) and is therefore called a full-admission turbine. It is also sometimes called a "pressure" turbine, which highlights the fact that there is a pressure change across the rotor of the turbine; the pressure/potential energy is converted to kinetic (i.e., mechanical) energy across the runner/blade/rotor. The direction of the water at the entrance to the turbine rotor is tangential (at the circumference of the runner) and at discharge it is parallel to the axis of rotation; this complex directional flow is the reason the Francis turbine is often called a "mixed flow" turbine. Despite its complex runner/rotor structure, its development was an empirical response to the need in rural areas of the 19th-century United States for higher speeds, greater power, and greater economy.

The *Nagler fixed-blade propeller turbine* is a full-admission reaction turbine with a direction of flow described as "axial," that is, parallel to the axis of rotation (Figure 42). It resembles a ship's propeller and indeed may be considered as the inversion of it; its rotor is therefore a much simpler structure than that of the Francis turbine. The low-head, high-speed characteristics of the Nagler turbine often make it possible to connect it directly to electrical generators. The Russian and American hydroelectric units mentioned later as available for developing countries use propeller turbines directly connected to electric generators.

The *Kaplan turbine* has propeller blades whose pitch can be varied during operation to keep the speed of the turbine constant, despite variations in load, so that the turbine will operate at maximum efficiency.

The *Pelton turbine/wheel* is often called an "impulse" turbine, though the principle of "reaction" plays an equally important role in its energy transformations. The Germans often designate it as a "pressureless" turbine, meaning that there is no pressure change across the runner. The pressure/potential energy of the water is converted completely to kinetic energy in a nozzle. The high-velocity stream of water is then directed onto hemispherical buckets to overcome the resistance of the load and transfer energy to the rotating turbine.

The Pelton turbine is usually found in mountainous areas where heads of hundreds of feet are available. The high heads compensate for the design's partial-admission limitation. In the New England area, there appears to be only one Pelton turbine still in place: a homemade unit used until 1972 to operate a woodworking mill. It operated under a head of 124 ft (36.8 m) and

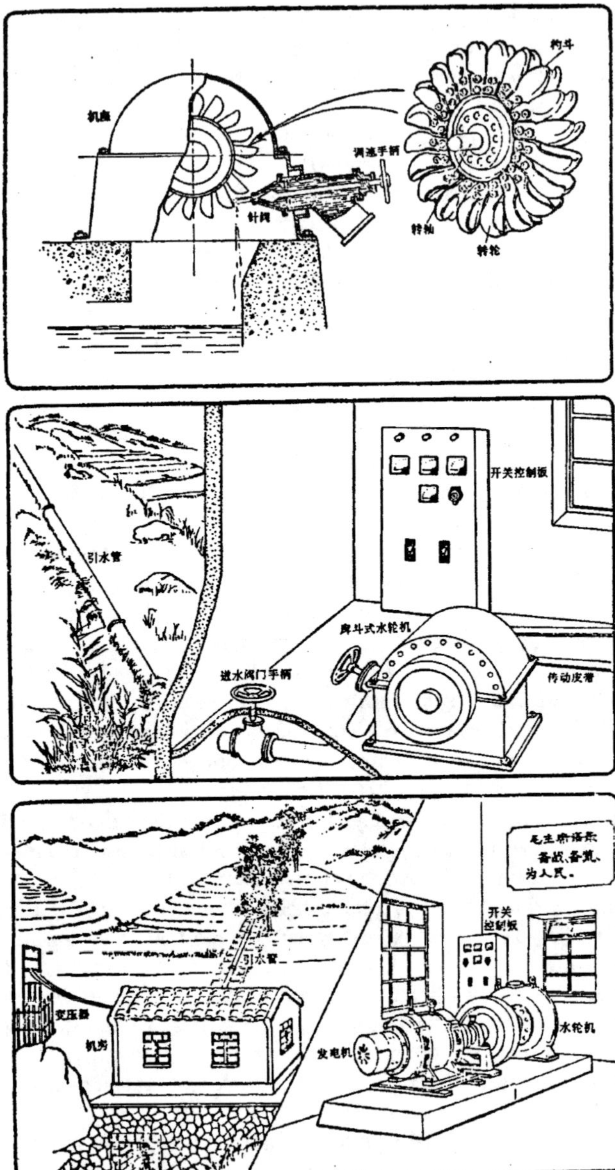

FIGURE 40 Pelton plant and details. [Source: see Ref. 2]

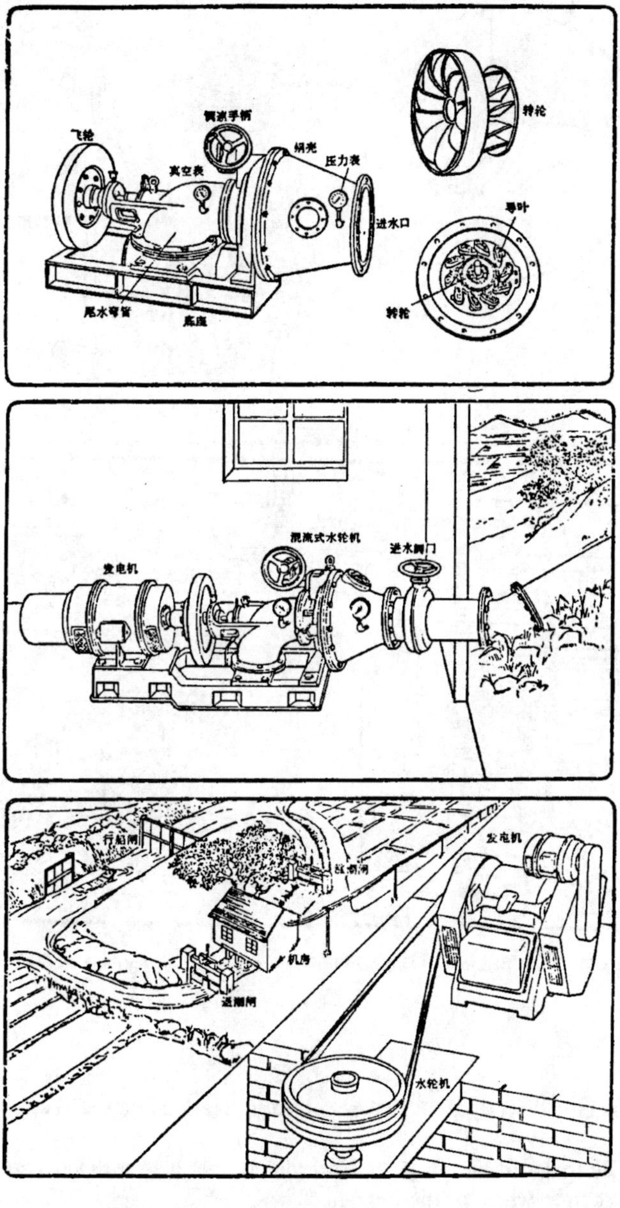

FIGURE 41 Francis turbine plant and details. [Source: see Ref. 2]

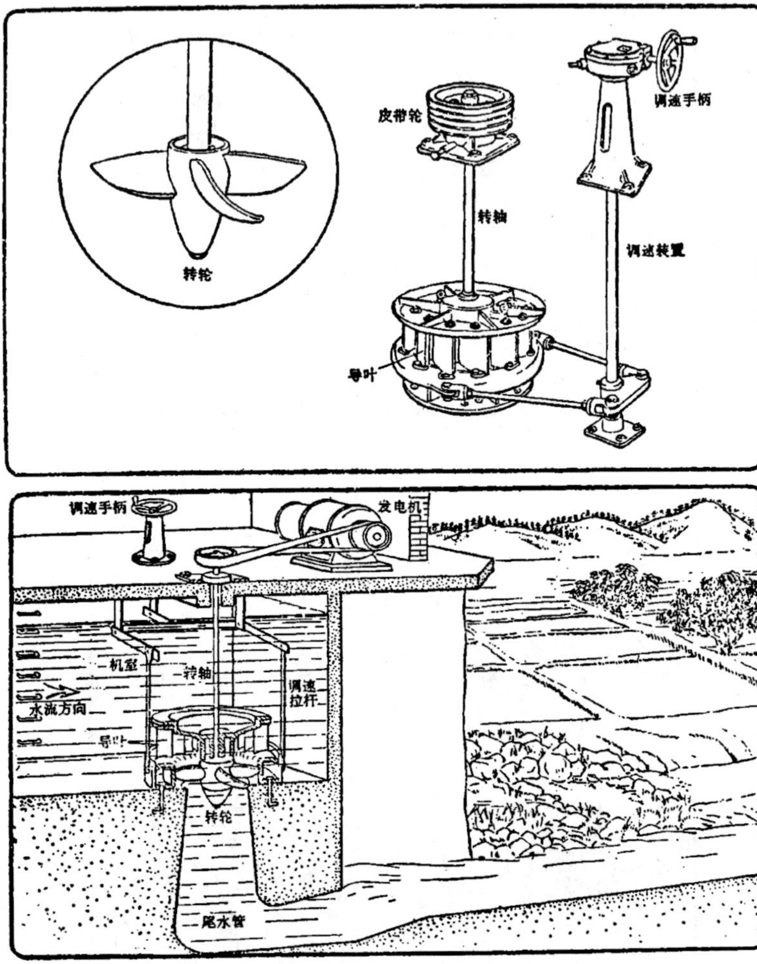

FIGURE 42 Fixed propeller (Nagler) plant and details. [Source: see Ref. 2]

was connected directly to a 5-kW generator operating at a speed of over 1,000 rpm.

The Pelton turbine may be considered the high-speed version of the undershot waterwheel or the tub wheel.

Water Supply

All turbines require that water be conveyed to them, under pressure, in large pipes or penstocks, from a source at some elevation above the turbine. The drop in elevation from the source to the turbine may be the result of natural conditions or may be artificially created by means of a dam. The dam not only serves to create the elevation difference (head), it also makes possible the storage of water to smooth out variations in stream flow. Ideally, the dam should be located at the point where a substantial drop in the stream bed allows maximum power to be developed with minimum conveyance of water to the turbine. However, for reasons of economy and structural simplicity, the most desirable dam of minimum length and height may have to be located upstream of this point and a long penstock used to convey water to the turbine. In New England, penstocks a mile (1.6 km) long or longer and 5 or 6 ft (1.53 or 1.83 m) in diameter, made of California redwood, have been a common sight for many years.

The selection of sites for hydropower development is dependent on the following factors:

- Topography (slopes, storage, evaporation);
- Geology (dam location, run-off); and
- Stream-flow (rainfall, watershed area).

Figures 43-48 illustrate some typical arrangements of small-scale hydroelectric plants.

The means of securing the necessary data for evaluating a site is a formalized procedure covered by engineering textbooks, some of which are listed later; the same is true of the power machinery.

Current Technology (Available within 5 Years)

A brief survey of turbine manufacturers in the United States has shown that there is only one, The James Leffel & Co. of Springfield, Ohio, offering suitable "packaged" hydroelectric units off the shelf. Electrical outputs (AC or DC) of 0.5, 1, 2, 3, 5, 7.5, and 10 kW (Figure 49) are available. For each power output, several turbine designs are offered, accommodating to variations in head requirements between 8 and 25 ft (2.5 and 7.6 m). (Details of the Leffel turbines and those of other manufacturers will be found in Appendix 8.)

The cost of these units ranges from $3,600 for the 0.5-kW unit to about $10,000 for the 10-kW size. A 10-kW unit developing 100-kWh/day, 350 days/year, with annual fixed and operating charges of 20 percent, would

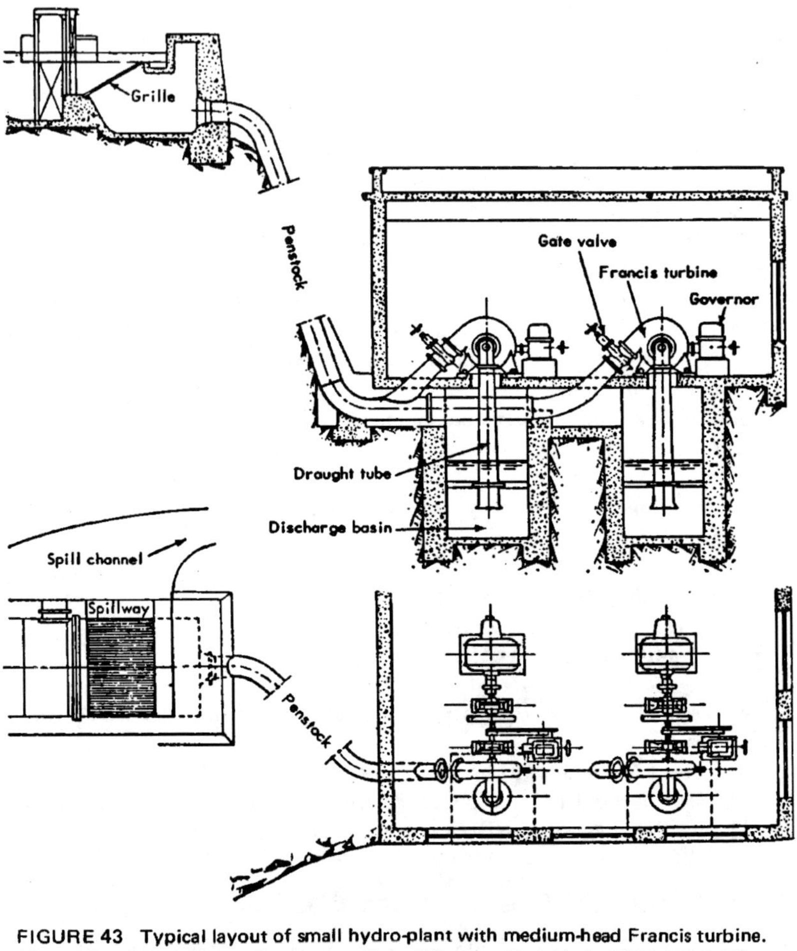

FIGURE 43 Typical layout of small hydro-plant with medium-head Francis turbine. [From: *Small-Scale Power Generation,* United Nations, New York, 1967. UN Publication Sales No. 67.II.B.7.]

supply power at a unit cost of about 6¢/kWh. This does not include the cost of the dam and penstock.

The packaged Leffel units employ propeller-type turbines with fixed blades and are connected directly to the electric generator, doing away with the inconvenience, expense, and reduced efficiencies of using belts or gears.

In addition to these packaged units, Leffel offers its traditional line of

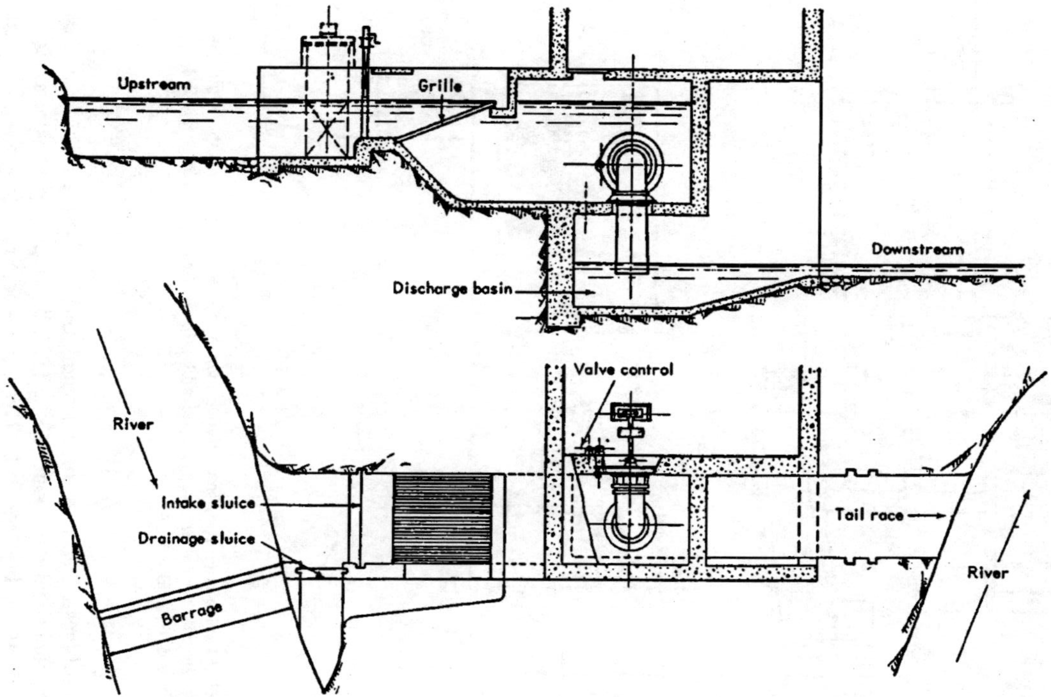

FIGURE 44 Typical layout of small hydro-plant with low-head Francis turbine. [From: *Small-Scale Power Generation*, United Nations, New York, 1967. UN Publication Sales No. 67.II.B.7.]

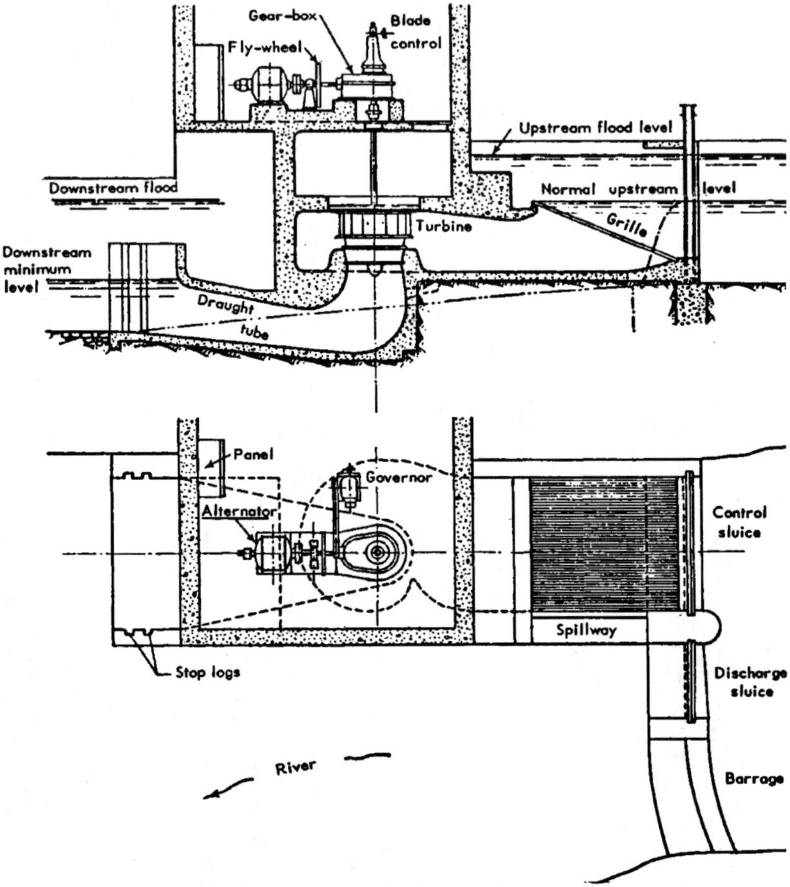

FIGURE 45 Typical layout of small hydro-plant with very low-head Kaplan turbine. [From: *Small-Scale Power Generation,* United Nations, New York, 1967. UN Publication Sales No. 67.II.B.7.]

Francis turbines. The cost of a Francis turbine having a 17-in. (43-cm) diameter runner and developing 20 hp (15 kW) at 294 rpm under a head of 10 ft (3 m) would be $7,900. Its low speed would require a belt or gear connection to an electrical generator.

The Soviet Union builds a fixed-propeller unit almost identical in size and form to the Leffel fixed-blade unit. It is called "Mikroges" and was designed by V. I. Gromov and Y. N. Fleksuer.

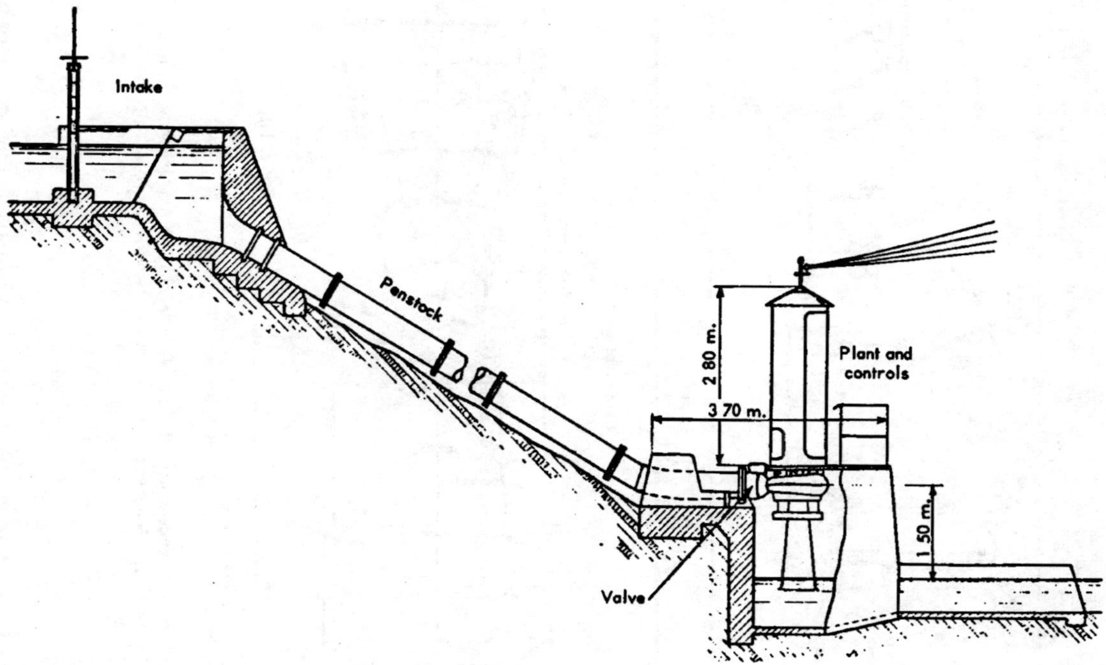

FIGURE 46 Typical layout of micro hydro-plant. [From: *Small-Scale Power Generation,* United Nations, New York, 1967. UN Publication Sales No. 67.II.B.7.]

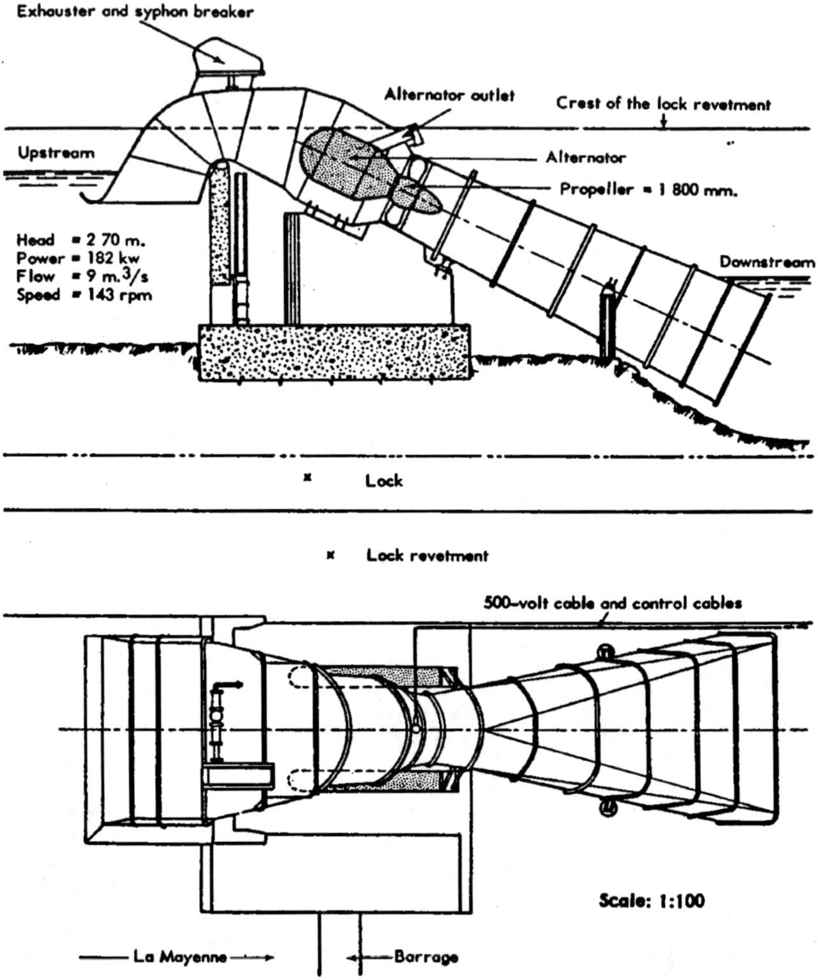

FIGURE 47 Syphon arrangement with bulb propeller turbine: La Mayenne, France. [From: *Small-Scale Power Generation,* United Nations, New York, 1967. UN Publication Sales No. 67.II.B.7.]

Gilbert Gilkes & Gordon Ltd., of Kendal, Westmorland, England, offers to design small 10- to 20-kW hydroelectric units for particular heads and flows. Their rough estimate of the cost of such hydroelectric units operating under a 10-ft (3-m) head would be about £17,000 ($34,000) and for a 15-ft (4.5-m)

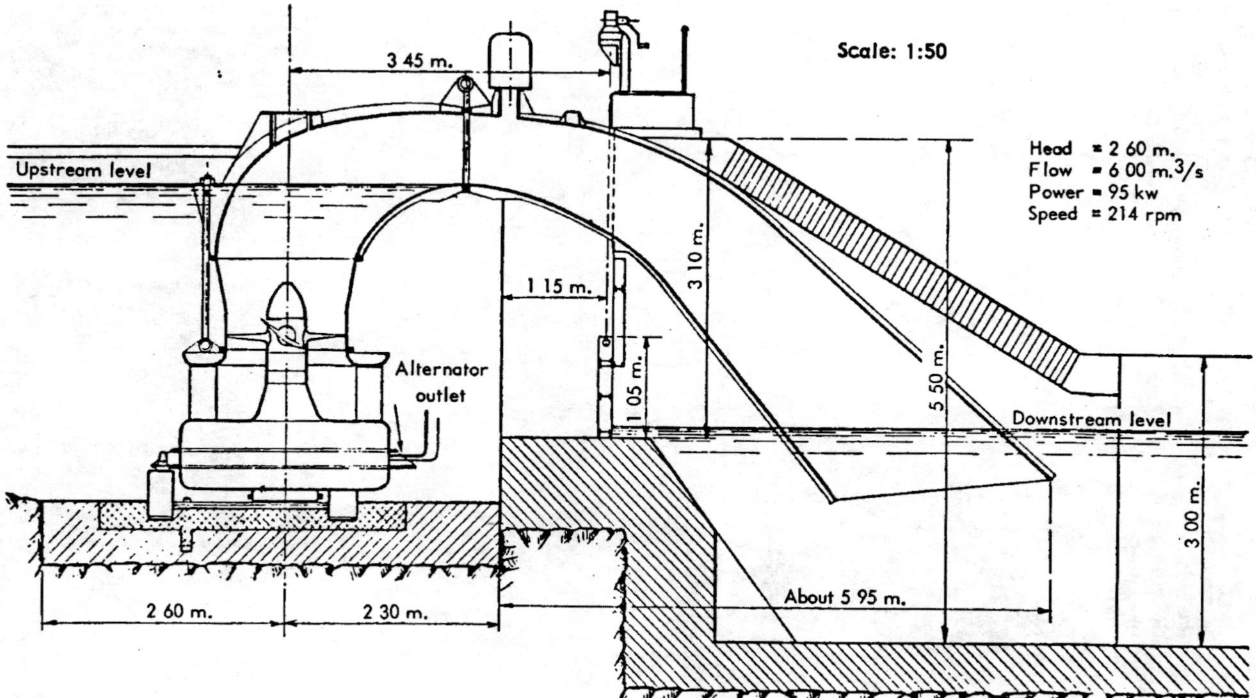

FIGURE 48 Submerged, upward flow micro turbine installation (sectional view): La Garonne, France. [From: *Small-Scale Power Generation*, United Nations, New York, 1967. UN Publication Sales No. 67.11.B.7.]

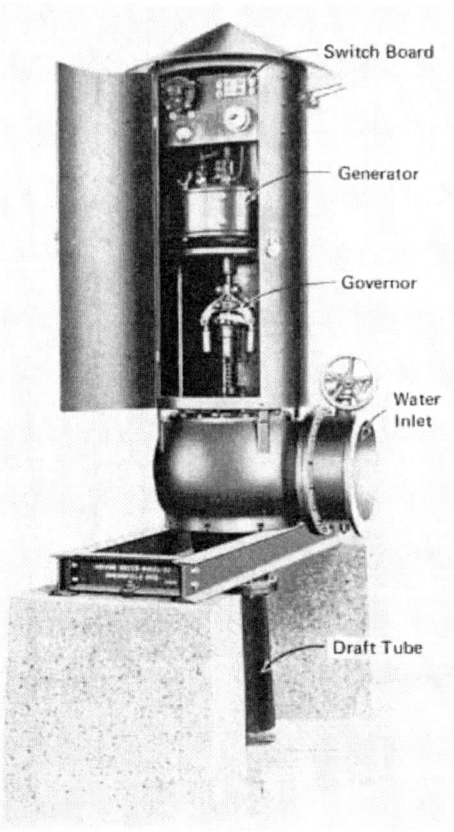

a

b

FIGURE 49 A 1-kW Leffel unit on a farm in Ohio. *a,* The 1-kW unit. *b,* Installation of the 1-kW unit. [Courtesy The James Leffel & Co.]

head about £14,000 ($28,000); this cost is exclusive of such factors as the cost of the dam, penstock, valves, and building.

The Canadian company Barber Hydraulic Turbine Ltd., Port Colborne, Ontario, has designed a small 18.5-kW (20-hp) hydroelectric unit for operational testing in 1976. The turbine is a fixed-blade type, with five blades, and a diameter of 10 in. (25 cm). This project is being partially financed by the Canadian Government.

The Drees Company of Werl, Westphalia, West Germany, manufactures a "micro" hydropower package power plant of about 20-kW capacity. In appearance it resembles the Leffel unit.

In Budapest, Hungary, the Ganz Company manufactures a "Mignon" hydroelectric package plant that develops 15 kW at heads ranging from 7 to 12 m (23 to 39 ft). It weighs about 1,300 kg (2,900 lb) and utilizes a propeller turbine of the Kaplan type.

The Ossberger-Turbinenfabrik of Weissenberg, Bavaria, West Germany, manufactures a hydroelectric package called the Hydro-light. It utilizes a turbine described as a "radial impulse" type, and is made in two sizes: the 3-kVa unit for heads ranging from 4 to 6 m (13 to 20 ft) and the 5-kVa unit for 6- to 9-m (20- to 29-ft) heads of water. These units have reportedly been installed in many developing countries.

Published information and news reports indicate that a rather widespread and reliable technology of small-scale, low-head hydroelectric devices exists in the People's Republic of China.[2,3,4,5] The available literature (in Chinese) describes the use of small hydroelectric units, including Pelton, Francis, and Kaplan turbines. News reports indicate their growing popularity in rural areas and describe the construction and installation of thousands of units of less than 100-kW capacity, including one wooden turbine for a 20-kW hydroelectric plant.[6] Information is also available on details of construction of wooden and reinforced-concrete turbines for small-scale hydroelectric units.[5] However, no details are available to this panel on either the specifications of such units or their commercial availability.

A particularly interesting use of an irrigation canal has been made in Lin Hsien, where a series of small hydroelectric plants has been built along the canal, taking advantage of the progressive drop in elevation. The first such plant, capable of generating 250 kW, is located at the bottom of a 15-m (50-ft) drop, at the top of which is a sedimentation basin to remove the sand and silt that would cause considerable erosion of the metal parts. Thereafter, at every 5-m drop in elevation a 40-kW turbine has been installed, each one preceded by a sedimentation basin. Figure 50 shows one of these 40-kW plants.

Wooden turbines for small hydroelectric units are also being manufactured in the Soviet Union, but again, the panel has been unable to obtain any details.[7]

a

b

FIGURE 50 Forty-kilowatt hydroelectric power station in Lin Hsien, People's Republic of China. This installation is one of a series of 26 placed at every 5-meter drop in elevation along the 12th channel of the No. 1 Branch Canal of the Red Flag Canal irrigation system. [These photographs courtesy of Amir U. Khan of the International Rice Research Institute. Dr. Khan visited these stations as a member of a delegation sponsored by the Committee on Scholarly Communication with the People's Republic of China, of the National Academy of Sciences' Commission on International Relations.] *a,* Housing and sluiceways. *b,* Generator and control mechanisms. (The generator is driven by multiple V-belts from the pulley on the turbine shaft.)

Reliability

Small units of the types described may be expected to have a useful life of about 25 years with minimal maintenance.

Local Implementation

Although most of the discussion has centered on manufactured devices, small-scale hydroelectric systems lend themselves to significant cost reductions through local enterprise.

- The dams may be made of earth, stone, or logs. Hundreds of such dams have been used in New England and last 25 years and more. Leffel's book on mill dams is excellent on the building of small dams for heads up to 15 ft (4.5 m).[8] (See also Figure 51.)
- Penstocks may be made of reinforced concrete or of wood planks strengthened by iron straps.
- Turbines may be made of wood, though if there is a rudimentary iron and steel industry, these more enduring materials may be used.

The units now available should be examined with the aim of making them simpler and cheaper. It may be possible to eliminate the governor and much of the switchboard equipment. The general problem of simplification should be explored with the companies mentioned above.

As mentioned previously, it has been reported that small wooden turbines are being made in the Soviet Union and China, although no firm information has been obtained from the Soviet Union. It may be worthwhile to pursue the idea of a wooden turbine design. This could become the subject of engineering theses at technical institutes or engineering colleges, preferably within a developing country, with funds provided for students to construct turbines for testing and demonstration. This could serve as a stepping-stone to formal instruction in wooden turbine building for artisans in developing nations.

HYDRAULIC RAMS

The hydraulic ram, the ingenious, automated water-pumping machine developed by Montgolfier in 1796, is largely forgotten today; at one time water was supplied to the fountains and lakes of the Taj Mahal by these simple devices.[9] Present listings of manufacturers making "hydraulic rams" refer to manufacturers of hydraulic pistons and not Montgolfier's ingenious pump.

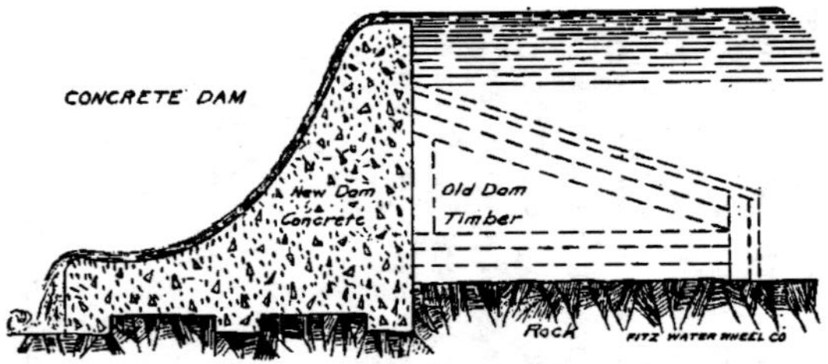

a To anyone familiar with the handling of concrete, this makes a very neat and permanent means of obtaining water storage. It is possibly the best and most shapely structure which can be used. The old wood dam need not be removed.

b A very good type of dam made from stone. The outer layers should be bound with concrete or concrete mortar to render it perfectly tight.

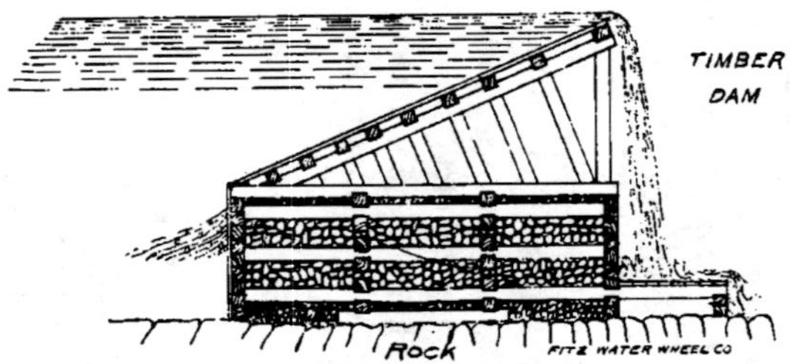

c Another method of building a wood dam where greater height is desired.

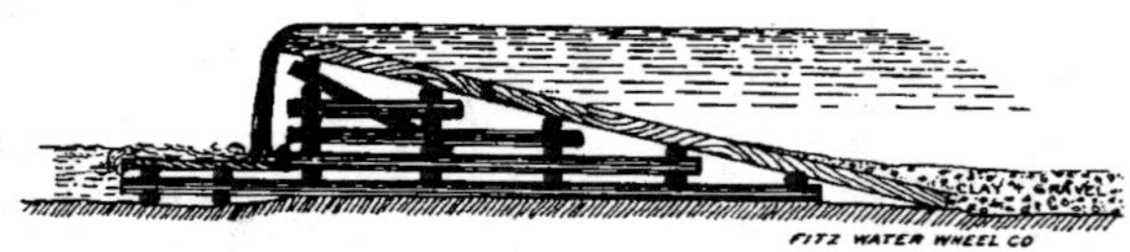

d In sections of the country where timber is easily obtained, this is a very inexpensive dam to build. Easy to construct and very satisfactory where only a small dam is required.

FIGURE 51 Dams for use with small heads. [Drawings and captions reprinted from: "Electric Light Plants Driven by Small Water Powers," Bulletin No. 16, Fitz Water Wheel Company, Hanover, Pennsylvania, 1916]

The basic principle of the hydraulic ram is that of "water hammer." In fact, it was the problem of water hammer that was responsible for its re-invention in England during the early part of the 19th century. What happened was this: a water line was installed in a multistory building for use at its lowest level, the head available being inadequate to serve the upper floors of the building. Whenever the faucet was closed quickly, however, the pipe would rupture near the faucet. A resourceful plumber solved the problem by connecting to the water pipe, just before the faucet, a vertical pipe connected to an open tank a floor or so above to relieve the sudden hammering pressure buildup. Eventually, it was necessary to locate the tank on the top floor, and it soon became obvious that with the supply of water reaching the top floor, a limited amount of water could be supplied to all floors. The idea then blossomed into the English version of the hydraulic ram.

The hydraulic ram, via the phenomenon of water hammer, uses the kinetic energy of water flowing in a pipe to elevate part of this water to a height greater than that causing the flow, with an efficiency of about 60 percent. Figure 52 shows a hydraulic ram and some typical installations. The principle of operation is illustrated in Figure 53. Water flows through the supply pipe D, and through the opening of outside valve F. As the water velocity increases, the dynamic pressure on the underside of the check valve F increases until it overcomes the weight of the valve. Then F moves rapidly upward, closing the opening. The pressure of the moving water causes valve B to open and water is driven into the air-cushion chamber A, increasing the air pressure and discharging through the delivery pipe I. As the momentum of the water in the ram decreases, valve B drops down and closes and the water rebounds somewhat. This creates a sudden drop in pressure under valve F, causing it to drop down and open quickly and the cycle begins again. When valve B is closed, of course, the pressure in A is determined by the head of water in the storage tank and delivery line. Thus, the incremental additions of water to the storage system are the result of the momentary increases in air pressure in A at each cycle. The frequency with which the cycle is repeated is regulated by adjustment of the counterweight C.

A ram using a vertical fall of water of 12 ft (3.7 m), for example, will deliver 5 or 6 percent of the flow to an elevation of 125 ft (38 m); if the lift is to be only 25 ft (7.6 m), about 22 percent of the water flowing will be elevated.

Though the hydraulic ram is customarily a device for pumping water, it has also been used to compress air for rock drills—during the building of the Mont Cenis Tunnel in Switzerland, for instance.[10] The surge of water into the chamber at A can act as a piston to compress the air normally used as an air cushion when the device is used for pumping water. The suction part of the cycle becomes the air-intake cycle in this application.

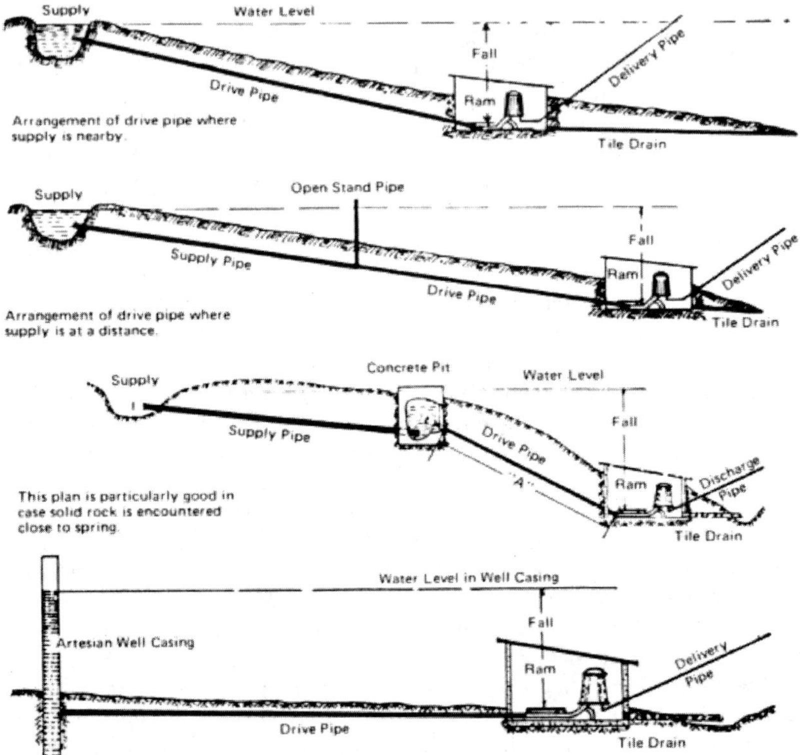

FIGURE 52 Hydraulic ram and examples of typical installations. [Courtesy Rife Hydraulic Engine Manufacturing Co.]

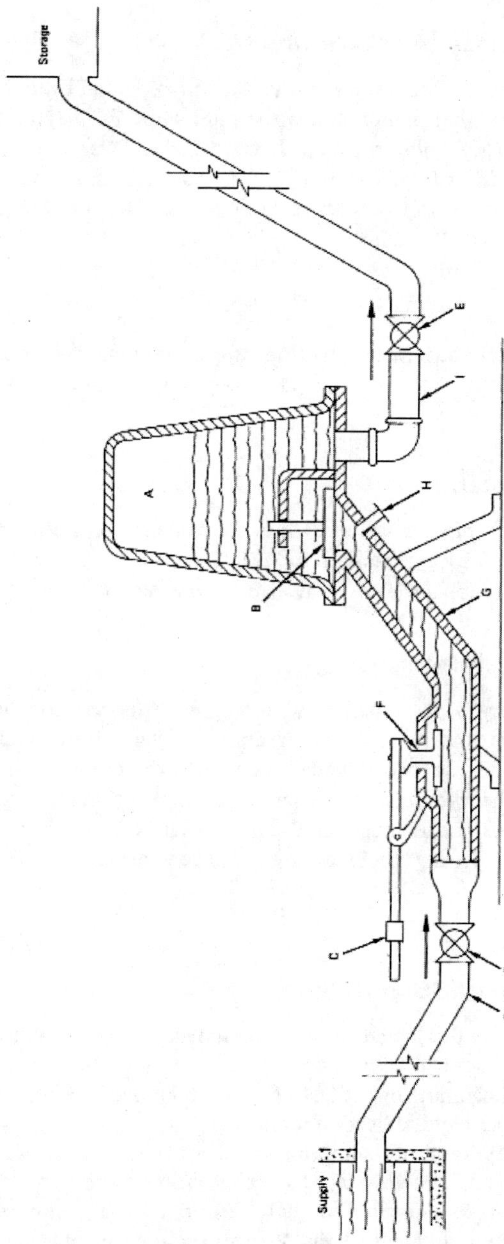

FIGURE 53 The hydraulic ram [Source: *Village Technology Handbook*, VITA, Mt. Rainier, Maryland, USA, 1970]

Current Technology (Available within 5 Years)

A canvass of about 20 hydraulic ram manufacturers in the United States has turned up only one that manufactures the pump: the Rife Hydraulic Engine Manufacturing Co. of Millburn, New Jersey 07041. This company supplies rams requiring minimal falls–about 2-4 ft (0.6-1.2 m)–ranging in cost, according to pumping capacity, from about $300 to about $2,600. (See Appendix 8 for details.)

Reliability

Rams are completely automated, will operate 24 hours a day with minimum attention for months, and may be used to pump water for irrigation and water supply.

Local Implementation

In areas where iron pipe and pipe fittings are available, hydraulic rams could be constructed with local materials and utilizing local skills. Complete details for constructing these devices are available from VITA.[11, 12] (See p. vii)

Research and Development

The hydraulic ram is so reliable and effective in its performance for the pumping of water that it does not require further development for this purpose. It would be worthwhile, however, to support research and development of the hydraulic ram as an air compressor. Such compressed air would be produced continuously and stored in tanks for use in reciprocating engines or turbines driving machines or electric generators.

HYDRAULIC AIR COMPRESSOR

In medieval Europe this machine with no moving parts was known as the "trompe" and it supplied compressed air to aid combustion in Catalan iron furnaces. At the beginning of the 20th century it enjoyed a brief vogue in the United States as the "hydraulic air compressor."

Basically, the hydraulic air compressor is the inversion of the air-lift pump. It is a simple device in which water at some elevation is allowed to flow into a vertical pipe through a circumferential Venturi (convergent-divergent) entrance. At the minimum areas of the Venturi section, openings are provided so that the water flowing past these openings will induce air to be sucked into them and become entrained in the water stream (Figure 54). If the water velocity is high enough, the entrained air will be carried to the bottom of the

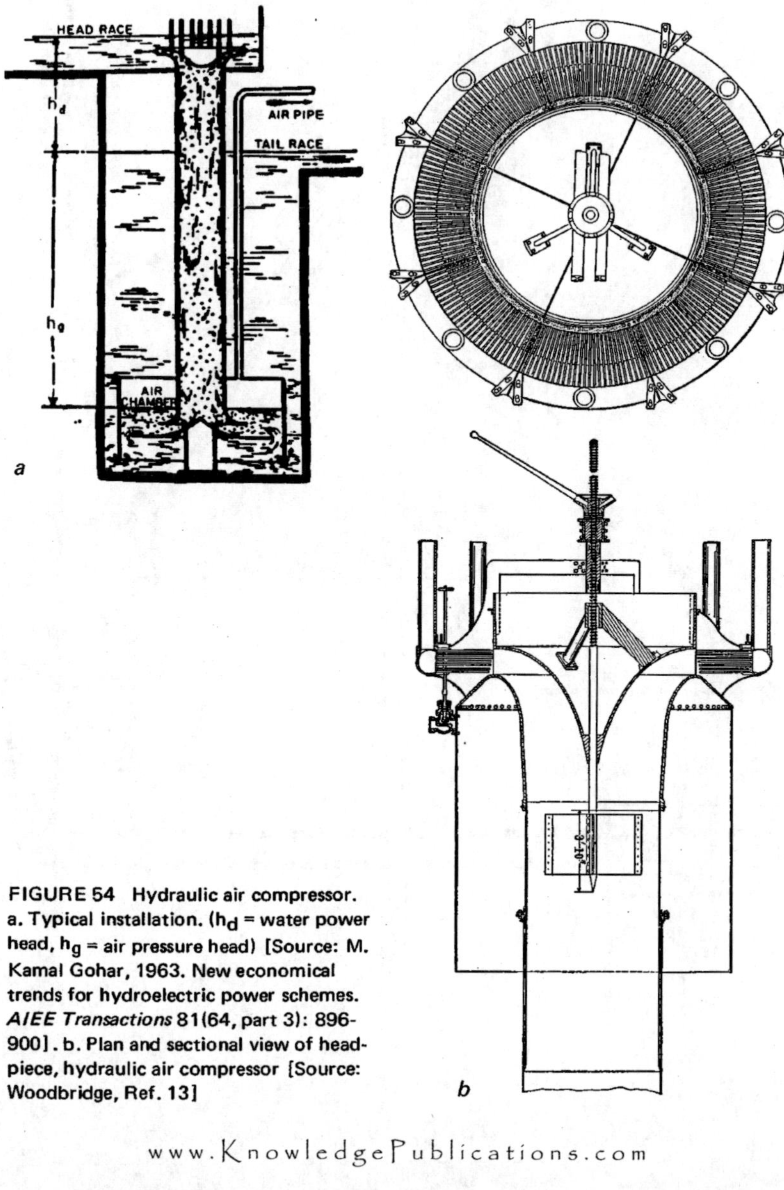

FIGURE 54 Hydraulic air compressor. a. Typical installation. (h_d = water power head, h_g = air pressure head) [Source: M. Kamal Gohar, 1963. New economical trends for hydroelectric power schemes. *AIEE Transactions* 81(64, part 3): 896-900]. b. Plan and sectional view of head-piece, hydraulic air compressor [Source: Woodbridge, Ref. 13]

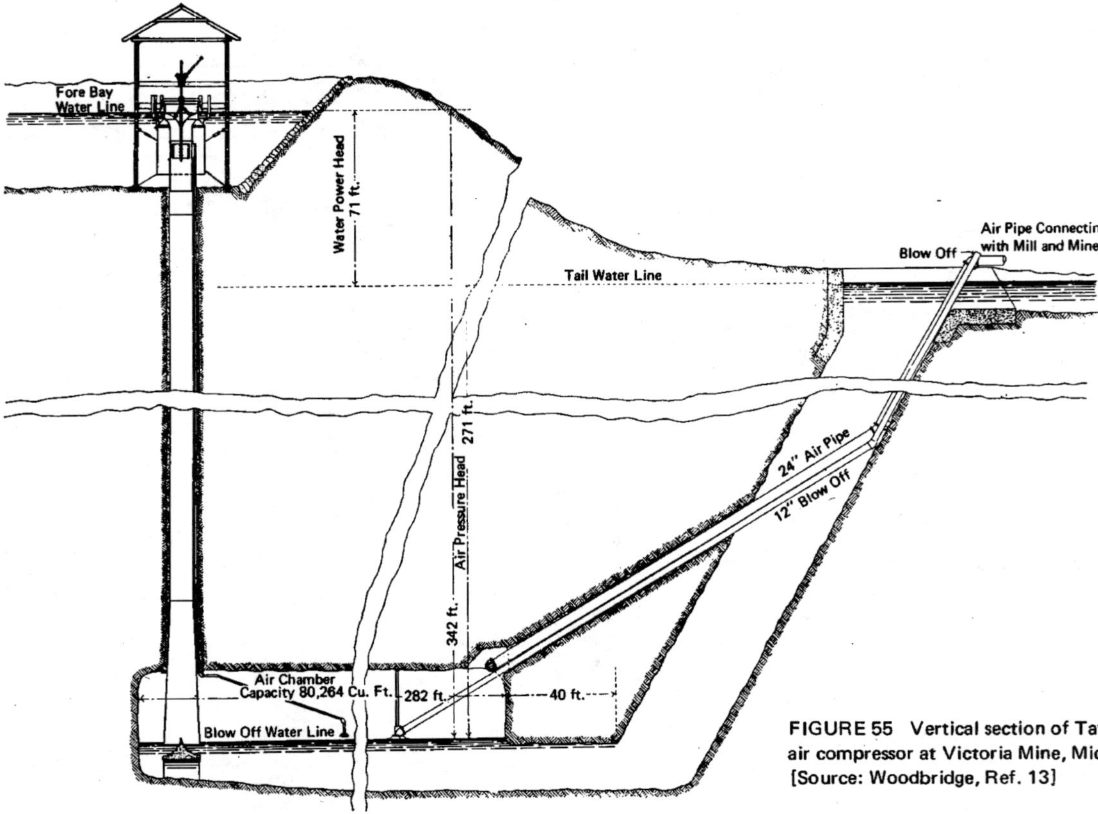

FIGURE 55 Vertical section of Taylor hydraulic air compressor at Victoria Mine, Michigan. [Source: Woodbridge, Ref. 13]

pipe and into a chamber where the air is trapped and compressed by the conversion of its kinetic energy to potential energy.

It has been reported that an installation utilizing a fall of water of 71 ft (21.6 m) with a pipe 5 ft (1.53 m) in diameter compressed enough air to a pressure of 117 lb per square in. (8.05 atmospheres) to develop about 1,000 hp (750 kW).[13] The efficiency of this installation was reported as 82 percent (Figure 55).

Current Technology (Available within 5 Years)

This is a technology that could be available now. But because it seems to have been forgotten for almost 70 years, there are no known operating installations or available manufactured Venturi devices designed for this application.

Research and Development

The simplicity of this device and its high performance suggest that it should be resurrected for further study and possible use in hilly terrain where ample water is available. The ability of the device to operate continually day and night, with its simple storage of energy in the form of compressed air in tanks or caves, makes it an interesting and potentially fruitful problem to investigate. The compressed air could be piped to sites to drive reciprocating engines or turbines that, in turn, could power production machines or electric generators.

REFERENCES

1. Mosonyi, E. 1960. *Water power development*, Vol. 2, pp. 837-886. 2 vols. Budapest: Hungarian Academy of Sciences.
2. *How to run small power stations efficiently*. 1971. People's Republic of China: Shanghai People's Press. (In Chinese)
3. Editorial Publishing Group, East China College of Water Conservancy, Nanking. 1973. *Rural small-scale hydro-electric stations*. People's Republic of China: Shanghai People's Press. (In Chinese)

   See especially Chapters 6 and 11. Chapter 6, "Self-made Turbines," is devoted to construction, in the village, of wooden, ferrocement, and welded turbines. Subheadings include: "Self-made wooden turbines," "Self-made ferrocement turbines," "Self-made steel plate welded turbines," "Self-made partial-admission ferrocement turbines," "Self-made axial-flow full-admission water turbines." Chapter 11 is devoted to construction of dams.
4. Canton Bureau of Water Conservancy and Electric Power. 1973. *Rural small-scale hydroelectric stations*. People's Republic of China: Canton People's Press. (In Chinese)

5. Editorial Publishing Group on Rural Small-Scale Hydropower Stations, Bureau of Water Conservancy and Electric Power, Revolutionary Committee of Hunan Province. 1974. *Rural Hydroelectric Stations*. 2 vols. People's Republic of China: Hunan People's Press. (In Chinese)

This book is concerned with installations of less than 400-kW capacity, using water heads under 30 meters. Part 2 of Volume I discusses construction of earth and stone dams, sluices, and penstocks (reinforced concrete, wood, ferrocement).

6. New China News Agency. 1970. Radio broadcast in English, 6:38 A.M., GMT, 11 August. *BBC Summary of World Broadcasts*, Part 3: The Far East weekly economic report. London: British Broadcasting Company.
7. Mosonyi, E. 1960. *Water power development*, Vol. 2, p. 869. Budapest: Hungarian Academy of Sciences.
8. *The construction of mill dams*. 1874. 1970 reprint. Springfield, Ohio: The James Leffel and Co.
9. Reeves, W. J. 1961. Country workshop. *Australian Country*, September: 32-33. Sydney, Australia.
10. Peele, Robert. 1930. *Compressed air plant*. 5th ed. New York: John Wiley & Sons.
11. *Hydraulic rams for village use*. 1970. Mt. Rainier, Maryland: VITA (Volunteers in Technical Assistance).
12. *Village technology handbook*. 1970. Mt. Rainier, Maryland: VITA (Volunteers in Technical Assistance).
13. Woodbridge, D. E. 1907. The hydraulic compressed-air power plant at the Victoria Mine. *Engineering and Mining Journal* 83(3):125-130.

SUGGESTED READING

HYDROELECTRIC PLANTS

Frizell, J. P. 1905. *Water power*. New York: John Wiley & Sons.

Koester, F. 1909. *Hydro-electric developments and engineering*. New York: Van Nostrand.

Daugherty, R. L. 1937. *Hydraulics: a text on practical fluid mechanics*, Chapters XIV-XVIII, pp. 343-418. 4th ed. New York: McGraw-Hill.

Reports on water power of the U.S. 1880. Washington, D.C.: U.S. Department of the Interior, Census Office.

Mead, D. W. 1950. *Hydrology*. New York: McGraw-Hill.

Wagner, E. G., and Lanoix, J. N. 1959. *Water supply for rural areas and small communities*. Geneva, Switzerland: World Health Organization.

HYDRAULIC RAM

Gibson, A. H. 1970. *Hydraulics and its application*, pp. 683-693. London: Constable.

Meade, D. 1933. *Hydraulic machinery*, pp. 358-383. New York: McGraw-Hill.

HYDRAULIC AIR COMPRESSOR

Gibson, A. H. 1970. *Hydraulics and its application*. London: Constable.

Schulze, Leroy E. 1954. *Hydraulic air compressors*. U.S. Bureau of Mines Information Circular, No. 7683. Washington, D.C.: Department of the Interior.

Photosynthesis

THE TERRESTRIAL BIOMASS AS A RESERVOIR

In the process of photosynthesis, solar energy absorbed by green plant tissue provides energy to reduce CO_2 and form carbohydrates, which are then utilized as energy sources and raw materials for all other synthetic reactions in the plant. Thus, solar energy is captured and stored in the form of chemical energy. Cellulosic plant material presents a vast, untapped supply of energy. It is renewable, has a low content of sulphur and other pollutants, and storage of energy from the season in which it is produced until it is needed is simple and inexpensive. When fossil fuels were abundant and low priced, there was little incentive to develop methods to utilize this ready energy supply. Now that traditional fuels are scarce and expensive, however, the vast potential of plants as an energy source merits serious and immediate attention.

Stored plant energy may be released by drying the material and burning it directly, or various processes may be utilized to obtain potential fuels such as ethanol, methane, or other gaseous or liquid fuels. The processes used to derive these fuels, however, may require different forms of the plant material. This section will deal only with the amount of energy involved and with factors that may affect the efficiency of energy capture, not with the processes by which the energy might be extracted or converted to use. (One energy-extraction technology, however—by microbial action—is considered in the next chapter.)

ENERGY CONTENT OF PLANT MATERIALS

The energy content of several different plant materials is shown in Table 1 as compiled by Alich and Hinman, 1974.[1] All of the materials listed were air- or oven-dry with the exception of the wet sugarcane bagasse. On a dry-weight (moisture-free) basis these materials have an average energy content of 8,700 Btu/lb (4,800 kcal/kg). Obviously, if combustible material is required for the energy-recovery process, the amount of water in the plant material will affect the recoverable energy; fresh leaves may contain 95 percent water. Thus, correction for water content must be made and the energy cost of drying must be considered when plant materials are evaluated as energy sources for combustion. All energy estimates and calculations in this report are based on dry weights.

Szego and Kemp, 1973,[6] discuss the problem of water in plant-energy conversion and use 6,500 Btu/lb (3,600 kcal/kg) as an average energy content for many different air-dry plant materials (10-15 percent moisture) which, on a moisture-free basis, would be about 7,500 Btu/lb (4,200 kcal/kg). As a rule of thumb, 7,500 seems a reasonable and conservative figure for dry plant material.

TABLE 1 Energy Content (Btu/lb) of Plant Biomass (Table Adapted from Alich and Hinman, 1974)[1]

	Plant Part	Percent H_2O	Btu/lb	kcal/kg
Sugarcane (*Saccharum officinarum*)[2]	Bagasse	12	7,281	3,860
Sugarcane (*Saccharum officinarum*)[3]	Bagasse	52	4,000	2,220
Bamboo (*Phyllostachys* spp.)[2]	Cane	10.5	7,398	3,925
Buckwheat (*Fagopyrum esculentum*)[2]	Hulls	10	7,594	4,030
Chamise (*Adenostoma fasiculatum*)[4]	Leaves	0	9,750	5,160
Chamise (*Adenostoma fasiculatum*)[4]	Stems	0	9,450	5,015
Coconut (*Cocos nucifera*)[2]	Shells	13	7,560	4,010
Beech (*Fagus* sp.)[2]	Wood	13	7,506	3,990
Birch (*Betula* sp.)[2]	Wood	12	7,578	4,015
Oak (*Quercus* sp.)[2]	Wood	13	7,182	3,810
Oak (*Quercus* sp.)[2]	Bark	7	8,139	4,310
Oak (*Quercus* sp.)[3]	Bark	0	8,370	4,450
Pine (*Pinus* sp.)[2]	Wood	12	7,956	4,230
Pine (*Pinus* sp.)[3]	Bark	0	9,030	4,790
Fir (*Abies* sp.)[3]	Bark	0	8,810	4,675
Spruce (*Picea* sp.)[3]	Bark	0	8,740	4,640
Redwood (*Sequoia sempervirens*)[3]	Bark	0	8,350	4,425
Oilseed crop[5]	Seed	–	9,000	4,775

ENERGY PRODUCTION BY PLANTS

Examples of the annual biomass yields and energy equivalents of aboveground dry matter from several types of plants are given in Table 2 (adapted from Table 1). The yields given in Table 2 show much variability within and among species. They are not intended to indicate the yield potential of the species; rather, they are examples of yields from various sites and under varying systems of management. Other examples are found in references 2, 6, 7, 8, and 9. Obviously, the energy accumulated by highly productive plants is massive.

We do not have the climatic data to permit us to estimate the efficiency of capture of solar energy represented by the yields in Table 2. For certain agricultural situations, however, sufficient climatic and yield information is available to make that calculation. For example, an Iowa maize field planted May 1 and harvested October 1 may produce 200 bu/a (180 hl/ha) of shelled grain (4,325 kg dry weight). This would be a high, but not unreasonable, yield for Iowa. Above ground, this crop would have an equal weight of plant material other than grain, giving a total yield of dry plant material of 21,400 kg/ha. This represents the capture of 9×10^7 kcal/ha. Between May 1 and October 1, a central Iowa maize field will receive about 7.4×10^9 kcal per ha insolation. Thus, a 180-hl/ha (200-bu/a) crop has captured 1.2 percent of the total incident insolation it receives from planting until harvest. When the seeds were germinating and when the plants were small and intercepting little radiation, however, the capture of solar energy would have been essentially 0. Thus, the figure of 1.2-percent capture is an average over the lifetime of this seeded annual crop. Energy capture would be greater if a ground cover of leaves could be attained more quickly. Such could be the case if species could be used that would regrow from root systems after harvest. Some highly productive species such as sugarcane and sorghum do produce successive crops from root systems and, in tropical or subtropical climates where year-round growth is possible, they appear to be very promising species to use for energy capture.

The daily growth rate of a 180-hl/ha (200-bu/a) Iowa maize crop averages 14 g/m^2 per day for the season. Under favorable conditions, growth rates of 15-25 g dry matter per m^2 per day have been observed for a number of crops, [1,7,8,9] and under ideal conditions sorghum, maize, sugarcane, and bullrush millet have produced 50 g dry matter per m^2 per day for a significant part of their growing season. This latter production rate represents a capture of about 3 percent of the incident solar radiation.

It is not realistic to expect to maintain a 3-percent capture rate over long periods of time, although fertile sites under ideal conditions may be able to approach that figure on a sustained basis. For the majority of situations,

TABLE 2 Aboveground, Dry Biomass Yields and Energy Equivalents of Selected Plant Species or Complexes (Adapted from Alich and Hinman, 1974)[1]

Species	Location	(tons/acre-year)	(metric tons/ hectare-year)	(10^7 Btu/ acre-year*)	(10^7 kcal/ hectare-year*)
Annuals					
Sunflower x Jerusalem artichoke	Russia	13.5	30.0	20	12.6
Sunflower hybrids (seeds only)	California	1.5	3.4	2	1.4
Exotic forage sorghum	Puerto Rico	30.6	68.7	46	29
Forage sorghum (irrigated)	New Mexico	10	22	15	9
Forage sorghum (irrigated)	Kansas	12	27	18	11
Sweet sorghum	Mississippi	9	20	14	8
Exotic corn (137-day season)	North Carolina	7.5	17	11	7
Silage corn	Georgia	7	16	10	7
Hybrid corn	Mississippi	6	13	9	6
Kenaf	Florida	20	45	30	19
Kenaf	Georgia	8	18	12	7
Perennials					
Water hyacinth	Florida	16	36	24	15
Sugarcane	Mississippi	20	45	30	19
Sugarcane (state average)	Florida	17.5	39.3	26	16.3
Sugarcane (best case)	Texas (south)	50	112	75	47
Sugarcane (10-year average)	Hawaii	26	58	39	24
Sugarcane (5-year average)	Louisiana	12.5	28.1	19	11.7
Sugarcane (5-year average)	Puerto Rico	15.3	34.3	23	14.3
Sugarcane (6-year average)	Philippines	12.1	27.2	18	11.3
Sugarcane (experimental)	California	32	72	48	30
Sugarcane (experimental)	California	30.5	68.5	46	28.5
Sudangrass	California	16	36	24	15
Alfalfa (surface irrigated)	New Mexico	6.5	15	10	6.1
Alfalfa	New Mexico	8	18	12	7
Bamboo	South East Asia	5	11	7	5
Bamboo (4-year stand)	Alabama	7	16	10	7
Abies saccharinensis (dominant species) & other species	Japan	6	13	9	6
Cinnamomum camphora (dominant species) & other species	Japan	6.8	15	10	6.3
Fagus sylvatica	Switzerland	4.3	9.7	6	4.0

Larix decidua	Switzerland	2.2	4.9	3	2.1
Picea abies (dominant species) & other species	Japan	5.5	12	8	5.1
Picea omorika (dominant species) & other species		6.4	14	10	6.0
Picea densiflora (dominant species) & other species	Japan	6.1	14	9	5.7
Castanopsis japonica (dominant species) & other species	Japan	8.3	19	12	7.7
Betula maximowicziniana (dominant species) & other species	Japan	3	7	4	3
Populus davidiana (dominant species) & other species	Japan	5.5	12	8	5.1
Hybrid poplar (short-rotation)					
Seedling crop (1 year old)	Pennsylvania	4	9	6	4
Stubble crop (1 year old)	Pennsylvania	8	18	12	7
Stubble crop (2 years old)	Pennsylvania	8	18	12	7
Stubble crop (3 years old)	Pennsylvania	8.7	20	13	8.1
American sycamore (short-rotation)					
Seedlings (2 years old)	Georgia	2.2	4.9	3	2.1
Seedlings (2 years old)	Georgia	4.1	9.2	6	3.8
Coppice crop (2 years old)	Georgia	3.7	8.3	6	3.5
Black cottonwood (2 years old)	Washington	4.5	10	7	4.2
Red alder (1-14 years old)	Washington	10	22	15	9.3
Eastern cottonwood (8 years old)		3	7	4	3
Eucalyptus sp.	California	13.4	30.1	20	12.5
Eucalyptus sp.	California	20.1	45.1	30	18.8
Eucalyptus sp.	Spain	8.9	20	13	8.3
Eucalyptus sp.	India	17.4	39.1	26	16.2
Eucalyptus sp.	Ethiopia	21.4	48.1	32	16.9
Eucalyptus sp.	Kenya	8.7	20	13	8.1
Eucalyptus sp.	South Africa	12.5	28.1	19	11.7
Eucalyptus sp.	Portugal	17.9	40.2	27	16.7
Miscellaneous					
Algae (fresh-water pond culture)	California	39	88	58	36
Tropical rainforest complex (average)		18.3	41.2	27	17.1
Subtropical deciduous forest complex (average)		10.9	24.5	16	10.2
Puckerbrush complexes (average)	North Carolina	2.2	4.9	3	2.1
Puckerbrush complexes (average)	Maine	4.4	9.9	7	4.1
World's oceans (primary productivity)		6	13	9	6

*Assuming 7500 Btu/lb.

however, it is probable that managed, productive sites could maintain production rates equivalent to the capture of 1 percent of the incident solar radiation during the crop-growing season. Thus, 1-percent capture of solar energy incident during a growing season could be used to estimate the potential energy supply obtainable by a plant system used as a solar collector. Obviously, in temperate climates this production would occur only during the warm months, but it could be continuous in fertile tropical and subtropical areas.

SOURCES OF PLANT MATERIAL

Plants that may be useful as energy sources may be found in one of three cultural schemes. First, the least disruptive source would be to utilize crop residues from land being used to grow food.* Only a part of the above-ground portions of plants is utilized as food. (As stated earlier, for many cereal crops only half of the above-ground dry weight is in the grain.) The energy in the crop residue is generally wasted. Even for plants that are fed to livestock, most of the energy is found in the animal wastes. Thus, crop residues and animal wastes are potential sources of energy for man.

The second source of plant material that could be utilized in an energy-recovery scheme is the native vegetation on non-farm lands. Most of this land will be untilled, but for good reason; it may have low productivity or be too rough or too wet to farm, or the yield per unit area may be low. Therefore, there are likely to be serious problems in gathering much plant material. Nevertheless, much uncultivated land could be a source of an occasional harvest of plants to be used for energy generation.

The third possibility could be the growing of special crops intended for use as energy sources—the so-called energy farm—which would allow the utilization of rapidly growing species without regard to the stringent quality requirements that often limit yields in food or fiber crops. The use of land for energy crops would compete, however, with the need for land on which to grow food. This may be a serious limitation to the establishment of energy farms, and decisions on how to use land must be made by the community involved. Here, we will consider factors that affect the productivity of plants, wherever they are grown.

*While not likely to be found in rural areas in developing countries, feed lots would be a source of residues that would belong in this category.

ENERGY YIELD

There are many factors that affect the energy yield of plant materials. These include climate, morphology (ground cover), insolation, availability of nutrients, soil conditions, and the efficiency of the photosynthetic reaction.

Climate

The climate of a region will exert a major controlling influence on the possibility of utilizing plants as a source of energy. In many developing countries the meager rainfall seriously inhibits plant growth. Dry countries have a limited capacity to produce plants that could be used as energy sources. Furthermore, the crops that traditionally supply the food in dry climates have little residue, and the potential for a significant source of energy from plants is nil.

Temperate-zone climates, with their seasonal variation, do not impose serious limitations on production to the point that it is unreasonable to think of plants as energy sources; rather the constraint in temperate climates is simply that production is seasonal. This seasonal constraint is compensated for in some cases, however, by the fact that high-latitude areas have long days during their summers, and the daily plant production per unit land area can be great.

In the tropics, the potential exists for a constant rate of production the year round; therefore, energy-recovery schemes could operate continuously, utilizing plant material in an identical form at all times. Even so, local rainfall patterns are likely to place constraints on the choice of plant species that could be utilized as energy sources at various times of year. Consequently, there is need to match the plant species to the climatic constraints of specific localities in order to provide a continuous supply of plant material for utilization as an energy source.

Ground Cover and Morphology

Apart from the obvious considerations of climate and soil, a primary condition that must be met before sizable productivity can be realized is that a large percentage of the available insolation must be absorbed by green plant tissue. This involves a number of considerations. First, harvests that leave only bare soil must be followed by rapid revegetation. Generally, plants growing from seed will be slow to provide an adequate ground cover compared to those that regrow from an established root system. Perennial species, or those that may regrow after harvest for several harvests, appear to have definite advantages over species that must be seeded. Thus, to maximize production, a species must be selected carefully. Several species that have a regrowth potential–and also possess the highly productive C_4 photosynthetic system (see below)–are adapted to the climate of the humid tropics.

The shape of the leaves and their orientation with respect to the sun are also important in determining the potential productivity of a species. Not only are there diverse genotypes from which to choose, but there is also the possibility of breeding specific morphological types to take best advantage of solar radiation. Thus, as far as plant morphology is concerned, there are both short-term and long-term possibilities.

Insolation

Plant productivity on a daily basis is highly correlated with the total daily insolation.[10] The long days in temperate climates or the cloudless days in semiarid tropics present unique possibilities in terms of daily production per unit land area. In estimating the local productivity potential, local insolation information would be essential.

Nutrient Elements

Plants as biological entities require large amounts of potassium, phosphorus, and especially nitrogen. Systems that recycle these elements as much as possible, by returning to the soil the sludge from anaerobic digestion or the ash residues from combustion schemes, would be highly desirable for minimizing the need for additional fertilizer. Where recycling is impractical, it is essential that a developing country, in order to maintain high plant productivity, have an economic system that will permit the manufacture or importation of fertilizer containing these elements. The problem will be exacerbated in areas of high rainfall where leaching by drainage water often depletes the soils of these elements.

A high rate of plant productivity is possible only if the soil has good tilth and water-holding capacity and is free of toxic substances. Generally, any soils meeting these requirements will already have been discovered and will be in use for agricultural crops. However, some areas having desirable soils may not be utilized in agriculture because they are virtually inaccessible or difficult to manage (they may be too steep or too wet, for example). These areas will have little value as energy-crop sites for the same reasons they are not being utilized by agriculture. It is likely, therefore, that energy farms will compete with food production. The need for food may preclude their utilization as sites for energy crops, and the utilization of crop or animal wastes may be the only practical way that agricultural land can provide alternative energy sources.

Photosynthetic Efficiency

In recent years, it has been discovered that plants utilize three different biochemical systems to convert sunlight to chemical energy.[11] All higher plant species utilize a biochemical system for the capture of light by chlorophyll and the transfer of this energy to stable chemical compounds. In some species, however, CO_2 is reduced to form a three-carbon compound and the biochemical system in which this is done is now referred to as the C_3 system for photosynthesis.

Other species that are especially adapted to high temperatures fix CO_2 in the light to form a four-carbon compound and are referred to as C_4 plants. Plant species that have the C_4 system are capable of high rates of CO_2 fixation and are generally more tolerant of high temperatures than are C_3 plants. They do poorly at cool temperatures, however.

Many species native to deserts have evolved a third system for fixing CO_2 at night, known as Crassulacean Acid Metabolism (CAM) from the fact that the system was discovered and studied in the *Crassulaceae*. The advantage of this system is that CAM plants can keep their stomata closed during periods of high temperatures and insolation and thereby drastically reduce their water loss. Therefore, they are particularly well adapted to arid climates. The system requires a large acid-storage capacity and all CAM plants have large, fleshy, photosynthetic organs. Unfortunately, most CAM species grow very slowly and their use in energy-recovery schemes would be confined to occasional harvests. In Table 3 are listed examples of C_3, C_4, and CAM species. In Table 4 we have listed some general environmental conditions and the plant type that is likely to display the greatest productivity under given conditions. An understanding of a species' photosynthetic system can be helpful in evaluating whether it should be used in a particular cropping situation. In some instances, productive energy schemes may well entail the importation of desirable non-native species suited to particular environmental conditions. Only a few of the world's plant species have been characterized as to their photosynthetic systems, however, and more work in characterization is needed.

MULTIPLE CROPPING AND INTERCROPPING

Because species differ in their reaction to light, temperature, and water stress, the highest yearly production in many climates will be obtained from a combination of species assembled to conform to seasonal weather. However, multiple crops present the problem of minimizing the inevitable periods of low productivity between the harvest of one crop and the establishment of an adequate leaf area on a succeeding crop. The most rapid establishment of a new ground cover after harvest would appear to be one of the criteria that should help define the optimum energy-crop sequence for a particular area.

Even in temperate zones, where often only a single crop is grown per year, it may be feasible to fit an energy crop into time slots not now being utilized. In Minnesota, for example, it is possible to plant winter rye late in the fall and raise it as an early spring crop, suitable as an energy feed stock, on farmland that will be seeded to maize in May. The rye germinates in the fall and develops a leaf cover that carries over into spring. The winter rye grows rapidly in cool spring weather and a harvest of 3 tons per a (7 metric tons per ha) or more of dry matter can be taken off the land before it is seeded to maize. The rye is of little value for forage and produces no grain, but it does contain energy that, presumably, could be recovered. Thus, an energy crop offers the potential of utilizing a part of the season now wasted for the 6 million acres on which maize is grown in Minnesota each year. Another example for Minnesota agriculture would be the possibility of utilizing land for an energy crop on which oats, barley, and wheat are grown after these crops are harvested in late July. Millions of acres lie idle in Minnesota during August and September after small grains are harvested.

These examples for an area in which there is strong tradition that a short

TABLE 3 List of some common C_3, C_4, and CAM species.

Typical C_3 plants	Typical C_4 plants	Typical CAM plants
Avena sativa L. (oats)	*Sorghum vulgare* pers. (sorghum)	*Agave americana* L. (century plant)
Oryza sativa L. (rice)	*Panicum miliaceum* L. (proso millet)	*Ananas comosus* Merr. (pineapple)
Hordeum vulgare L. (barley)	*Saccharum officinarum* L. (sugarcane)	*Aloe* sp.
Triticum aestivum L. (wheat)	*Zea Mays* (maize)	*Cactaceae* sp.
Gossypium hirsutum L. (cotton)	*Amaranthus retroflexus* L. (redroot pigweed)	
Medicago sativa L. (alfalfa)	*Cynodon dactylon* (L.) pers. (bermudagrass)	
Glycine max (L.) merr. (soybeans)	*Digitaria sanguinalis* (L.) (crabgrass)	
Arachis hypogeae L. (peanuts)		
Helianthus annuus L. (sunflower)		
Dactylis glomerata L. (orchardgrass)		

TABLE 4 General Environmental Features That Favor Growth of Certain Plant Types.

General Environmental Conditions	Favored plant type
High light intensity and high day and night temperatures	C_4
Low (25°C) day temperatures	C_3
High to medium light intensity low night temperature (20°C)	CAM
High light intensity and temperature, low fertility	C_4
Low and/or sporadic rainfall	CAM
Low annual rainfall, but seasonal high light, high temperature	C_4

growing season permits only a single crop illustrate the adaptability of "energy" crops (if energy recovery is the principal criterion and the quality or form of the product is unimportant). It is highly probable that similar seasonal niches will be found and could be utilized in most agricultural systems.

In addition, since individual crop species often make unique demands on space and the environment, two carefully selected productive species grown simultaneously not only will not interfere with each other, but can give greater yields from a plot they share than either species would if grown in isolation.[12] This is especially true in less intensively managed situations.[13,14,15] Relatively little experimental work has been done on intercropping, but it is a practice that should be evaluated in greater depth as part of the attempt to utilize plants to capture solar energy.

TECHNOLOGY OF PLANT PRODUCTION

The production of the base biomass for the capture of solar energy involves many of the same plant-production techniques and many of the same problems that are encountered in present-day agriculture.

Cost

The cost of producing biomass for energy capture will approximate the costs of crop production. To this must be added the cost of delivery of the plant material to the energy generator sites and the handling of the material during conversion. These costs will depend to some extent on the conversion scheme selected to recover the energy from the plant material.

High energy yields per acre may well depend on the availability of an ample supply of fertilizer. This will always represent a direct cost for energy production. Some saving can be made by recycling residues from the energy recovery facilities. Nevertheless, distribution problems and fluctuations in fertilizer production (witness the recent worldwide shortage) may present serious problems of supply and foreign exchange that could limit energy farming in some countries. Even so, the available agricultural residues alone present an attractive source of potential energy in most humid climates.

Another major "cost" that must be considered is that energy crops or energy farms may compete directly with food production for both land and fertilizer. The need for food may limit the use of choice land for energy farms. Perhaps crop residues and occasional harvests from untilled land may be the only practical sources of plant material for energy feed stocks. These materials will often be available to communities without direct cost other than that of collection and handling.

Reliability

The knowledge and experience for growing plants is available in all countries. On-site advisors may be able to increase the efficiency of the production process, but the reliability has been established, over many centuries, for nearly all local environments. In general, the reliability will depend more on water supply than on any other factor.

Local Implementation

The work involved in growing, harvesting, and collecting plant material and even conducting necessary short-term research can be done locally with local technology (except, perhaps, for providing chemical fertilizers in situations where they are needed). The choice of management practices and of species to be used in energy-capture systems will probably require, at the start, the availability of trained consultants, not to mention the personnel needs that will ultimately arise in connection with the energy-conversion factory.

RESEARCH AND DEVELOPMENT

An engineering plan for recovering energy from plant material has not been specified. Once that is done and the acceptable form of the plant material has been specified, the general limits on production schemes can be established and local areas can be classified as to their acceptability for use in producing energy. Factors such as distance from the recovery plant, suitability for different cropping schemes, and the decision on the need to use land to grow food or to collect solar energy must be made. Trained personnel will likely be needed to advise the local inhabitants.

On-site research will be required to determine the species to be used and how they are to be managed. Non-native species may be required to maintain the highest productivity. Agronomic advisors will be necessary to suggest efficient plants for capturing energy. Local people could be trained as technical assistants. Important aspects of this research will include the efficient use of fertilizer and an analysis of the energy consumed in growing, harvesting, drying, and handling as against the energy produced.

POTENTIAL FROM UNTILLED LAND

Many communities have land, not now used for farming, that could be used for energy crops. Scrub woodland, swampy areas, and land too steep or too infertile for agricultural use may support plants that could be used as fuel for energy-recovery schemes. Some of these sites, such as swamp land, may be highly productive. Others may support an occasional harvest of an energy crop. Every community would need to conduct a local survey of such areas to evaluate the potential of local energy sources. Trained personnel should be involved in those local surveys. Care must be exercised in utilizing such areas to prevent possible long-lasting ecological damage. Some local experimentation may be required to determine the effects of harvesting plants from wasteland.

TIME SCALE OF RESULTS

Four types of research are involved in establishing energy-crop schemes. First, the principles of maintaining highly productive systems should be outlined in a detailed training manual. The information is available in current literature and the task could be completed within 6 months after being commissioned. The principles that have been touched on in this report should be included, as well as directions for conducting field experiments, sampling populations, and evaluating results. This manual would be useful for on-site agronomic advisors (who perhaps may be familiar only with food crops, but should not exclude any species from consideration) and in training technical people to assist with the research.

The second type of research would involve inventories of local resources including crop residues and potential harvests from untilled land. This should take no longer than about 2 weeks for each community. It will require trained agronomists who are good botanists and ecologists as well, or small, mobile, interdisciplinary teams.

The third area involves local or regional research on cropping schemes. This will take a minimum of 2 years and should be relatively complete in 5. It will involve factors such as the identification of optimum species, selection of combinations for multiple-cropping schemes, evolution of management schemes for maximum productivity and efficient use of fertilizers, and determination of energy and food crop combinations. However, it need not be completed in order to begin utilizing plant materials readily available in many communities.

The fourth type of research will address long-term agronomic questions and such questions as the ecological management of waste lands. Management of low-fertility sites for energy crops, adaptation of major technological developments, and long-term questions such as development of irrigation potential should be investigated. Results are unlikely before 10 years at the earliest.

The basic research to support these suggested applications cannot be divorced from the plans for energy recovery from plant material; therefore, it should involve both engineers and ecologically oriented agronomists. Appropriate research groups could be assembled from personnel at a few universities and institutes that have specialized in agricultural research. The research could be supported by foundations or governments as independent efforts. The problem will be to identify groups who have an interest in addressing these questions. Announced availability of support from foundations, governments, or international agencies, on a contract or grant basis, should help draw responses from interested and qualified groups.

REFERENCES

1. Ahlich, J. A., Jr., and Hinman, Robert E. 1974. *Effective utilization of solar energy to produce clean fuel*. Report to the National Science Foundation. Menlo Park, California: Stanford Research Institute.
2. Hodgman, C. D., ed. 1945. *Handbook of chemistry and physics*. 29th ed. Cleveland, Ohio: The Chemical Rubber Publishing Co.
3. *Steam: its generation and use*. 1963. New York: Babcock and Wilcox Co.
4. Rothermel, R. C., and Philpot, C. W. 1973. Predicting changes in chaparral flammability. *Journal of Forestry* 71(10):640-643.
5. Loomis, R. S. *et al.* 1971. Agricultural productivity. *Annual Review of Plant Physiology* 22:431-468.
6. Szego, G. C., and Kemp, C. C. 1973. Energy forests and fuel plantations. *Chemical Technology* (May): 275-284.

7. Loomis, R. S., and Williams, W. A. 1963. Maximum crop productivity: an estimate. *Crop Science* 3:67-72.
8. deWit, C. T. 1959. Potential photosynthesis of crop surfaces. Reprint. Wageningen, Holland: Institute of Biological and Chemical Research on Field Crops and Herbage.
9. Zelitch, I. 1971. *Photosynthesis, photorespiration and plant productivity*. New York: Academic Press.
10. Moss, D. N.; Musgrave, R. B.; and Lemon, E. R. 1961. Photosynthesis under field conditions. III. Some effects of light, carbon dioxide, temperature, and soil moisture on photosynthesis, respiration and transpiration of corn. *Crop Science* 1:83-87.
11. Hatch, M. D. 1971. Mechanism and function of the C_4 pathway of photosynthesis. In *Photosynthesis and photorespiration*, Hatch et al., eds., pp. 139-152. New York: Wiley-Interscience.
12. Sprague, H. B. 1936. *Bulletin. New Jersey. Agricultural Experiment Station* 609:1.
13. Andrews, D. J. 1972. Intercropping with sorghum in Nigeria. *Experimental Agriculture*, London 8:139-150.
14. Multiple cropping. 1974. In *IRRI 1974: annual report for 1973*, pp. 15-34. Los Baños, Philippines: IRRI (International Rice Research Institute).
15. Kelly, Isabel T., and Palerm, Angel. 1952. *Tajin Totonac: history, subsistence, shelter and technology*. Institute of Social Anthropology, Publication No. 13, pp. 99-149. Washington, D.C.: Smithsonian Institution.

SUGGESTED READING:

Begg, J. E. 1965. High photosynthetic efficiency in a low-latitude environment. *Nature* 105:1025-1026.

———, and Jarvis, P. G. 1968. Photosynthesis in Townsville lucerne (*Stylosanthes humilis*). *Agricultural Meteorology*, Amsterdam 5:91-109.

Black, C. C.; Chen, T. M.; and Brown, R. H. 1969. Biochemical basis for plant competition. *Weed Science* 17(3):338-344.

Bonner, J. 1962. The upper limit of crop yield. *Science* 137:11-15.

Chandler, R. F., Jr. 1969. Plant morphology and stand geometry in relation to nitrogen. In *Physiological aspects of crop yield*, J. D. Eastin *et al*., eds., pp. 265-285. Madison, Wisconsin: American Society of Agronomy (ASA) and Crop Science Society of America (CSSA).

Field crops. 1974. Washington, D.C.: U.S. Department of Agriculture, Crop Reporting Board, Statistical Reporting Service.

Heichel, G. H. 1973. Comparative efficiency of energy use in crop production. *Bulletin of the Connecticut Agricultural Experiment Station*, New Haven, Connecticut 739:1-26.

Pimentel, David *et al*. 1973. Food production and the energy crisis. *Science* 182:443-449.

Monteith, J. L. 1965. Light distribution and photosynthesis in field crops. *Annals of Botany New Series*, London 29:(113)17-38.

Moss, D. N. 1965. Capture of radiant energy by plants. *Meteorological Monographs* 6:90-108.

———. 1967. Solar energy in photosynthesis. *Solar Energy* 11:173-179.

Westlake, D. F. 1963. Comparison of plant productivity. *Biological Reviews of the Cambridge Philosophical Society* 38:385-425.

Microbiological Conversion of Plant Materials to Liquid Fuels

The microbiological approach to the identification of alternative energy sources falls into two categories, gaseous- and liquid-fuel generating systems, both of which, in general, utilize anaerobic microbial populations. The generation of methane fuel gas by means of the classical "anaerobic digestion," the best known of these anaerobic processes, is the subject of a separate, concurrent study and will not be discussed here.[1] The focus in this chapter is on the microbiological production of other gaseous and liquid fuels—more specifically, production by microbial action of the following fuels and fuel systems:

- ethanol;
- acetone/butanol + hydrogen;
- butanol/isopropanol + hydrogen; and
- acetone/ethanol + hydrogen.

In addition, the raw materials for the production of these fuel systems will be discussed, as will the present technology and the requirements for the translation of this technology to developing countries.

A TYPICAL PROCESS TECHNOLOGY

Virtually all of the chemicals that can be considered as potential sources of energy have been produced microbiologically in commercial-scale operations in the past. Most of these materials were produced during the critical years of the 1940s, when nations on both sides in World War II were seeking supplementary fuel sources, and to a lesser extent in the early 1950s.

From a chemical-process point of view, the methods of microbiological production of the variety of fuel systems outlined above are quite like one another. Since current industrial production relies on synthesis from petrochemical sources, however, with the exception of ethanol none of these chemicals is being commercially produced by microbiological processes at the present time. (A typical schematic flow diagram, in this case illustrating the production of a mixture of acetone, ethanol, and butanol, with minor amounts of ethanol and hydrogen, is shown in Figure 56.)

The proper selection of the raw material (substrate) is critical to the economic viability of all microbiological processes. The example in Figure 56 is based on the uses of molasses as the raw material, though we shall discuss other raw materials as well in considering the transferability of this technology to developing nations.

In the industrial process for solvent production by microbial action, the general procedure calls for the sequential development of inoculum necessary for the final production fermentor. The inoculum buildup is shown in the upper left quadrant in Figure 56; required quantities of the desired microbial cultures are produced, starting with test-tube batches and moving to successively larger culture vessels.

The final fermentors for the production of chemicals may vary in size; this example employs fermentors ranging from 60,000 to 500,000 gal (227 to 1,900 m^3). As the nutrients for the production fermentors undergo processing, the medium must be sterilized to eliminate interfering or otherwise undesirable organisms. The nutrient needs of the microorganisms must also be defined.

The conversion of the raw materials (carbon, nitrogen, salts, etc.) into the microbial biomass takes place in the fermentor. The organisms act here as "bio-catalysts" capable, generally in the absence of oxygen (anaerobically), of producing the desired products during the conversion process. Sufficient stirring is required to maintain a homogeneous suspension as well as to facilitate the transport of nutrients to the microorganisms and residual and metabolic products away from the microorganisms. In addition, since in these processes the overall reaction is exothermic, the liberated heat must be removed through convective heat transfer. The growth of the microorganisms and the subsequent production of chemicals are generally performed in batch-wise fashion; typical times of operation for this segment of the process range from 40 to 60 hours.

The main products from the example cited are acetone, ethanol, and butanol. These materials are produced extracellularly and are recovered from

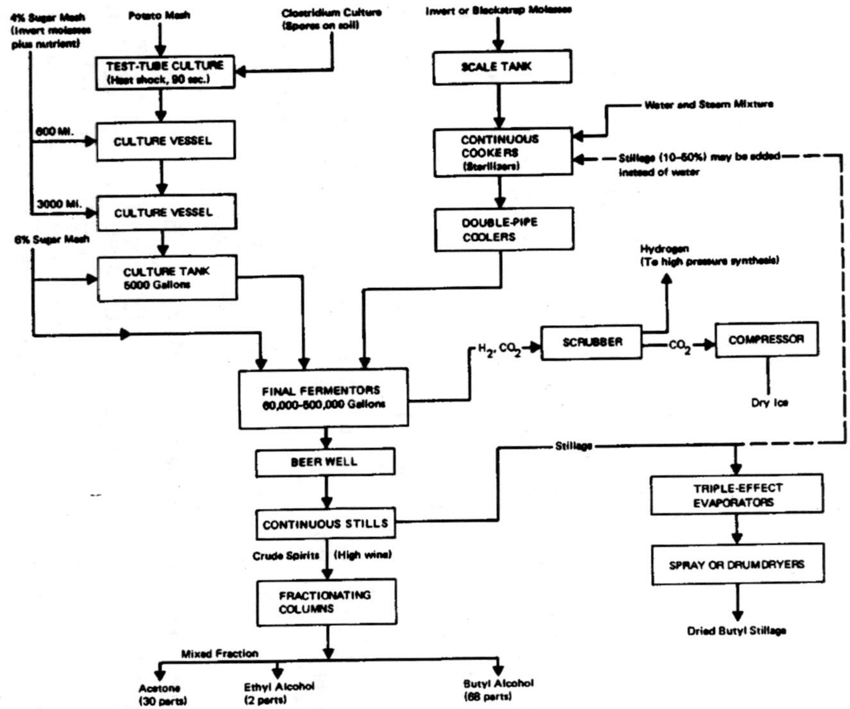

FIGURE 56 Flow diagram of typical modern acetone-butanol fermentation using sugar products. [Source: Beesch, Ref. 2. Reprinted with permission from *Industrial and Engineering Chemistry*. Copyright by the American Chemical Society.]

the reaction mixture by conventional distillation. In addition, there are several by-products that have potential value. During fermentation, for instance, hydrogen and carbon dioxide are formed as gaseous by-products. The carbon dioxide can be removed by scrubbing and recovered as dry ice. The hydrogen gas can also be recovered as a raw material for other syntheses or, alternatively, as an energy source. The stillage (or bottoms) from the distillation is composed of the microbial biomass as well as nonvolatile products formed during fermentation. Portions of these materials can be recycled directly to the fermentor as nutrients for microbial growth, or, dewatered by evaporation and then dried, these materials can become an excellent animal-feed supplement or fertilizer.

The production of chemicals that might be considered as energy sources must be evaluated from the point of view of material and energy balances. In Table 1 are shown the results of this analysis with energy balance estimated for the purpose of this report, based on materials balance information obtained from Beesch, 1952[2] (Figure 56). The material balance was based on raw-material input of 100 kg of molasses containing 57 kg of fermentable sugars, 3.1 kg of protein, and 6.2 kg of ash. The other material added (15.2 kg) was presumably recycled stillage. The caloric values associated with the different raw materials are calculated on the basis of average values of heats

TABLE 1 Typical Material and Energy Balances of Acetone-Butanol-Ethanol Fermentation

Starting Material–100 kg Blackstrap Molasses

Component	Quantity (kg)	Energy (kcal. × 10^{-3})
Materials and Energy Input:		
Sucrose and Invert Sugar	57.0	212
Protein	3.1	12
Ash	6.2	0
Other	15.2	56
Total	81.5	280
Materials and Energy Output:		
Butanol	11.5	100
Acetone	4.9	36
Ethanol	0.5	3
Carbon Dioxide	32.1	0
Hydrogen	0.8	24
Dry Feed (6 kg prot/6 kg ash)	28.6	106
Total	78.4	269

of combustion. The primary products were butanol, acetone, and ethanol. By-products included carbon dioxide, hydrogen, and animal-feed proteins.

In terms of the energy available from the combustible products (i.e., the solvents and the gaseous hydrogen), this system achieves a conversion of 77 percent of the input sugars. Although in this example the primary product was butanol, it is technically feasible to manipulate the product solvent mixture to any desired ratio, without altering significantly the conversion efficiency.

CURRENT TECHNOLOGY (AVAILABLE WITHIN 5 YEARS)

The usefulness of microbiological production of energy in developing nations must be considered in the light of the technical and economic constraints facing those nations. These constraints are quite different from those operating in an industrialized nation, and include:

- The availability and appropriateness of raw materials for energy conversion;
- The technical challenge—village-scale concept; and
- The type of research and development required and projected outcome.

RAW MATERIALS FOR ENERGY CONVERSION

It has been stated previously that the selection of the raw material for the production of chemicals is extremely important. In an industrialized country this selection is generally based on economics. In rural areas where a cash economy may not even exist, the raw materials needed for production must be freely available. Furthermore, the types of raw materials available frequently differ from one locality to another. The raw material requirements of the microbiological production process are somewhat less critical than those of most chemical processes, since a variety of primary raw materials can be considered.

Table 2 lists various primary raw materials that can serve as carbon and energy sources for the production of the various output chemicals. In general, the soluble carbohydrates can be used directly for fermentation. However, waste products and the cellulosic type of substrates including wood, hulls, and citrus waste, are likely to be more abundant and less important to the economy of developing nations. Unfortunately, the technology for their utilization is less advanced than for soluble and easily fermentable substrates such as molasses, sugars, and starches.

In addition to the energy or carbon source needed for the microbial process, there are other raw-materials needs. The most important of these are the inorganic nutrients such as ammonia, phosphorus, magnesium, and other minerals in trace amounts. The fact that these necessary chemicals must be purchased could constitute an economic barrier to use of the microbial process in some places.

THE TECHNICAL CHALLENGE: VILLAGE-SCALE CONCEPT

There is also a serious question as to whether, even with suitable raw materials available, the technology exists at the village level in a developing country to make the microbiological production of energy practical; the developments in this technology have been achieved mainly in industrialized nations, where all of the necessary advanced technological capabilities are readily available. The degree of technological advancement necessary for success will be illustrated by considering the various segments of the microbial conversion process.

Microbial Culture Maintenance and Inoculum Preparation

The microbial cultures that are potent producers of the desired chemicals must be maintained according to established procedures. In the industrialized nations, this involves the availability of trained technical personnel and proper physical facilities. However, the usual criterion of maximum product yield, on the basis of which high-technology industries operate, may not be

TABLE 2 Types of Raw Materials Potentially Useful for Microbial Conversion to Fuels in LDCs*

Ethanol	Acetone-Butanol	Butanol-Isopropanol	Acetone-Ethanol
Molasses	Molasses	Molasses	Molasses
Sulfite Liquors	Sulfite Liquors	Sugar Cane	Potatoes
Cellulose Pulp	Corn Cobs	Raw Sugar	Corn
Pineapple Juice	Wood Sugars	Wood Sugars	Peanut Hulls
Potatoes and Potato Products	Cassava	Sulfite Liquors	Oat Hulls
Citrus Waste		Starchy Products	Corn Cobs
Sweet Potato			Wood Sugars
Manioca Meal			

*Production of the chemicals listed has been demonstrated from all of the raw materials listed under each chemical.

essential in a village operation. One could therefore relax the stringent culture maintenance and development programs that commercial competition necessitates.

At the same time, since the general practice for the microbiological production of chemicals has been to employ batch rather than continuous operations, the preparation of inoculum for seeding subsequent fermentors becomes an integral part of the overall scheme. This would require laboratory-scale and pilot-scale inoculation fermentors as well as trained personnel for their successful operation. It is difficult to conceive that technical talents of this nature are readily available at the village level.

Production Fermentor Operations

The heart of the process for solvent production is the fermentor, in which microbial propagation and fermentation of the raw materials take place. These involve sterilization of nutrients as well as maintenance of aseptic conditions during fermentation. The process control for such operations is minimal, but it must be reliable. On the other hand, it is probable that in most villages—certainly within village clusters—some kind of beer is routinely brewed, which surely indicates some degree of technical competence.

Product Recovery

The other critical technical challenge for the village would involve the unit operations for product recovery that are necessary if the solvents produced are to be used as fuel. Recovery of gaseous products from fermentors has become a common and routine practice at the farm and village level in many developing countries[1] and, apart from considerations of safety, the low-pressure storage of a combustible fuel such as hydrogen might be useful in this situation. In a typical example, off-gas from a fermentor might contain 40 percent H_2 and 60 percent CO_2. The utility of this type of fuel must first be determined, however.

The most critical aspects of the product recovery process lie in the distillation requirement to recover a useful solvent. To be useful as a fuel to operate machinery—or even to use as a heat source for cooking or other processes—the solvent must be recovered in a sufficiently pure form that the concentration of water will not be high enough either to interfere with combustion or to reduce the heat value (Btu/lb or kcal/kg) to an uneconomic level. To do this for the solvents involved requires distillation equipment and modes of operation easily available in an industrial complex, but not within the capabilities of even the village producers of distilled alcoholic liquors. Lastly, the recovery of stillage as a potential animal feed or fertilizer would require dewatering technology. Here again, one would have some reservation as to the capabilities of achieving this successfully at the village level.

All of these considerations, then, from availability of raw materials to the technological prerequisites, appear to dictate the following conclusion: the present state of technology in the microbiological production of fuels from agricultural products does not make this a practical approach.

RESEARCH AND DEVELOPMENT

The probability of microbiological production of fuels at a village level could be enhanced provided research and development efforts are initiated. Research needs include:

- Examination of the type and nature of available raw materials, e.g., starches, soluble carbohydrates, waste products (cellulosics) in candidate locales (see chapter on photosynthesis);
- Exploration of the potential of microorganisms known to perform the desired conversions, with raw materials already proven useful;
- Isolation and screening programs for new microorganisms capable of performing the desired conversion using either known or new and unproven raw materials, e.g., indigenous carbohydrate wastes and cellulosics;
- Development of fermentation systems and product recovery schemes in which minimum technical skills will not prevent high probability of success, e.g., non-aseptic fermentations, minimum process-control requirements, reliable operations;
- Development of equipment for fermentation and recovery operations that is less capital-intensive (e.g., economical and reliable dewatering techniques, non-metallic fermentors made with cheap materials and construction techniques*); and
- Delineation of the exact requirements and specifications of the microbial products sought from locale to locale or application to application.

These suggestions are all directed to adapting a known technology to the constraints characteristic of rural areas in a developing country. The potential

*For example, the report, *Ferrocement: Application in Developing Countries* (National Academy of Sciences, Washington, D.C., February 1973), recommended that consideration be given to "the use of ferrocement to replace steel–particularly stainless steel–in manufacturing at least some units of basic food-processing equipment." Among the possibilities suggested were fermentation vats, storage vats or tanks, and driers of various sorts.

already exists for much of this research and development to be performed by institutions in developing countries. For example, for countries located in Central America, the Institute for Central America Industries and Technology (ICAITI) located in Guatemala City has the equipment, the scientific personnel, and the perspective for the task. When research is conducted using equipment and physical facilities incompatible with situations in developing countries, the results are often totally irrelevant or unacceptable for other, less-tangible reasons; therefore it is important that research institutes in industrialized nations play a limited—perhaps only a consulting—role. Other research institutes similar to ICAITI in other parts of the world (such as the Central Food Technology Research Institute in Mysore, India) should also be considered.

The funding of these activities could be part of the programs of the aid-donor nations, international agencies, or perhaps private foundations. Although it would be useful to make one agency, e.g., AID, responsible for technical coordination and monitoring of progress, it is strongly recommended that scientific personnel from universities and industry play an active role.

REFERENCES

1. National Academy of Sciences. *Methane generation from human, animal, and agricultural wastes.* Report of an *ad hoc* panel of the Advisory Committee on Technology Innovation, Board on Science and Technology for International Development, Commission on International Relations. (In preparation. See p. 306.)
2. Beesch, S. C. 1952. Acetone-butanol fermentation of sugars. *Industrial and Engineering Chemistry* 44(7):1677-1682.

SUGGESTED READING

Ethanol Production

Agar, H. D., and McCarthy, J. L. 1951. Fermentation to alcohol of concentrated calcium, magnesium, and ammonium sulfite waste liquors. *TAPPI* (Technical Association of the Pulp and Paper Industry) 34(3):126-131.

Barron, H. 1944. *Distillation of alcohol.* Louisville, Kentucky: Joseph E. Seagram and Sons, Inc., Division of Education.

Bilford, H. B.; Scalf, R. E.; Stark, W. H.; and Kolachov, P. J. 1942. Alcoholic fermentation of molasses, rapid continuous fermentation process. *Industrial and Engineering Chemistry* 34:1406-1410.

Boruff, C. S., and Van Lanen, J. M. 1947. The fermentation industry during World War II. *Industrial and Engineering Chemistry* 39:934-937.

Borzani, W., and Falcone, M. 1953. Control of starter preparation in fermentation of molasses. *Journal of Agricultural and Food Chemistry* 1:1070-1071.

Dunn, P. M. 1945. Industrial alcohol from wood waste. *Chemical Products and the Chemical News* 9(1/2):15-17.

Jacobs, P. B. 1947. *Alcohol from agricultural commodities.* U.S. Department of Agriculture, Report AIC-95. Washington, D.C.: U.S. Department of Agriculture.

Joseph, H. G. 1947. Industrial alcohol production from sulfite waste liquor by the Ontario Paper Co., Ltd. *Sewage and Industrial Wastes* 19(1):60-69.

Leonard, R. H., and Hajny, G. J. 1945. Fermentation of wood sugars to ethyl alcohol. *Industrial and Engineering Chemistry* 37:390-395.

Lichty, L. C., and Phelps, C. W. 1938. Gasoline-alcohol blends in internal-combustion engines. *Industrial and Engineering Chemistry* 20:222.

Liebmann, A. J. 1942. Alcohol in war. *Wallerstein Laboratories Communications* 5(15):131-137.

Miller, E. S. 1943. Industrial alcohol from wheat. *American Miller* 71:39-42.

Thorne, C. B.; Emerson, R. L.; Olson, W. J.; and Peterson, W. H. 1945. Evaluation of malts for production of alcohol from wheat. *Industrial and Engineering Chemistry* 37:1142-1144.

Acetone–Butanol Production

Arzberger, C. F.; Peterson, W. H.; and Fred, E. B. 1920. Certain factors that influence acetone production by *Bacillus acetoethylicum. Journal of Biological Chemistry* 44:465-479.

Banzon, J. R. 1941. Fermentative utilization of cassava. *Iowa State College Journal of Science* 16:15-18.

Beesch, S. C. 1952. Acetone–butanol fermentation of sugars. *Industrial and Engineering Chemistry* 44:1677-1682.

————. 1953. Acetone–butanol fermentation of starches. *Applied Microbiology* 1(2):85-95.

Bryner, L. C.; Christensen, L. M.; and Fulmer, E. I. 1936. Hydrolysis of oat hulls with hydrochloric acid. *Industrial and Engineering Chemistry* 28:206-208.

Fred, E. B.; Peterson, W. H.; and Anderson, J. A. 1923. Production of acetone, alcohol and acids from oat and peanut hulls. *Industrial and Engineering Chemistry* 15:126.

Gabriel, C. L. 1928. Butanol fermentation process. *Industrial and Engineering Chemistry* 20:1063-1067.

———, and Crawford, F. M. 1930. Development of the butyl-acetone. fermentation industry. *Industrial and Engineering Chemistry* 22:1163-1165.

Gill, A. 1919. The production of normal butyl alcohol and acetone by fermentation of horse chestnuts. *Society of Chemical Industry Journal*, London 38:411-412T.

Leonard, R. H.; Peterson, W. H.; and Ritter, G. J. 1947. Butanol-acetone fermentation of wood sugar. *Industrial and Engineering Chemistry* 39:1443-1445.

Morikawa, K., and Prescott, S. C. 1928. A new fermentation yielding butyl and isopropyl alcohols. *Journal of Bacteriology* 13:58.

Nathan, F. 1919. The manufacture of acetone. *Society of Chemical Industry Journal*, London 38:271-273T.

Peterson, W. H., and Fred, E. B. 1932. Butyl-acetone fermentation of corn meal. *Industrial and Engineering Chemistry* 24:237.

Prescott, S. C., and Morikawa, K. Production of butyl and isopropyl alcohols. U.S. Patent 1,933,683 (Cl. 195-20), 7 November 1933, Appl. 26 January 1927; 3 pp.

Robinson, G. C. 1922. A study of the acetone and butyl alcohol fermentation of various carbohydrates. *Journal of Biological Chemistry* 53:125-154.

Waksman, S., and Kirsh, D. 1933. Butyric acid and butyl alcohol fermentation of hemicellulose- and starch-rich materials. *Industrial and Engineering Chemistry* 25:1036-1041.

Wendland, R. T.; Fulmer, E. I.; and Underkofler, L. A. 1941. Butyl-acetonic fermentation of Jerusalem artichokes. *Industrial and Engineering Chemistry* 33: 1078-1081.

Wiley, A. J.; Johnson, M. J.; McCoy, E.; and Peterson, W. H. 1941. Acetone-butyl alcohol fermentation of waste sulfite liquor. *Industrial and Engineering Chemistry* 35:606-610.

Butanol—Isopropanol Production

Baba, T. 1952. On the industrial production of butanol, isopropanol, and acetone by a fermentation process. *Bulletin. Faculty of Engineering, Hiroshima University* 1: 101-110.

Osburn, Oiven Lloyd. 1935. The production of butyl and isopropyl alcohols by fermentative processes. *Iowa State College Journal of Science* 10:97-98.

——; Brown, R. W.; and Werkman, C. H. 1937. The butyl alcohol-isopropyl alcohol fermentation. *Journal of Biological Chemistry* 121:685-695.

——; ——; and, ——. 1938. Dissimilation of intermediary compounds in the butyl-isopropyl alcohol fermentation. *Iowa State College Journal of Science* 12:275-284.

——, and Werkman, C. H. 1935. Utilization of agricultural wastes. II. Influence of nitrogenous substrate on production of butyl and isopropyl alcohols by *Clostridium butylicum*. *Industrial and Engineering Chemistry* 27:416-419.

PART III
OTHER TECHNOLOGIES

Geothermal Energy

Geothermal resources have been known since ancient times when they were widely utilized to support establishments for therapeutic hot bathing. For hundreds of years this was their only use, particularly in Europe, where cities grew up as spas based upon the therapeutic use of hot springs. At the beginning of the 20th century, a new application of geothermal energy was found at Larderello, in Italy, where geothermal steam was used for the generation of electricity.

Exploitable geothermal resources are most likely to exist in those parts of the world that have experienced geologically recent volcanism (in the past million years), though they may exist in other areas also. Known geothermal areas include the circum-Pacific "circle of fire," which is a region of active earthquakes and volcanoes. This comprises the western part of South America and the Andes, most of Central America, large parts of the western United States and the Rocky Mountains, the Aleutian Islands, the Kamchatka peninsula, Japan, Taiwan, Indonesia, the Phillipines, New Zealand, and many South Pacific Islands. A large part of East Africa, associated with the Rift Valley System, also has geothermal activity in Ethiopia, Kenya, Uganda, Tanzania, and Zaire. In Asia and Europe hot springs extend from Nepal, India, Afghanistan, Iran, and Turkey to Greece and Italy.

Many other countries and geological provinces possess geothermal resources, although they are devoid of hot springs or other evidence of hot water at depth. For example, it has been estimated that as much as one-third of the Soviet Union is underlain by exploitable geothermal energy.

Geothermal energy is the natural heat contained within the earth.

Temperatures in the earth increase with depth. At a depth of 20 km the temperature is close to 750°C. However, most of this heat is too deeply buried to be exploited. The depth from which heat may be extracted economically is unlikely to exceed 10 km; even in this outer 10 km most natural geothermal heat cannot easily be extracted and put to practical use. Geothermal energy does, however, have economic significance where the hot rock at depth is fractured or has pore spaces that permit water or steam to circulate and carry heat from the rock to the surface if a natural channel exists or if a geothermal well is drilled. It has been proposed that in order to exploit the heat contained in rocks that are not penetrated by fractures or pore spaces, fractures could be induced artificially, using explosives or hydraulic fracturing techniques. These proposals are now under active practical investigation in the United States, but it seems unlikely that this technology will be available for practical use in the near future.

PAST AND CURRENT USES

The main uses of geothermal energy have been to generate electricity and to heat buildings and greenhouses. Geothermal energy is used for space heating and in greenhouses in Iceland, Hungary, the Soviet Union, and the United States. It is used for air-conditioning buildings, product-processing in a paper-pulp plant in New Zealand, drying diatomite in Iceland, recovering salt from sea water in Japan, and for heating soil for agriculture in Israel. Some chemical products may be associated with geothermal energy utilization, such as the production of dry ice from geothermal carbon dioxide in the United States, the production of borax associated with geothermal steam in Italy, and the production of calcium chloride from geothermal brines in the Imperial Valley, California.

CURRENT TECHNOLOGY (AVAILABLE WITHIN 5 YEARS)

Applications of current technology to exploit geothermal sources are relatively simple or complex, depending on the temperature of the source–hot water sources that are below the normal boiling point, or sources of steam or superheated water.

Simple Applications

Applications involving water at a temperature below the boiling point–in general, those using water for heating or bathing–do not involve a complex technology. All that is required is an ability to dig shallow wells, to lift water through small vertical distances and to construct channels or pipes to conduct water to the point of use. When applied on a small scale where hot springs occur naturally, this type of thermal energy use can be had at virtually zero cost. There are few locations in the world, however, where opportunities for this type of development occur that have not already been exploited.

The development of geothermal resources in other circumstances (where high-temperature water and steam are required or where the resource lies at a considerable depth) presents special problems in technology and finance.

Complex Applications

Geothermal resources have attracted considerable interest among energy users in the recent past as a possible substitute for oil in the generation of electricity and heat.

In many places in the world, there is high probability of finding hot water for space heating and process heating. High-temperature geothermal water and steam have been exploited in countries such as Iceland, Italy, Indonesia, Japan, the Philippines, South America, the western United States, and Central America, where volcanic activity has occurred within the geologically recent past. In general, high-temperature geothermal water and steam are suitable for electricity production at delivered costs that are significantly lower than those for electricity produced from fuel oil at current prices. Plant requirements are similar to those of standard thermal power stations, with the difference being that neither boiler nor fuel is required to produce the steam. The capital cost of a geothermal electricity-generating plant, therefore, is lower than that of a fossil-fueled thermal plant, and there are, of course, no fuel costs. There is, however, an element of financial risk and expense in exploring and drilling for water and steam, since it cannot be guaranteed that they will be found in adequate quantity and quality for electricity generation. This must be considered part of the total cost of production.

Some high-temperature geothermal water contains dissolved mineral salts that, when present in high concentration, make it difficult to discard necessarily substantial quantities of geothermal water without causing unacceptable pollution of surface or ground water. Disposal by channel to the sea has been adopted in the Ahuachapan Field in El Salvador, where alternative reinjection beneath the surface has also been tested. Apart from being an environmental problem, this is also a cost item.

Geothermal exploration makes use of advanced technology; geothermal drilling requires the use of equipment and techniques similar to those in use in the oil and gas industry, though it requires the use of drilling personnel with geothermal experience. It follows, therefore, that there is a minimum scale of activity and investment of risk capital that is required if geothermal exploration is to be carried out with good prospects of success. Exploration that is economically and technically justified normally must be undertaken where there is a requirement for electricity production of several megawatts or several tens of megawatts. When the requirement for electricity is in the kilowatt range—as, for example, in some rural applications—geothermal sources will not be economically viable unless steam-production costs can be shared among several communities.

EXPLORATION

Exploration and prospecting are field activities that must be carried out before drilling if the risk of unproductive drilling is to be minimized. Exploration for geothermal resources in many countries is likely to take place in at least two phases: an initial phase of reconnaissance, and a second phase of more-detailed investigations and possibly exploratory drilling.

Surveys commonly undertaken as part of the reconnaissance phase include: a) airborne infra-red scanning surveys, in situations where little information on the location of hot springs is available; b) regional studies of the hydrogeochemistry of known hot springs and associated cold-water springs; and c) studies of the regional geology, particularly of the principal tectonic features and the relation of these to known hydrothermal activitiy.

A variety of surveys is undertaken during the second phase of investigation. Detailed hydrogeochemical investigations of the prospect area are conducted to locate and evaluate those anomalies in mineral or gas content that may be related to the existence of hot water or steam beneath the surface. Geophysical surveys—most commonly electrical resistivity surveys—are used to determine the sites of the initial "wild-cat" or exploration wells. Exploration experience indicates that steam-producing areas are commonly associated with indications of low electrical resistivity at a depth of several hundred meters beneath the surface. Other geophysical exploration techniques, such as micro-earthquake or ground-noise surveys, are still in the process of evaluation. Still others, such as seismic reflection surveys and gravity surveys, may be useful where previous experience indicates an association of the geothermal resource with a structural feature exhibiting a contrast in seismic velocity or in density.

Cost of Exploration Using Current Technology

The cost of geothermal reconnaissance surveys ranges between several cents and several dollars per square kilometer; the actual cost depends on whether airborne infra-red, hydrochemical, and reconnaissance geological surveys are all carried out and whether the survey covers all potentially productive areas. Costs can be minimized in the case of rural projects if financial considerations require that some limit be set to the distance over which any electricity developed may be transmitted. If this distance is, say, 50 km, then the reconnaissance can be limited to an area within a 50-km radius of the communities to be served.

Detailed geophysical and geochemical surveys each may cost up to $1,000-$2,000 per day. Typical expenditures for detailed investigations of a prospect several tens of square kilometers in area may amount to $100,000.

AVAILABILITY OF TURBO-GENERATING EQUIPMENT

Steam turbines in sizes suitable for use in rural geothermal applications are readily available. Single-stage back-pressure turbines are available in power ranges up to 400 kW and small turbines with 2 or 3 stages, usable as back-pressure or as condensing machines, are available in power ranges up to 4 or 5 MW.

Cost of Geothermal Development Using Current Technology

A geothermal plant using a condensing turbo-generator of 30-MW capacity was commissioned in El Salvador in 1975. Installed costs, estimated at $257/kW in 1972, actually were approximately $500/kW; however, almost one-half of this figure represents the cost of installing the plant for disposal of the geothermal brine produced.

The cost of steam in geothermal electricity-generating plants is a function of the ratio of dry wells to productive wells drilled and of the productivity of the wells. If, for example, each well costs $300,000, if 4 dry wells are drilled for each productive well, if the productive well yields steam equivalent to 5 MW, if the life of the well is 10 years and the interest is 10 percent, then the cost of steam will be $0.006/kWh at a load factor of 0.9. Analysis of operating-plant economics indicates that if the life of the generating plant is 25 years and a 10-percent interest rate is used, then fixed charges, including maintenance, will be $0.004/kWh and the total generating cost (that is, steam charges plus fixed charges) will be $0.010/kWh for geothermal-generated electricity when the plant is used for base-load operation (at a load factor of 0.9).

This kind of installation—consisting of a condensing turbo-generator of a few megawatts capacity—would seem to be economically feasible in rural applications only where a number of sizable rural communities cluster around a geothermal field so that electricity generated at a central location near the geothermal field could be distributed to the surrounding communities by relatively short transmission lines. In rural communities the load factor would not usually be 90 percent; costs in rural applications would, therefore, be higher than $0.010/kWh, depending on the load factor.

An alternative approach that may be economically feasible where geothermal resources are located in areas with few and relatively small communities, would be to make use of non-condensing turbo-generators. These would minimize capital costs as well as maintenance and operating costs. Communities in such an area might also rely upon some central organization, external to the rural community being served, to provide the steam wells required. This organization could, of course, be financed either from public funds or from private venture capital.

Geothermal wells drilled in such scattered rural areas would be more expensive than wells drilled within a geothermal field. If the cost per well in a rural area is, therefore, taken to be $300,000, the steam produced is equivalent to 2 MW, and other factors are the same as those given above, then the cost of steam will be $0.015/kWh. If, as a reasonable approximation, we take the installed cost of a rural non-condensing generating plant at 50 percent of a condensing turbo-generator, then fixed charges will be roughly $0.002/kWh and the total cost of electricity $0.017/kWh if the load factor is 0.9. This cost is still significantly lower than the cost of electricity generated using oil-fired thermal or diesel generators that would range from $0.02/kWh to much higher values in a rural area in a developing country, but two factors must be considered. First, the costs depend on the actual costs of fuel transport and maintenance. The additional costs that may be involved in transporting the fuel from the port to the point of use in a rural area in a developing country may result in very high fuel costs at the point of use. Second, at lower load factors the cost would rise in inverse proportion to the load factor; for example, in rural areas and in small communities where the main use of electricity is for lighting, load factors may be in the range of 0.2-0.5.

DIRECT APPLICATION OF GEOTHERMAL ENERGY FOR HEATING

Lower-temperature geothermal water resources that are less suitable for electricity production are highly suitable for many applications requiring thermal energy, such as district heating and greenhouse heating. Such resources are found not only in volcanic areas, but also in areas with thick accumulations of permeable sediments, for example, the Gulf of Mexico in the United States, and the Paris Basin and the Hungarian Basin in Europe, all of which are areas without recent volcanic activity. Geothermal water is already in use in district-heating and greenhouse-heating applications in Iceland, Hungary, Japan, New Zealand, and the Soviet Union, and on a somewhat smaller scale in the United States, France, Czechoslovakia, and Rumania.

In rural areas, the use of low-temperature geothermal water as a source of heat in agricultural applications appears to be promising in those developing countries with a cold season that ordinarily prevents crop production and processing. Experiments are taking place in Israel, for example.

The cost of thermal energy from geothermal wells producing water below the boiling point at atmospheric pressure depends upon such factors as seasonal variation in demand and the actual temperatures of the water before and after use, as well as upon the productivity of the wells. Representative costs in Europe fall within the range of $1 to $4/Gcal, which is about one-half to one-eighth the cost of such thermal energy at current (1976) local prices for imported oil. Furthermore, low-temperature geothermal waters are less commonly highly charged with mineral salts and therefore present fewer problems of disposal of waste water. The deposition of carbonate scale in pipes carrying low-temperature geothermal water has occurred in some cases, but control measures have proved effective in dealing with the problem.

In situations where information on the presence of geothermal waters is available as a by-product of the search for oil and gas, the element of financial risk is largely absent from subsequent geothermal development. This will occur where low-temperature geothermal waters occur in deep sedimentary basins that have been investigated for the presence of petroleum.

CURRENT RESEARCH AND DEVELOPMENT

Geothermal energy can be used in many places in the world where power requirements are small. This is true provided adequate equipment is available for utilizing hot water/steam mixtures directly from one or two hot-water wells of moderate size capable of producing mixtures of steam and hot water. However, conventional steam/water separators and steam turbines of the type available at present may not function efficiently when used in conjunction with geothermal water that deposits mineral scale, because of the mechanical and thermal problems caused by these deposits. During the past 3 years, engineering tests have been made on new equipment that has been proposed as a possible way to overcome these problems.[1] Tests performed in Mexico on brine from a well at the Cerro Prieto field, using a helical rotary screw expander as a prime mover, demonstrated that this device can operate with such brines and produce power at reasonable efficiencies. These results have been supplemented by further tests, with larger units, using brine from a well in the East Mesa field of the Imperial Valley, California. Appendix 9 contains a description of the helical rotary screw expander, a discussion of its efficiency, and details of the tests referred to above.

GEOTHERMAL RESOURCES DEVELOPMENT POLICY

Those countries that have not yet begun to evaluate or develop their geothermal-resource potential may be assisted in determining policy if they take the following points into consideration:

- Base-load electricity produced from high-temperature geothermal resources is likely to be substantially cheaper than electricity produced at solid-fuel-fired or oil-fired thermal generating stations, considering the present world prices of oil and coal;
- Most countries having geologically recent volcanic activity are likely to possess high-temperature geothermal resources;
- Exceptionally great thicknesses of sediment and an exceptionally high regional heat flux are not necessary for these geothermal resources to be exploitable economically;
- Exploration and development of high-temperature geothermal resources involve an element of financial risk;
- Low-temperature geothermal sources can be developed to satisfy a need for heating in winter, for example, and thus substitute for oil or coal; and
- In many locations, where petroleum exploration has taken place, low-temperature geothermal resources can be developed at less financial risk.

REFERENCE

1. McKay, R. A., and Sprankle, R. S. 1974. Helical rotary screw expander power system. In *Proceedings of the Conference on Research for the Development of Geothermal Energy Resources, Pasadena, California, 23-25 September*, pp. 301-307. Pasadena: California Institute of Technology, Jet Propulsion Laboratory.

SUGGESTED READING

1. Proceedings of the United Nations Symposium on the Development and Utilization of Geothermal Resources, 1970, Pisa, Italy. *Geothermics*, Special Issue 2, 2 vols. Vol. 1 contains the rapporteurs' summaries. Vol. 2, in two parts, contains all the technical papers presented at the symposium.

2. *Abstracts, Second United Nations Symposium on the Development and Use of Geothermal Resources, San Francisco, California 20-29 May 1975.* Berkeley, California: University of California, Lawrence Berkeley Laboratory.

   The abstract volume, in English, French, and Spanish, was distributed at the conference. The full proceedings are to be published in 1976.

3. Armstead, H. C., ed. 1973. *Geothermal energy: a review of research and development.* Paris: United Nations Educational, Scientific, and Cultural Organization.

Energy Storage

The use of intermittent or variable sources of energy, such as solar and wind energy and some of the forms derived from moving water, often becomes practical only if some means of energy storage is possible. Specifically, we are concerned with storage methods feasible within the constraints under which rural communities in developing nations operate: there must be minimal requirements for capital investment, sophisticated equipment, or complicated maintenance.

TECHNOLOGY OF STORAGE

There are numerous ways to store energy in almost any form; some are more attractive economically in particular situations than others. Many are well known, particularly in industrialized nations. Their suitability to village applications in developing nations will vary with local conditions.

Energy is manifested in several important forms, including mechanical, electrical, chemical, and thermal—the latter being a special form of mechanical energy. Energy may be converted from any of these forms to any other form by one or more intermediate processes, always, however, entailing some loss of energy. The loss usually, but not always, manifests itself as heat. For example, chemical energy is converted to electrical energy in an ordinary power plant. Fuel (chemical energy) is burned (thermal energy) to produce steam that turns a turbine (mechanical energy) coupled to an electrical generator (electrical energy). The overall process is, on the average, about 33 percent efficient. There are ways, using fuel cells, to turn fuel (chemical energy) directly into electricity (electrical energy) without all the intermediate steps. Theoretically, this latter technique should be more efficient and less costly than the chemical-thermal-mechanical-electrical route; for several practical reasons, mostly having to do with materials problems, this is not yet true.

The important point is that conventional fuels are simply cases of sun-derived energy converted to, and stored in, chemical form. It is true that for oil, natural gas, and coal, nature has done the storage for us. But the fact remains that they are merely chemically stored energy forms, and that we shall have to devise shorter processes to obtain the same results if we are to make the use of intermittent energy sources practical.

The basic energy-storage problem is to devise ways to take energy, whenever and in whatever form available, convert it (if necessary) to forms best suited for storage, and then reconvert it, with minimal loss, to a useful form at the time it is needed. The principal physical possibilities include mechanical, chemical, and thermal storage.

CURRENT TECHNOLOGY (AVAILABLE WITHIN 5 YEARS)

With current technology there are practical ways to store energy in mechanical, chemical, and thermal forms.

Storage of Energy in Mechanical Form

Mechanical energy may be stored in the form of kinetic or potential energy. Although the use of the flywheel as a kinetic-energy storage device is well known, there is little prospect that this technique would find significant use in the immediate future in rural situations at power levels that would be useful; it depends too heavily on sophisticated materials and maintenance capabilities.

Mechanical energy may be stored in the form of potential energy by pumping water to a reservoir at a higher level and allowing it to flow down to a lower reservoir whenever the energy is required. This is conventionally called a pump-back system. The kinetic energy of the flowing water is usually converted to electrical energy, although it could, of course, be used directly as mechanical energy. Simple water wheels provide direct mechanical energy whenever stationary mechanical power can be used; when the energy must be moved from place to place, electric generators are probably the best possibility.

A difficulty of the pump-back storage system is that relatively large reservoirs are required. Generally, the best possibility in developing areas is to try to find a reasonably large basin for an upper reservoir located near an existing lake or river. The latter then becomes the lower reservoir. With current technology in pump-back systems, generally about two-thirds of the energy is available for reuse after storage. The feasibility of such a system, however, usually depends on an energy market sufficient to justify the installation of the primary pumping system and its energy supply. Then the pumping from lower to upper reservoir can be accomplished economically during off-peak periods, making supplementary energy available during periods of peak demand. Wind-energy systems providing electricity and/or mechanical services during the day and pumping at night might be one example.

Still another method for storing mechanical energy is through compressed air. Two basic methods are of interest: direct compression and isothermal hydraulic compression. Direct compression, to be practical, requires the availability of natural or man-made underground caverns that can be sealed and pressurized. Reuse of the stored energy requires the availability of low-pressure turbines or piston engines similar to the old-style steam engines. Generally, such systems provide a relatively low overall efficiency. About one-third of the initial energy can finally be reutilized. This figure can be improved considerably when compressed-air systems are combined with gas-turbine cycles or with refrigeration cycles, but the latter are probably not immediately applicable to our major interest.

Of much greater interest to small communities and rural areas in developing nations is isothermal hydraulic compression. In this process (which is at least 3,000 years old), a vertical tube is placed in front of a dam that is creating a head of water, typically 5-10 m (15-30 ft) or more. Water flowing into the inlet entraps air bubbles that are carried to the bottom with the water. The water then passes through the bottom of a sealed tank or cavern where the air bubbles rise. Ultimately, the air in the tank or cavern will be compressed to a pressure slightly less than that of the hydraulic head of the vertical (Venturi) tube. That compressed air can then be used in a simple turbine or other suitable expander to obtain usable mechanical energy.

Such isothermal hydraulic compressors were used from ancient times until the latter part of the 19th century when they were gradually replaced by electrical, steam, or internal-combustion devices. (See chapter on hydropower.)

Storage of Energy in Chemical Form

Although the input and output energy for a battery is electrical, the actual energy storage is in chemical form. The most widely known device is the conventional lead-acid battery; the technology is mature and the costs are fairly well known. The storage efficiency is about two-thirds, or roughly the same as for pump-back storage. The battery life is fairly long, perhaps 10 years or more when properly maintained, but the cost is high if one wishes to store enough energy for significant use. In some cases, however, it is the only thing available. In such circumstances, it may be used to operate a few lights, television sets, and similar low-demand devices, but it will be too expensive for light industrial or vehicular applications.

Electrolysis of water represents another technique for storing energy in chemical form. The stored product is hydrogen, which then can be used as a simple fuel. To minimize storage requirements, it must be stored at relatively high pressures; even so, the energy density is fairly low. (Greater energy densities could be achieved by liquefaction, but the mechanical and cryogenic technologies required seem out of place in this discussion.) Nevertheless, one of the major attractions of this approach is that the basic raw material is distilled water, and it is possible to develop simple devices to run on hydrogen directly. These include simple heaters and internal-combustion engines. The former could be used for space heat, grain drying, cooking, etc.; the latter could be adapted to small, rather short-range vehicles such as "mini" tractors and "mini" trucks. In a rural, small-village situation, these devices could be extremely useful and would not require imported fuel. Development of simple electrolysis systems, and simple heaters and engines using hydrogen as a fuel, should have high priority.

One interesting facet of electrolysis is that it can be performed with very high efficiency at very high pressures, provided the temperature is elevated to about 200°C. Efficiencies of 92-93 percent can be achieved at pressures of 150-250 atmospheres at these temperatures. These pressures can be created by the gas generated, if the system is properly designed, and this would eliminate the need for compressor pumps. Generally, the elevated temperatures can be achieved by insulating the pressure vessels and using the internal (I^2R) losses to supply the necessary heat. Thus, the whole electrolysis system is essentially a static (no moving parts) system with the exception of manual valves. The input energy must be electricity, which for the time being could be obtained from wind-generation systems (Figure 57).

Energy can also be stored chemically in the form of combustible hydrocarbons and alcohols synthesized from hydrogen and organic materials. Many such processes are under study throughout the world, but none is yet known to have short-term applicability to the rural community in a developing nation. Anaerobic bacterial processes may be an important exception to this statement; this subject is covered in the chapter on microbial conversion of plant materials to liquid fuels, and is treated in detail, with respect to generation of methane, in a separate Academy study.[1]

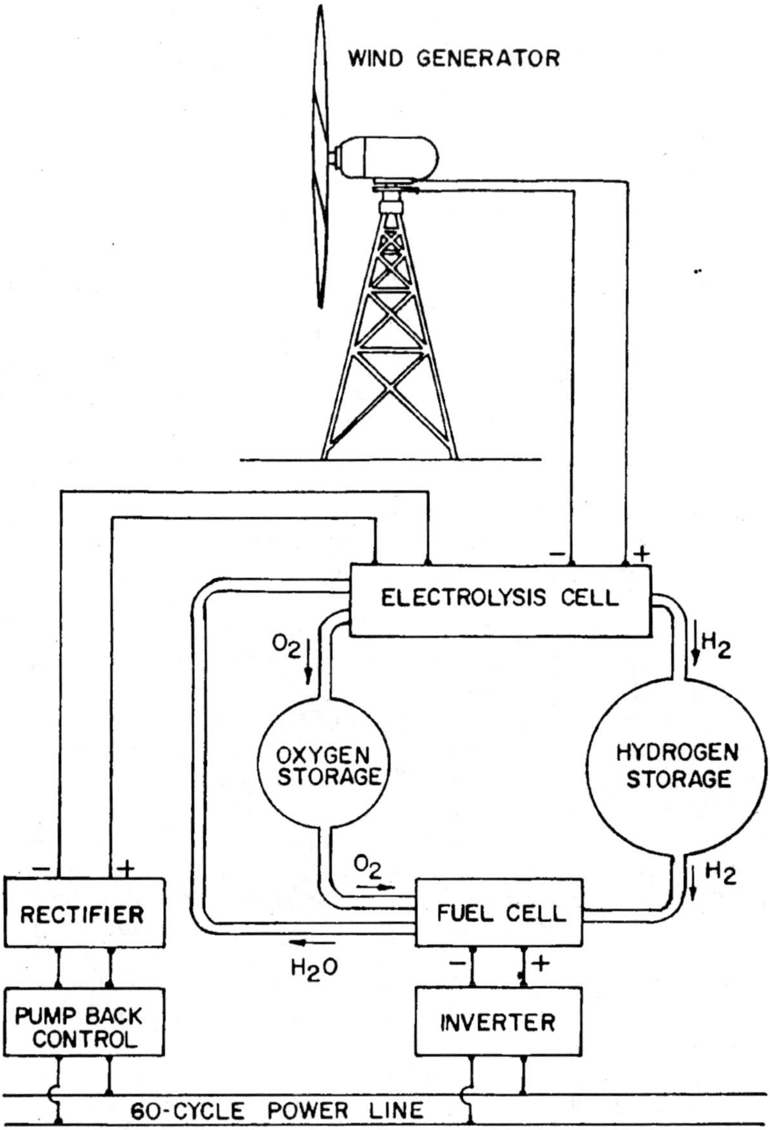

FIGURE 57 Wind as power source. [Source: Utilization of Resources. The Role of Energy in the Development of Human Settlements in Africa. United Nations Economic and Social Council, Document E/CN.14/HUS/6, 17 April 1975.]

Storage of Energy in Thermal Form

Energy can obviously be stored as heat in appropriate materials, provided suitable thermal insulation surrounds the storage substance. Water and rocks, materials of high-heat capacity that are generally available, are commonly used as storage media. Sources of energy such as solar heaters or wind generators can be used. The major difficulty is that the energy can be effectively reutilized only as heat; reconverting the heat energy to electrical or mechanical energy always involves significant losses because of fundamental thermodynamic limitations. Storage temperatures generally would be low, and maximum reconversion efficiency (to mechanical or electrical form) would be on the order of a few percent. Nevertheless, direct reuse as heat can be practical, and this is done in some places; more information on this point is provided in the chapter on heating, cooling, distillation, crop drying, and power generation.

RESEARCH AND DEVELOPMENT

Although other means of storing energy have been suggested, they require further research and development before they can be put to practical use.

Storage of Energy in Mechanical Form

There is much current interest in the development of "super flywheels," which are proposed devices constructed from exotic materials of high tensile strength. Long-term development possibilities look rather good, but because the manufacture of the required sophisticated materials will call for high technological capability and large capital investments, the applicability of this technology to rural areas in developing nations is really a long way off.

The possible use of isothermal hydraulic compression, while not appearing to depend significantly on technical research and development, would benefit from construction and operation of a few prototypical installations to determine economic feasibility.

Storage of Energy in Chemical Form

Much research has been conducted on batteries other than the conventional lead-acid battery, and that research is continuing. It is primarily directed at reducing the weight per unit of energy stored so that electrically driven vehicles might become practical. The most promising work has been with batteries operating at rather high temperatures, or with batteries using expensive materials such as silver. As far as we know, there is no new battery on the horizon that shows more promise for rural areas than the conventional lead-acid cell. The use of hydrogen as a fuel will depend on results of further research on storage and transportation as well as on developments leading to simplification of the electrolysis process. However, there is little prospect that this process will become practical for use in rural areas of developing countries within 10 years or so. For similar reasons, current research and development on fuel cells should not be counted on to make these devices available in villages any time soon.

Storage of Energy in Electromagnetic Form

The only known techniques of storing electrical energy directly without changing form are in very high-intensity electric fields or very high magnetic fields. While some work is being done with these techniques, the technology requirements and costs are extremely high, and the possible energy densities achievable are, so far, very low. Unless some breakthrough occurs in techniques or materials not now foreseen, these methods will probably not be useful for long-term storage in either industrialized or developing nations in the foreseeable future.

REFERENCE

1. National Academy of Sciences. *Methane generation from human, animal, and agricultural wastes.* Report of an *ad hoc* panel of the Advisory Committee on Technology Innovation, Board on Science and Technology for International Development, Commission on International Relations. (In preparation. See p. 306.)

APPENDIXES

APPENDIX 1

Energy Research and Development with Potential for Small-Scale Application

[The information in this list was compiled by NAS staff, primarily from responses to inquiries sent to organizations and individuals in the countries listed. It is not necessarily complete, but represents the best information available as of early 1976.] *

*Since the compilation of this material, a report prepared for the U.S. Energy Research and Development Administration has been published that supplements the information in this appendix: de Winter, F., and de Winter, J. W., eds. 1976. *Description of the Solar Energy R&D Programs in Many Nations.* Available from the National Technical Information Service, U.S. Department of Commerce, Springfield, Virginia 22161 U.S.A.

Country	Institution/Organization (Location)	Contact	Activities
Argentina	Observatorio Nacional de Física Cósmica Comisión Nacional de Estudios Geoheliofísicos Departamento de Radiación Solar Programa de Helioenergética Av. Mitre 3100, San Miguel, Pcia. de Buenos Aires	Lic. Jorge Luis Guerrero Ing. Alfredo Tomás Rapallini	Thermal conversion of solar energy; solar distillation; solar heating of fluids
	Observatorio Nacional de Física Cósmica Departamento de Radiación Solar Meteorología Av. Mitre 3100, San Miguel, Pcia. de Buenos Aires	Lic. Ernesto Santino Crivelli Lic. Manuel Agustin Rela	Radiation measurements and calibration of instruments; radiation climatology
	Universidad Nacional de Salta Departamento de Ciencias Exactas Buenos Aires 177, Salta	Dr. Luis Saravia	Solar crop dryers, heaters, collectors; selective surfaces
	Universidad Nacional de San Luis Departamento de Física y Química Chacabuco y Pedernera San Luis	Lic. Mario Diaz	Solar water heating; solar distillation
	Universidad Nacional de Rosario Departamento Física Av. Pellegrini 250 Rosario, Pcia. Santa Fe	Dr. Carlos Garibotti	Low-loss structures; thermal conversion of solar energy
	Universidad Nacional de La Pampa Instituto de Estudios Regionales Mansilla 178, Santa Rosa, La Pampa	Prof. Juan Elias Zabala	Solar stills for local communities
Australia	CSIRO [Commonwealth Scientific and Industrial Research Organization] Division of Mechanical Engineering P.O. Box 26, Highett, Melbourne	Dr. B. Rawlings, Chief of Division	Thermal conversion; collector design and development; water heating and low-pressure-steam generation; building heating and cooling
	Division of Mechanical Engineering Highett, Victoria 3190	P.I. Cooper, Research Scientist W.R.W. Read, Principal Research Scientist	Stills; timber kiln

CSIRO Solar Energy Studies P.O. Box 89, East Melbourne, Victoria 3002	Mr. Roger N. Morse, Director of Solar Energy Studies	Solar energy feasibility studies; energy utilization in industry; contact with solar-energy research workers in Australia and overseas.
CSIRO Division of Atmospheric Physics Melbourne [Radiation Centre for Region V, World Meteorological Organization]	G. W. Paltridge	Radiation measurements
CSIRO Division of Building Research Melbourne		Radiation measurements in Papua
CSIRO Division of Irrigation Research Griffith	K. V. Garzoli	Thermal characteristics of glass and plastic greenhouses
CSIRO Division of Land Use Research P.O. Box 1666 Canberra City, A.C.T. 2601	Dr. J. D. Kalma Dr. R. J. Millington	Radiation measurements; insolation; energy use in agriculture and in urban environments; climatological/meteorological aspects of energy use
CSIRO Division of Food Research P.O. Box 52, North Ryde, Sydney, N.S.W. 2113	Mrs. W. Szulmayer	Sun-drying of fruit
University of Melbourne Department of Mechanical Engineering Melbourne	W. W. S. Charters, Reader in Thermodynamics, Chairman of Department	Thermal properties of solar air heaters; solar water heaters, collector surfaces; solar dwellings; solar collectors, system performance
University of Queensland Department of Mechanical Engineering St. Lucia, Brisbane, 4067	Norman R. Sheridan, Reader in Mechanical Engineering	Heating and cooling of buildings; photovoltaic conversion; heat engines; energy storage and transmission; publishes "Solar Research Notes"

Country	Institution/Organization (Location)	Contact	Activities
	University of New South Wales School of Mechanical and Industrial Engineering P.O. Box 1, Kensington 2033, New South Wales	C. M. Sapsford	Radiation measurements; residential heating and cooling; performance of flat-plate collectors
	International Solar Energy Society Australian and New Zealand Section c/o CSIRO Division of Mechanical Engineering P.O. Box 26, Highett, Victoria 3190	Robert V. Dunkle, Editor	Publishes annually the "Solar Energy Progress in Australia and New Zealand"
	The University of Western Australia Department of Mechanical Engineering Nedlands, W. A. 6009	R. S. Minchin, Senior Lecturer	Low-cost heater–design, construction, installation (total materials cost A$15)
		J. A. Appleyard, Senior Lecturer	Cylindrical solar water heater combining absorber and storage tank in one unit–design, construction, testing
Belgium	Ministère de l'Agriculture Bruxelles 1000	J. F. M. Ronchaine, Ingenieur Agronome	Greenhouses, stills
	Université Catholique de Louvain Faculté des Sciences Appliquées Laboratoire de Physico-Chimie et de Physique de l'État solide Batiment Boltzmann Place Croix du Sud, 1 B 1348 Louvain-La-Neuve	Prof. J. P. Issi J. P. Michenaud	Thermoelectric conversion
Brazil	Universidade Federal da Paraíba Laboratório de Energía Solar João Pessoa, 58.000 Paraíba	Cleantho Torres, Head Antonio MacDowell Pio Caetano Lobo Julio Goldfarb Rogério Klüppel	Water distillation; drying of fruits; solar furnaces; solar engines for irrigation. Publishes a biannual bulletin (in Portuguese).

	Instituto Tecnológico de Aeronáutica Division of Mechanical Engineering 12.200 São José dos Campos, São Paulo	Dr. Sérgio Nelo Vannucci, Assistant Professor	Sea water distillation
	Centro de Pesquisas do Cacau Agricultural Engineering Division Caixa Postal 7, Itabuna - B.A.	Dr. Biswa Nath Ghosh, Head [Current Address: Gulf and Western Advanced Development and Engineering Center, 101 Chester Road, Swarthmore, Pa. 19081, U.S.A.]	Crop dryer
Burma	Central Research Organization Rangoon	U Maung Maung	Solar cookers; stills; hot-water heaters; crop drying; salt production
Canada	The University of Western Ontario Faculty of Engineering Science London 72, Ontario	Prof. R. K. Swartman	Solar collectors; solar-powered refrigeration and ice making
	Brace Research Institute Macdonald College McGill University Ste. Anne de Bellevue Quebec	Thomas A. Lawand, Director of Field Operations	Radiation measurements; crop drying; cookers; distillation; hot-water heaters; solar still/greenhouse combinations; environmentally adapted greenhouses; solar-powered engines; solar ponds; windmills for electrical generation and water pumping; use of solar and wind energy in heating buildings, etc. (A variety of practical publications available.)
Chile	Universidad del Norte Solar Energy Laboratory Antofagasta	Prof. Carlos Espinosa	Radiation measurements; hot-water heaters

Country	Institution/Organization (Location)	Contact	Activities
	Universidad de Chile Department of Mechanical Engineering Casilla 2777, Santiago	Sergio Alvarado Felipe Wainer	Radiation cooling
	Universidad Técnica Federico Santa María Solar Energy Research Center Casillas 110-V. - 132-V., Valparaiso	Dr. Julio R. Hirschmann Director, Solar Energy Research Center Dr. Bernardo Seifert, Director, Solar Energy Laboratory Prof. Germán Frick, Director, Lab. for Horticulture Technology	Radiation measurements; cookers; hot water heaters; stills; solar ponds; publishes Revista "Scientia"; evaporative cooling of greenhouses
		Dr. Arnold Keller	Solar collectors; surface treatments
	Universidad Técnica Federico Santa María Solar Energy Research Center Viña del Mar	Dr. Bernardo Seifert	Experimental station for hot-water heaters, stills, solar furnaces
	Universidad Técnica Federico Santa María Solar Energy Research Center Calama	Prof. Germán Frick, Director	Experimental station for solar ponds and for horticulture in controlled environment (plastic greenhouses)
	Universidad Técnica Federico Santa María Solar Energy Research Center Quillagua	–	Experimental station with solar saline water distillation plant
Egypt	Ain Shams University Heliopolis, Cairo	M. K. Elnesr	Radiation measurements; solar collectors; stills

	National Research Centre Solar Energy Laboratory Dokki, Cairo	Dr. Ing. Ibrahim A. Sakr, Dir.	Solar cookers; stills
France	Université de Nancy Faculté des Sciences Laboratoire de Physique de Dépots Métalliques Boulevard des Aiguillettes 54 - Nancy	Prof. Jean Fléchòn	Solar refrigeration; air conditioning; photovoltaics
	Centre National de la Recherche Scientifique [CNRS] Laboratoire de l'Énergie Solaire Odeillo	Dr. Felix Trombe, Director	Solar furnace; refrigeration; residential heating; radiation measurements
	Université de Provence Centre de Saint Jerome Département d'Héliophysique Marseilles	J. Millet, Mâitre Assistant M. Bazan, Assistant B. Imbert, Assistant J. P. Legre, Assistant J. Gervais, Assistant	Radiation measurements
		L. Aiache, Assistant G. Durr, Chercheur J. Le Troquer, Chercheur B. Imbert, Assistant	Helio-electrochemical conversion
	Société Française d'Études Thermiques et d'Énergie Solaire [SOFRETES] Zone Industrielle d'Amilly B.P. 163, 45203 Montargis	J. P. Girardier	Heat engines using solar energy for water pumping
	PROMETHEE (joint venture between SOFRETES and French Atomic Energy Agency) Zone Industrielle d'Amilly B. P. 163, 45203 Montargis	M. Clemot J. P. Girardier	R and D in solar energy. First application: solar pumps for irrigation purposes
	Coopération Méditerranéenne pour l'Énergie Solaire (COMPLES) Siège Social: Palais de la Bourse 13001 Marseille	Prof. M. Perrot, President	Publishes semiannual bulletin, "Revue Internationale d'Héliotechnique"

Country	Institution/Organization (Location)	Contact	Activities
	Association Française pour l'Étude et le Développement des Applications de l'Énergie Solaire (A.F.E.D.E.S.) 28, rue de la Source, 75016 Paris	P. Girard, Délégué General	Publishes *Cahiers A.F.E.D.E.S.*; organizes study meetings with the reports sent to members; documentation center for members' use
Germany (Federal Republic of)	Forschungsinstitut für Energietechnik 7441 Wolfsschlugen Hauffstr. 14	Prof. Dr.-Ing. G. Schöll	Solar collectors
	Forschungsinstitut für Windenergietechnik 7 Stuttgart 80 Pfaffenwaldring 31	Prof. Dr. V. Hütter	Wind energy
	Technische Universität Braunschweig 33 Braunschweig Mendelssohnstr. 16	Prof. E. W. Justi	Solar cells
	Institut für Systemtechnik und Innovationsforschung (ISI) 75 Karlsruhe Breslauerstr. 98	–	Solar energy
	Institut für Wasserbau und Wasserwirtschaft TH Aachen 51 Aachen Mies-van-der-Rohe-Strasse	–	Hydropower
	Rheinische-Westfälische Technische Hochschule Institute for Applied Thermodynamics Aachen	Dr. Ing. A. Beckers	Bilateral program with Universidad Técnica Federico Santa María Chile
	Kernforschungsanlage Jülich 517 Jülich	Dipl.-Ing. F. J. Friedrich	Collectors, solar cells, solar thermal gas processes
	Deutsche Forschungs- und Versuchsanstalt für Luft- und Raumfahrt E. V.	–	Solar and wind energy

	Philips Forschungslaboratorium 51 Aachen Postfach 1980, Weisshausstrasse	Dr. Hörster	Solar energy
Greece	Hellenic Industrial Development Bank Amalias Ave. 20, Athens 118	P.P. Kokkaliaris, Chemical Engineer	Solar distillation; desalination
	Greek Atomic Energy Commission Nuclear Research Center "Demokritos" Aghia Paraskevi - Attiki	Dr. Argyrios-Demetrios Spyridonos Dr. (Mrs.) E. Delyannis	Solar stills; water heaters; collectors; concentrators; space-heating
India	Regional Engineering College Warangal (A.P.) 506004	Dr. T. L. Sitharama Rao Mr. S. Subramanyam	Solar water heater; multipurpose solar collector
	Aurobindo Centre for Environmental Studies Pondicherry	Dr. C. L. Gupta	Solar water heaters
	Central Arid Zone Research Institute Division of Wind Power and Solar Energy Utilization Jodhpur, Rajasthan	H. P. Garg Dr. A. Krishnan	Radiation measurements; flat-plate collector design; wind power
	Birla Institute of Technology and Science Pilani, Rajasthan	A. K. Jain T.L.S. Rao, Assistant Prof.	Solar water heaters
	National Physical Laboratory Hillside Road, New Delhi	Dr. G. D. Sootha Dr. V. G. Bhide	Has developed solar cookers, solar water heaters, solar-heated Stirling engine, solar-distillation systems; current activity not certain
	Indian Meteorological Department New Delhi	Dr. A. Mani, Deputy Director	Radiation measurements
	Ghandian Institute of Studies Appropriate Technology Development Unit P.O. Box 116 Rajghat, Varanasi 221001	M. M. Hoda, Head of Unit	Appropriate technology documentation directory; encouraging research on appropriate technology in rural areas and its adoption in India

Country	Institution/Organization (Location)	Contact	Activities
	Defence Research Laboratory Jodhpur	Dr. J.P. Gupta, Senior Scientific Officer R. K. Chopra, Jr., Scientific Officer	Solar space heating
	Central Building Research Institute Roorkee	R. Ganguli J. S. Pursi	Radiation measurements; flat-plate collectors; water heaters; residential heating
	Central Salt and Marine Chemical Research Institute Bhavnagar, Gujarat	Dr. R. L. Datta, Assistant Director	Radiation measurements; stills
	Indian Institute of Technology Department of Mechanical Engineering Kanpur 208016, U.P.	Prof. H. C. Agrawal	Solar collectors for rural use; solar drying of agricultural products; water heaters
	Department of Chemical Engineering	Dr. C. V. Sheshadri Dr. A. V. S. Prabhakara Rao Dr. G. D. Agrawal	Cultivation and use of algae for cattle feed and digestion to produce biogas.
	Department of Civil Engineering	Dr. S. Ramaseshan	Effects of solar energy on water resources (temperature, precipitation, use)
		Dr. G. D. Agrawal Dr. Y. C. Das	Solar-radiation measurements as possible index of environmental deterioration; solar thermal stresses in structures
	Indian Institute of Technology Department of Mechanical Engineering Madras 600036	Dr. M. C. Gupta, Head, Solar Energy Lab	Solar space cooling; solar drying; solar water heaters
	Motilal Nehru Regional Engineering College Mechanical Engineering Department Allahabad 211004	Dr. B. K. Gupta	Cooling of industrial and public buildings; flat-plate collectors; Fresnel lenses

		R. K. Bhardwaj, Reader in Mechanical Engineering	Solar water heaters
	All India Solar Energy Working Group c/o R. L. Datta, Central Salt and Marine Chemical Research Institute Bhavnagar, Gujarat	Dr. R. L. Datta, Chairman	Publishes proceedings; devoted to promotion of solar energy research and development and dissemination of knowledge in field of solar energy
Iran	Pahlavi University Department of Mechanical Engineering Shiraz	Prof. Mehdi N. Bahadori	Solar residential heating; solar cooking; solar sterilization; solar desalination; solar dehydration of fruits and vegetables
	Institute of Natural Resources Karaj	-	Hot-water heaters
	Teheran University Agricultural Engineering Department Karaj	Dr. Javed Maghsood	Greenhouses
Iraq	Building Research Center Scientific Research Foundation Baghdad	A.E.A. Almaini, Research Asst. A.E.A. Alhadithi, Research Asst.	Solar water heaters; space heating
	Mustansiriya University Faculty of Science Physics Department Baghdad	M. A. Abbas	Radiation measurements
Israel	Scientific Research Foundation Dan Danziger Building Hebrew University Campus P.O. Box 3745, Jerusalem	Dr. Harry Z. Tabor, Scientific Director	Solar-energy systems; collectors and materials; climatic control of buildings, including fenestration studies; energy recovery; solar ponds for heat and power; power units

Country	Institution/Organization (Location)	Contact	Activities
	The National Physical Laboratory of Israel Hebrew University Campus Danziger Building A Jerusalem	Avram Kalisky, Director	Hot-water heaters; hot-water heating systems; solar ponds; photovoltaics; research on materials and design to maximize natural heating and cooling of buildings; radiation measurements; heat exchangers
	The Hebrew University of Jerusalem Human Sciences Laboratory School of Applied Science and Technology Jerusalem	Prof. Hillel I. Shuval Dr. Georges Belfort Dr. Inka Dor	Use of solar energy to grow algae in wastewater as a low-cost protein animal feed; anaerobic fermentation of agricultural wastes for methane production
	The Hebrew University of Jerusalem Authority for Research and Development Department of Physics Mount Scopus P.O. Box 24100, Jerusalem	Prof. E. Harnik Prof. M. Schieber	Solar cell materials and fabrication technology
		Prof. D. Shaltiel Prof. Y. Jacoby	Hydrogen storage; use of light guides for the transport of solar energy
	Department of Physical Chemistry	Prof. G. Stein	Photochemical fuel production; hydrogen evolution; hydrogen storage
	Department of Botany	Prof. B. Z. Ginzburg Dr. B. Z. Dorfman	Algae/microbial conversion to liquid fuels
	Department of Organic Chemistry	Prof. S. Patai	Pyrolysis of cellulose and wastes
		Dr. S. Brenner Prof. J. Klein	Hydrogenation of cellulose and agricultural wastes
	The Hebrew University of Jerusalem Center for Research in Energy, Fuel and Petrochemistry	Prof. J. Klein Prof. M. Rabinowitz Prof. J. Blum Prof. I. Agranat	Coal liquefaction; coal reactions

Country	Institution	Researchers	Research area
	Ben-Gurion University of the Negev Research and Development Authority P.O. Box 1025 Beer-Sheva, 84110	Dr. D. Pasternak Dr. M. Twersky	Geothermal heating of soil (growing crops in cold season, protection from freezing, accelerating growth); geothermal heating of greenhouses, nurseries, and poultry and other small-animal facilities
		E. Rappaport Dr. D. Pasternak Dr. A. Richmond	Controlled environmental systems–greenhouses and algae-producing units
		Dr. I. Yaron Dr. I. Borde	Solar cookers; absorption refrigeration for preserving agricultural produce
		Dr. C. Forgacs	Desalination of brackish water; design of small field units for drinking water
	Department of Mechanical Engineering	Dr. O. Igra	Shrouded windmills for water pumping and generating electricity
		Dr. A. Bar-Cohen Dr. K. Preiss	Solar heating; hot-water heaters
Italy	European Community Joint Research Centre 21020 Centro Euratom di Ispra Varese	Dr. Joachim Gretz, Scientific Directorate	Collectors for residential heating; quantum conversion problems; large-scale electricity production
	Instituto Chimico "G. Ciàmician" dell' Università Bologna	V. Balzani F. Bolletta L. Moggi	Photochemical conversion
	Laboratorio Fotochimica e Radiazone d'Alta Energia del CNR Bologna	M. Gleria	Photochemical conversion
Japan	Keio University Tokyo	Prof. I. Tanishita	Hot-water heaters

Country	Institution/Organization (Location)	Contact	Activities
	Waseda University Department of Architecture Nishiokubo, Shinjuku-ku, Tokyo 160	Prof. K. Kimura M. Udagawa, Research Engineer	Experimental solar house; solar space-heating; hot water; solar collectors (theory and experiment); absorption cooling; calculation of heating and cooling loads of buildings
	Tokai University	Prof. I. Tanishita	Hot-water heaters
	Tokai University Department of Architecture Shibuya-ku, Tokyo	Prof. S. Tanaka	Basic study on solar heating and cooling systems
	Kogakuin University Department of Architecture Shinjuku-ku, Tokyo	Prof. Y. Nakajima	Heat storage tank, theory and experiment
	Osaka Institute of Technology Department of Mechanical Engineering Asahi-ku, Osaka	Prof. Y. Saito	Experimental combined flat-plate and focusing collector
	Tokyo Metropolitan University Department of Architectural Engineering Setagaya-ku, Tokyo	Prof. N. Ito	Bibliography on solar heating, cooling and hot-water supply
	University of Tokyo Sogoshikenjo School of Engineering Bunkyo-ku, Tokyo	Prof. Y. Matsuo	Solar-radiation measurements; measurement and theory of heating and cooling loads of buildings
Netherlands	Eindhoven University of Technology Laboratory for Fluid Mechanics Eindhoven	Ir. P. T. Smulders	Windpower for developing countries

	Mechanical Engineering Department	Prof. Ir. C. W. J. von Koppen	Production of hydroelectric power by means of Banki-type radial-flow turbine generators for small head, for developing countries; solar cooling
	Chemical Engineering Department	Ir. S. P. Bertram	Fermentation of organic materials to produce methane (for developing countries)
	Koninklijk Institut voor de Tropen Department of Product Research Mauritskade Amsterdam	Ir. F. W. Korthals Altes	Solar dryers
	Ingenieursbureau Dwars Hederick en Verhey Laan 1914 35, Amersfoort	Ir. H. Deibel	Solar stills
	Organization for Industrial Research TNO P.O. Box 406 Delft	Ir. P. van Staveren	Utilization of windpower
	University of Technology Enschede	E. Odijk	Windpower, solar energy
		Prof. Dr. Ir. W. P. M. van Swaay	Wood gasification
	Foundation TOOL (Technische Ontwikkeling Ontwikkelingslanden) P.O. Box 525 Eindhoven	-	Coordination of projects involving development of small-scale energy technology for developing countries
New Zealand	Dominion Physical Laboratory Department of Scientific and Industrial Research Physics and Engineering Laboratory Private Bag, Lower Hutt	Dr. R. F. Benseman	Water heaters and solar-radiation distribution

Country	Institution/Organization (Location)	Contact	Activities
Niger	Office de l'Énergie Solaire Organization Nigerienne de l'Énergie Solaire Niamey	Dr. Abdou Moumouni, Directeur Abdoulaye Hima, Technicien	Water pumps; water heaters
Nigeria	University of Ife Ile-Ife	Dr. V. A. Williams	Photovoltaics
	University of Lagos Department of Mechanical Engineering Lagos	Dr. S. O. Adenubi Dr. V. A. Akinsete	Solar-radiation measurements; flat-plate collectors; solar water heaters; solar crop dryers
Pakistan	Pakistan Council of Scientific and Industrial Research (P.C.S.I.R.) Labs. Kemal Ataturk Road, PPI Building Karachi	Director, Solar and Wind Energy Program	Solar water pumping; crop drying; integrated food/water/power complex; windmills; heating of buildings
Papua New Guinea	The Papua New Guinea University of Technology Department of Electrical Engineering Faculty of Engineering P.O. Box 793 Lae	Prof. J. L. Woodward	Simple pumps and hydraulic rams; water wheels; small-scale electric power generation; solar water heaters; solar refrigeration
		Prof. J.C.V. Chinappa	Solar refrigeration
Peru	Instituto de Investigaciones de Aplicaciones de la Energía Solar (INAES)	Dr. Maximiliano Duran Dr. Casio Ore Ing. Alfredo Oliveros Prof. Alejandro Duran	Cooker (current activity uncertain)
Republic of China (Taiwan)	National Tsing Hua University Power Mechanical Engineering Department 855 Kuang Fu Road Hsinchu, Taiwan 300	Dr. Ching-Kwei Kang	Design and construction of 20-kW windmill, also used for water pumping. Design and construction of 15-kW windmill

République du Mali	Solar Energy Laboratory Bamako	I. Toure, Director	Radiation measurements; absorption cooling
Saudi Arabia	College of Petroleum and Minerals Dhahran	Prof. M. Ali Kettani Hussein K. Abdel-Aal	Solar-radiation measurements; helio-hydroelectricity; direct energy conversion; hydrogen production; mineral production from the sea
	University of Riyadh College of Engineering Riyadh	Dr. J.A. Sabbagh, Dean Dr. A. A. M. Sayigh, Dr. E. M. A. El-Salam	Solar water heaters, stills, boilers; solar house; effect of radiation on the habitat
Senegal	Université de Dakar Institut de Physique Météorologique H. MASSON Faculté des Sciences Dakar-Fann	Prof. Djibril Fall	Flat-plate collector; solar motor; water heater; dryer
	Centre National de la Planification de la Recherche Scientifique et Technologique	Dr. Assad Takla	Solar pumps; cost reduction
South Korea	Korean Atomic Energy Institute Seoul	Dr. Jong Hee Cha	Flat-plate collector
Sri Lanka	University of Ceylon Peradeniya	Dr. N.S. Wijesundere, Lecturer in Mechanical Engineering	Solar water heaters
	Central Electricity Board	Dr. K. Ghanalingam	CdS cells for water pumping
	University of Ceylon Colombo Campus 61 Abdul Caffoor, Mawatha, Colombo-3	Prof. K. Kularatnam	Salt and minerals from the sea
Sweden	Chalmers Institute of Technology Department of Geology Fack S-402 20 Goeteborg 5	Prof. K. G. Eriksson	Geothermal heating

Country	Institution/Organization (Location)	Contact	Activities
	The Lund Institute of Technology Division for Building Construction Fack S-220 07 Lund	Prof. Bo Adamson	Solar heating and cooling
	State Power Board Fack S-162 87 Vaellingby	Benat Nordstroem	Hydroelectric power generation; windpower generation
	Swedish Board for Technical Development Fack S-100 72 Stockholm 43	Olle Ljungstroem	Windpower generation
	Svensk Kaernbraenslefoersoerjuings AB Fack S-102 40 Stockholm	Ingemar Lindholm	Solar heating and cooling
	Royal Institute of Technology Department of Building Technology Fack S-100 44 Stockholm 70	Prof. Ingemar Hoeglund	Solar heating and cooling
	Department of Chemical Technology	Prof. Olle Lindstroem	Pyrolysis of waste and production of methanol as a fuel
Thailand	Asian Institute of Technology Division of Community and Regional Development P.O. Box 2754 Bangkok	Dr. R.H.B. Exell	Solar radiation in Thailand, solar distillation of water; solar-powered water pump; solar refrigeration
		Dr. P. A. Cowell	Solar-powered water pump
	Division of Environmental Engineering	Dr. Maung Nay Htun	Solar distillation
	King Mongkut Institute of Technology Mechanical Engineering Department 91 Pracha Uthit Road,	Dr. Prida Wibulswas Ken Cooper	Solar water heaters

Country	Institution	Contact	Activities
Trinidad	University of the West Indies Department of Mechanical Engineering Faculty of Engineering St. Augustine	Dr. Suppramaniam Satcunanathan, Head of Department	Solar stills; air heaters; water heaters; crop drying; refrigeration/air conditioning
	University of the West Indies St. Augustine	Oliver St. C. Headley, Dept. of Chemistry	Solar crop dryers; flat-plate collectors; solar stills
Turkey	Middle East Technical University Ismet Inonu Bulvari Ankara	Dr. M. Kudret Selçuk (current address: Jet Propulsion Laboratory, Bldg. 277, Rm. 202, Pasadena, Calif. 91103, U.S.A.) Prof. M. Akyrut	Radiation measurements; thermal properties of controlled-environment greenhouses; heat engines; crop drying
United Kingdom	University College Solar Energy Unit Department of Mechanical Engineering P.O. Box 97 Cardiff	Dr. B. J. Brinkworth	Utilization of solar energy for domestic, commercial, agricultural and industrial purposes
	College of Technology Department of Mechanical Engineering Brighton, Sussex	Dr. J. C. McVeigh	Flat-plate collectors
	Electrical Research Association Cleve Road Leatherhead, Surry KT22 7SA	A. H. Stodart, General Mgr., Industrial Services Division	Wind power
	University of Reading Department of Engineering and Cybernetics Whiteknights, Reading RG6 2AY	Prof. P. D. Dunn	Water-piston pumps; Stirling engines; vertical-axis windmills; solar heaters; methane generation
	University of Sheffield Department of Building Science Sheffield	Prof. J. K. Page, Head of Department	Estimation of direct and diffuse insolation on vertical and inclined surfaces from radiation received on horizontal surfaces.

Country	Institution/Organization (Location)	Contact	Activities
	Intermediate Technology Development Group Ltd. Parnell House, 25 Wilton Road London SW1V 1JS	George McRobie, Director (Communication)	Source of information on water, wind & solar power; publishes *Appropriate Technology* quarterly.
Union of Soviet Socialist Republics	Ministry of the Electrical Engineering Industry Moscow	A. P. Landsmann	Photoelectric generators
	Power Engineering Institute *imeni* G.M. Krzhizhanovskiy Solar Energy Laboratory Leninskiy Prospekt 19, Moscow	Dr. Yuriy Nikolayevich Malevskiy, Director Dr. Sergey M. Lukomskiy Dr. Rafael Rafaelevich Aparisi Dr. Dmitriy Ivanovich Teplyakov Dr. Boris Arnol'dovich Garf	Solar stills; absorption cooling; thermoelectric generators; heat pumps; collectors and concentrators; cookers
	All Union Institute for Research in Solar Technology Leninskiy Prospekt 19, Moscow	Dr. B. V. Petukhov	Solar pumps
	All Union Scientific Research Institute of Sources of Current Solar Energy Laboratory Moscow	Dr. Nikolay Stepanovich Lidorenko, Director Dr. B. V. Tarnizhevskiy	Solar pumps; photovoltaics; surface coatings
	Turkmen Physico-Technical Institute Ashkhabad, Turkmen SSR	Dr. Aman Kh. Khanberdiyev Dr. R. Bairamov	Solar stills; refrigeration; thermoelectric generators; photovoltaics
United States	University of California (Berkeley) Sea Water Conversion Laboratory 1301 South 46th Street Richmond, California 94804	Prof. Alan D. K. Laird, Director Badawi W. Tleimat, Sr. Development Engr. Prof. Everett D. Howe, Emeritus Director	Solar stills; desalination

University of California (Richmond) Sanitary Engineering Research Laboratory Richmond Field Station 1301 South 46th Street Richmond, California 94804	Prof. William J. Oswald Prof. Clarence G. Golueke	Waste-water treatment; algae; methane generation
University of Delaware Institute of Energy Conversion Newark, Delaware 19711	Dr. Allen M. Barnett, Director	Photovoltaics (CdS/Cu_2S solar cells); solar heating and cooling of buildings; solar collectors; heat storage; energy management; distributed conversion systems in conjunction with utility grid; techno-economical analysis
University of California Los Angeles, California	Prof. Richard Schoen	Heating of buildings
Colorado State University Fort Collins, Colorado	Prof. George O. G. Löf	Cookers; residential heating and cooling; water heaters
Southern Methodist University Institute of Technology Dallas, Texas	Prof. Harold A. Blum	Flat-plate collectors—design
Oklahoma State University School of Engineering Stillwater, Oklahoma 74075	Prof. William L. Hughes, Prof. H. J. Allison Prof. R. Ramakumar	Windmill turbines, pumps, generators (variable speed, constant frequency); energy storage (hydrogen from electrolysis of water)
University of Wisconsin Solar Energy Laboratory Engineering Research Laboratory 1500 Johnson Drive Madison, Wisconsin 53706	Dr. John A. Duffie, Director	Collectors; residential heating; refrigeration
Battelle Memorial Institute Columbus Laboratories 505 King Avenue Columbus, Ohio 43201	Dr. James A. Eibling	Solar heating and cooling; solar stills; Stirling engines; energy storage; flat-plate collectors; concentrators

Country	Institution/Organization (Location)	Contact	Activities
	University of Minnesota Department of Mechanical Engineering 125 Mechanical Engineering Building Minneapolis, Minnesota 55455	Dr. Richard C. Jordan, Head Dr. Ernst Eckert Dr. Ephraim Sparrow Floyd Larsen	Solar thermal electric power; heating and air conditioning; flat-plate collectors; energy storage and heat-pump systems; selective surfaces
	University of Massachusetts Department of Civil Engineering Amherst, Massachusetts 01002	Prof. William Heronemeus	Wind energy conversion
	University of Maryland Department of Mechanical Engineering College of Engineering College Park, Maryland 20742	Prof. Stephen L. Sargent Prof. Redfield Allen	Organic Rankine-cycle engines; absorption cooling
	National Aeronautics and Space Administration (NASA) Goddard Space Flight Center Greenbelt, Maryland	Dr. M. P. Thekaekara, Code 912	Radiation measurements
	Arizona State University College of Architecture Tempe, Arizona	Prof. John I. Yellott, Chairman, Solar Energy Program	Residential and institutional heating and cooling
	Energy Research and Development Administration (ERDA) Division of Solar Energy 20 Massachusetts Avenue, N.W. Washington, D.C. 20545	Dr. Lloyd O. Herwig, Scientific Advisor Dr. Frederick H. Morse, (Heating and Cooling) Louis V. Divone, (Wind Energy Conversion) W. R. Cherry, (Crop Drying)	Supports and coordinates research on solar energy in U.S. institutions and development of commercial technology

	Volunteers in Technical Assistance (VITA) 3706 Rhode Island Avenue Mt. Rainier, Maryland 20822	Dr. Michael P. Greene	Technical inquiry service provides assistance on request in most areas; publishes plans and booklets on devices and implements; participates in projects on appropriate technology
	Route 4, Box 258 Mt. Airy, Maryland 21771	C. J. Swet	Solar cooker, pump; liquid heating solar collectors; decentralized hydrogen production from sunlight and water; decentralized ammonia fertilizer production from sunlight, water and air
	University of Florida Department of Mechanical Engineering Solar Energy and Conversion Laboratory Gainesville, Florida 32611	Prof. E. A. Farber, Director	Residential heating and cooling; distillation; refrigeration; cookers; heat engines; solar-powered automobile
	University of California Department of Materials Science and Engineering Hearst Mining Building Berkeley, California 94720	Prof. Marshall F. Merriam	Low-cost pyranometer; solar ponds; low-cost collectors; low-pressure solar steam
	Ohio University Mechanical Engineering Department Athens, Ohio 45701	Prof. William T. Beale	Low- and high-temperature Stirling engines; flat-plate collectors; windmills
Zambia	University of Zambia P.O. Box 2379 Lusaka	Prof. J.M.K. Dake	Solar ponds

APPENDIX 2

Manufacturers of Solar Heating and Cooling Devices

[This compilation is not necessarily complete, but represents the best information available to NAS staff at the time of publication. Although the list is meant to include only those organizations currently engaged in commercial marketing of solar devices, it is possible that some of those listed are not yet at that stage.]

Country	Company (Location)	Contact	Product or Activity
Australia	Braemar Engineering Co., Ltd. Bilsen and Zillmere Roads Geebung 4034 Queensland	–	Flat-plate copper collector and water-heater systems
	Turbon Engineering Pty., Ltd. 12 Bailey Street West End, Brisbane 4101 Queensland	–	Flat-plate copper collector and water-heater systems
	Sola-ray Appliances 6 Boag Road Morley 6062 Perth	–	Flat-plate copper collector and water-heater systems
	S. W. Hart and Co., Pty., Ltd. 112 Pilbara Street Welshpool, W.A. 6106	–	Flat-plate copper collector and water-heater systems
	Smalls Sola Heeta Co., Pty., Ltd. 10 Goongarrie Street Bayswater, W.A. 6053	–	Flat-plate copper collector and water-heater systems
	Beasly Industries Pty., Limited Bolton Avenue Devon Park, South Australia 5008	–	Flat-plate copper collector and water-heater systems
Chile	Compañía Química y Minera de Chile Departamento de Investigaciones Científicas Santiago	–	Solar evaporation ponds
	Compañía Anglo-Lautaro Santiago	–	Solar evaporation ponds; production of sodium sulfate

Country	Company (Location)	Contact	Product or Activity
Germany (Federal Republic of)	Dornier System GmbH Postfach 1360 7990 Friedrichshafen	K. Kögler, Head Thermodynamics Dept. Dr. H. Rieche, Head Marketing	Stills for sea-water desalination by solar-heated heat pipes; heating and cooling hot water; electricity generation
	AEG-Telefunken Abteilung Raumfahrttechnik und Neue Technologien 2000 Wedel, Industriestrasse	S. Korius H. Manthey	Small solar batteries
India	The Tata Iron and Steel Co., Ltd. 220 Outer Circle Road Jamshedpur	M. K. Ghosh	Solar cooker
Japan	Yazaki Buhin Co., Ltd. Kosai-shi, Shizuoka-ken	T. Ishibashi, Technical Director	Solar absorption-refrigeration; Yazaki solar house—cooling, heating hot water; selective surface manufacture
	Yanagimachi Laboratory Meguro-ku, Tokyo	M. Yanagimachi, President	Heat-pump-assisted solar heating system; heat-storage tank
	Nihon Kogyo Co., Ltd. Minato-ku, Tokyo	H. Igarashi, Technical Director	Selective copper absorber
	Toshiba Electric Co., Ltd. Kaden Research Laboratory Kawasaki-shi,Kanagawa-ken	M. Sakamoto, Chief Res. Engineer H. Koizumi, Chief Res. Engineer	Domestic hot-water unit
	Showa Aluminum Co., Ltd. 480 Imuzuka, Oyama City	Y. Asano, Technical Director	Solar hot-water heater, Roll-Bond® absorber
	Kawasaki Juko Co., Ltd. Minato-ku, Tokyo	S. Fijidai, Research Engr.	Double-effect absorption-refrigeration machine
	Ishikawajima-Harima Juko Co., Ltd. Chiyoda-ku, Tokyo	S. Ichikawa, Solar Energy Project, Manager	Freon turbine solar refrigerating machine

	Sanyo Electric Co., Ltd. Hirakata-shi, Osaka	K. Hinotani, Chief Res. Engineer	Solar absorption-refrigerating unit; selective surfaces
	Takenaka Research Laboratory Koto-ku, Tokyo	K. Maekawa, Technical Dir. Y. Tanaka, Chief Res. Engr.	Solar collectors for heating and cooling
Mexico	Circuito Novelistas 59-A, Cd-Satelite (Ed. de Mexico)	Maximo Chauvet, Ingenieur E.C.P.	Heaters
United Kingdom	Alcoa Ltd. Gowerton, Swansea	–	Roll-Bond® heat exchangers
	Air Distribution Equipment (Midlands and West) Ltd. 64 Whitebarn Road Llanishan, Cardiff Wales	–	Collectors, control equipment
	Commercial Solar Energy 163 Pelham Road Sherwood, Nottingham	–	Collectors
	A. T. Marston & Co., Ltd. Lancaster Road Shrewsbury, Salop	–	"Calorsol" collectors
	Pilkington Brothers, Limited Latham, Ormskirk, Lancashire	H. C. Agrawal, Senior Technologist, R&D Labs	Solar dryers
	Production Methods, Ltd. West Arthurlie Works Barrhead, Glasgow Scotland	–	"Raystore" collectors
	Rolls-Royce Technical Institute Derby Engines Division Rolls-Royce Anti-Poverty Group P. O. Box 31 Derby	–	Low-cost water heater

Country	Company (Location)	Contact	Product or Activity
	Solar Centre 171 Ifidel Road, Chelsea London, SW10 9AF	–	"Sunstore" collectors
	Solar Heat, Ltd. 99 Middleton Hall Road Kings Norton Birmingham B30 1AG	–	"Suntrap" collectors
	Solar Water Heaters, Ltd. Pillar House 21 South Parade Doncaster, York	–	Collectors
	Stellar Heat Systems, Ltd. Upper York House 30 Upper York Street Bristol BS2 8Q3	–	Collectors; consultancy services
United States	Thermo Electron Corporation Waltham, Massachusetts	Jerry P. Davis	Organic Rankine cycle engine for low-temperature heat sources
	Sunwater Company 1112 Pioneer Way El Cajon, California 92020	Horace McCracken, Pres. Ed Smith	Solar water heaters, stills, space heaters
	SES, Inc. 70 South Chapel Street Newark, Delaware	Dr. Steven DiZio, Pres.	CdS/Cu_2S solar cells
	PPG Industries, Inc. One Gateway Center Pittsburgh, Pennsylvania 15222	–	Flat-plate collector. Double-glazed with aluminum Roll-Bond® panel.
	Kalwall Corporation 1111 Candia Road	–	Flat-plate collector with fiberglass-reinforced plastic covers

P. O. Box 237. Manchester, New Hampshire 03105		
Olin Brass/Roll-Bond Products East Alton, Illinois 62024	–	Flat-plate aluminum alloy absorber for use in collectors
Tranter Manufacturing, Inc. 735 East Hazel Street Lansing, Michigan 48909	–	"Econocoil," a heat exchanger currently being sold as absorber for flat-plate collector
Energex Corporation 5115 South Industrial Road Suite 513 Las Vegas, Nevada 89118	–	Copper flat-plate collectors
U. S. Solar Corporation 6407 Ager Road West Hyattsville, Maryland 20782	–	Copper flat-plate collectors
Solaron Corporation 4850 Olive Street Commerce City, Colorado 80022	John Bayless	Complete solar heating and cooling systems
Copper Development Associates 405 Lexington Avenue New York, New York 10017	Paul Anderson	Collectors
Sunworks, Inc. 669 Boston Post Road Guilford, Connecticut 06437	Everett M. Barber, Jr.	Copper flat-plate collectors
E & K Service Co. 16824 74th Avenue, N.E. Bothell, Washington 98011	–	Flat-plate collector and home space-heating system
Free Heat P. O. Box 8934 Boston, Massachusetts 02114	–	Flat-plate collector and home space-heating system

Country	Company (Location)	Contact	Product or Activity
	Garden Way Laboratories Charlotte, Vermont 05445	–	Water-heating system
	Hitachi America Ltd. 437 Madison Avenue New York, New York 10022	–	Collector for domestic hot water
	Reynolds Metals Co. Building E-A 2315 Dominguez Street Torranse, California 90509	Plant Engineer	Flat-plate collectors, heat exchangers, header systems
	Solar Power Corporation 930 Clocktower Parkway Village Square New Port Richey, Florida 33552	–	Flat-plate collector
	Sunsource 9606 Santa Monica Boulevard Beverly Hills, California 90210	–	Flat-plate collector
	CSI Solar Systems Division 12400 49th Street North St. Petersburg, Florida 33730	–	Domestic hot-water-heating systems
	Energy Systems, Inc. 634 Crest Drive El Cajon, California 92021	–	Flat-plate collector
	Helio-Dynamics Inc. 518 South VanNess Avenue Los Angeles, California 90020	–	Flat-plate collectors, domestic hot-water, space-heating, air-conditioning systems; solar refrigeration and freezing
	Raypak, Inc. 31111 Agoura Road West Lake Village, California 91360	–	Flat-plate collectors

Piper Hydro 2895 East La Palma Avenue Anaheim, California 92806	–	Flat-plate collectors, domestic hot-water and space-heating systems; heat pumps
Rodgers and MacDonald 3003 N.E. 19th Drive Gainesville, Florida 32601	–	Flat-plate collectors, domestic hot-water systems
Solar Water Heater Co. P. O. Box 341872 Coral Gables, Florida 33134	–	Domestic water heaters, space-heating systems
Wilcon Corporation 3310 S.W. 7th Street Ocala, Florida 32670	–	Hot-water heating systems using collector of vinyl tubing with dyed heat-transfer fluid
W. R. Robbins and Son 1401 N.W. 20th Street Miami, Florida 33142	–	Flat-plate collectors
Youngblood Co., Inc. 1085 N.W. 36th Street Miami, Florida 33127	–	Flat-plate collectors for domestic hot water

---

NOTE: The United States Energy Research and Development Administration and the Department of Housing and Urban Development are currently compiling information for a forthcoming catalog of solar-energy products. The catalog is designed to:

"–Disseminate information to the general public and to the solar industry about the types of hardware available under development.

–Provide a general reference base of "state-of-the-art" technology immediately available through commercial suppliers.

–Enhance capabilities of individuals and organizations to collaborate in the assembly, development, and marketing of complete systems and subsystems.

–Establish a standard basis for collection and dissemination of information within the solar energy industry."

Information about this project may be obtained from either of the following addresses:

ERDA Technical Information Center
P. O. Box 62
Oak Ridge, Tennessee 37830

or

Division of Energy, Building Technology and Standards
Room 8158, Department of Housing and Urban Development
Washington, D.C. 20410

APPENDIX 3

Manufacturers of Terrestrial Solar Photovoltaic Devices

[This list includes only those manufacturers known to be producing commercially available terrestrial solar arrays at the time of writing. It may not be complete, but it represents the best information available to the panel at this time.]

Manufacturer	Cable, TWX, or Telex Address	Telephone
SOLAREX 1335 Piccard Drive Rockville, Maryland 20850 U.S.A.	TWX: 710 828 7909 Cable Address: SOLAREX	(301) 948-0202
Spectrolab, Inc. Division of Hughes Corporation 12484 Gladstone Avenue Sylmar, California 91342 U.S.A.	TWX: 910 496 1488	(213) 365-4611
Iran Solar Energy Company (Affiliate of Spectrolab, Inc.) 100 Hafez Avenue, Ehansi Building Teheran IRAN	Telex: 2516 Cable Address: 668 Teheran	43530-669775-8
Ferranti Limited Electronic Components Division Gem Mill Chadderton, Oldham OL9 8NP ENGLAND	Telex: 668038	061-624-6661
Centralab Semiconductor Division 4501 Arden Drive El Monte, California 91734 U.S.A.	TWX: 910 587 3429	(213) 448-4001
Solar Power Corporation 23 North Avenue Wakefield, Massachusetts 02185 U.S.A.	TWX: 710 348 7674	(617) 246-2355
Sharp Corporation Industrial Instruments Group Yamatokooriyama, Nara 639-11 JAPAN	Telex: J63428 or J63429 Cable Address: LABOMETSANKI OSAKA	
R.T.C. La Radiotechnique-Compelec 130, Avenue Ledru-Rollin 75540 Paris FRANCE	Telex: 68495	355-44-99

SPECTROLAB:
12-VOLT LIGHT ENERGY CONVERTER ARRAYS MODEL LECA

SPECIFICATIONS

Nominal Output Voltage
12 VDC

Minimum Operating Voltage
14.3 VDC @ LEC™ operating temperature of +60°C. (+140°F.).
Charges typical 12-volt lead acid storage battery at 2.25 volts per cell and allows for normal 0.8 volt blocking diode drop.

Nominal Output Current
From 0.3 to 7.2 amperes depending on number and configuration of standard LEC™ modules in the array. See table. Complete arrays can be paralleled for larger requirements.

Construction

Support Frame:
Extruded aluminum members. Baked epoxy acrylic coating—black.

Modules:
Individual modules easily replaceable. Array is self-cleaning in normal tilted installation.

Output Power/Array Total Area
6.0 Watts/ft²

Output Power/Array Total Weight
2.0 Watts/lb nominal.

Dimensions and Weights
See table.

ENVIRONMENTAL CONDITIONS

ELECTRICAL PERFORMANCE CHARACTERISTICS TYPICAL CURRENT/VOLTAGE CURVES

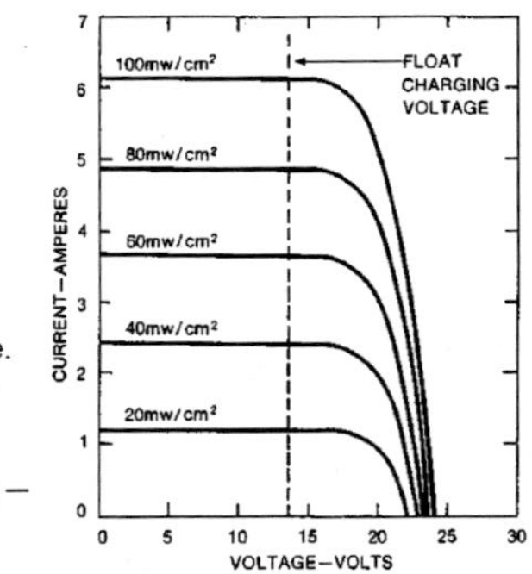

Array current as a function of voltage for various sunlight intensities. 12V 6A Array. Array temperature 60 C.

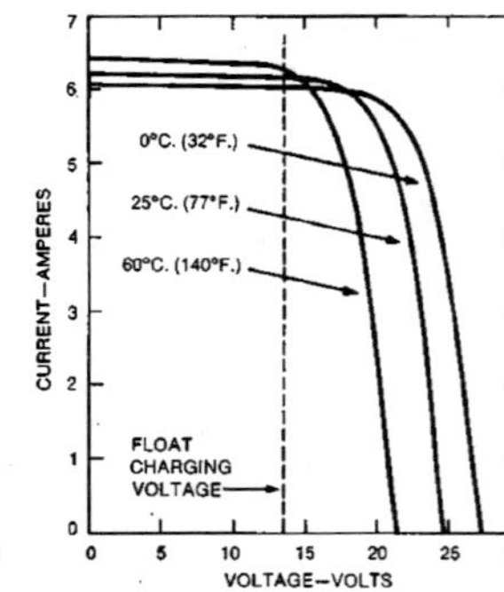

Voltage-Current Characteristics vs. LEC™ Temperature. 12V 6A Array, 100 mw/cm² intensity.

These graphs illustrate several important facts.
Output voltage of the LEC rises almost immediately to the float charging voltage . . . charging begins at first light and continues until darkness.
In practical application, temperature sensitivity prevents reliable charging at the maximum power point . . . the array must be operated in the flat portion of the characteristic curve.
Current output must be specified at a 60°C. (140°F.) cell temperature

Humidity:
100%

Altitude:
Sea level to 30,000 feet.

Submersion:
Full performance after prolonged submersion.

Wind:
175 knots without damage or loss of performance.

Ice:
6 inch buildup without structural damage.

ACCESSORIES

Solid-state voltage regulator.

Pole mount hardware for small arrays.

Low-self-discharge lead-acid and nickel-cadmium batteries.

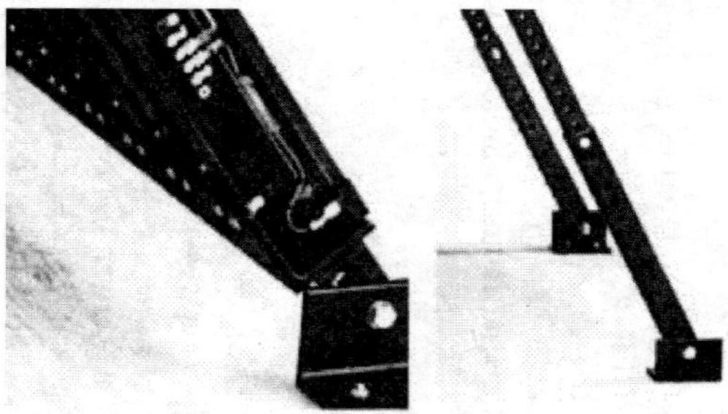

Left: *Tubes are protected and supported by aluminum extrusions. Modules are bolted at both ends to L-shaped cross members. Wiring is fast and simple. Redundant AMP connectors can be pulled and reconnected quickly.*
Right: *Telescoping "Telespar" struts and pivoted feet mean easy mounting to any surface.*

STANDARD 12-VOLT[1] LEC™ ARRAYS

Array Model No.	Approximate Dimensions (inches)	Current Output (Amps)[1]	Amp-Hours Generated Under Various Per Day Solar Conditions[2]			Approx. Weight (lbs.)
			Poor	Average	Good	
12V .3A	37 x 3 x 3	.3A	1.1	1.3	1.6	3.8
12V .6A	37 x 6 x 3	.6A	2.2	2.6	3.1	9.3
12V .9A	37 x 9 x 3	.9A	3.3	3.9	4.7	14.0
12V 1.2A	37 x 12 x 3	1.2A	4.4	5.2	6.3	18.6
12V 1.5A	37 x 15 x 3	1.5A	5.5	6.5	7.9	23.3
12V 1.8A	37 x 18 x 3	1.8A	6.6	7.8	9.4	28.0
12V 2.1A	37 x 21 x 3	2.1A	7.7	9.0	11.0	32.6
12V 2.4A	37 x 24 x 3	2.4A	8.9	10.4	12.6	37.3
12V 2.7A	37 x 27 x 3	2.7A	10.0	11.7	14.1	41.9
12V 3.0A	37 x 30 x 3	3.0A	11.1	13.0	15.7	46.6
12V 3.6A	37 x 37 x 3	3.6A	13.3	15.6	18.9	55.9
12V 4.2A	37 x 43 x 3	4.2A	15.5	18.2	22.0	65.2
12V 4.8A	37 x 49 x 3	4.8A	17.7	20.8	25.2	74.5
12V 5.4A	37 x 55 x 3	5.4A	20.0	23.4	28.3	83.8
12V 6.0A	37 x 61 x 3	6.0A	22.2	26.0	31.5	93.1
12V 6.6A	37 x 67 x 3	6.6A	24.4	28.6	34.7	102.4
12V 7.2A	37 x 73 x 3	7.2A	26.6	31.2	37.8	111.7

1. Minimum current output under Standard Test Conditions (STC) Intensity = 100 mw/cm^2; Temperature = 0°C. to +60°C.
2. Usable energy generated for use in a solar power supply system with lead acid storage batteries under typical conditions and based on annual mean solar radiation data for various locations in the contiguous United States.

FIGURE 58 Typical data sheet for commercial photovoltaic arrays. [Courtesy Spectrolab Inc.]

APPENDIX 4

External-Combustion Engines — Rankine and Stirling Engines as Small-Scale Power Sources for Developing Countries

Two types of external-combustion engines adaptable to a variety of heat sources are the Stirling- and Rankine-cycle engines. Both are based on modifications of the Carnot cycle, as illustrated in Figures 59 and 64, but thermal efficiencies are such that for operating temperatures below about 300-350°C, the Rankine engine is more efficient.

RANKINE ENGINES

The traditional example of a Rankine engine is the familiar steam engine, with thermal efficiencies of some 5-15 percent. This discussion, however, will be restricted to engines that are more useful on a smaller scale and that operate at temperatures in the range that can be achieved with solar collectors. For these lower temperatures, Rankine-cycle engines have been designed that use organic materials instead of water as the working fluid. The analysis that follows illustrates the type of system that would be associated with a solar-powered Rankine-engine-operated electric power plant.[1,2] It is based on current manufacturing capabilities for the individual components.

Introduction

Figure 60 shows a schematic of a solar-powered organic Rankine-cycle engine loop. The system of the schematic indicates the engine driving an electric generator. However, the engine could just as well be driving a water pump or the compressor of a cooling system.

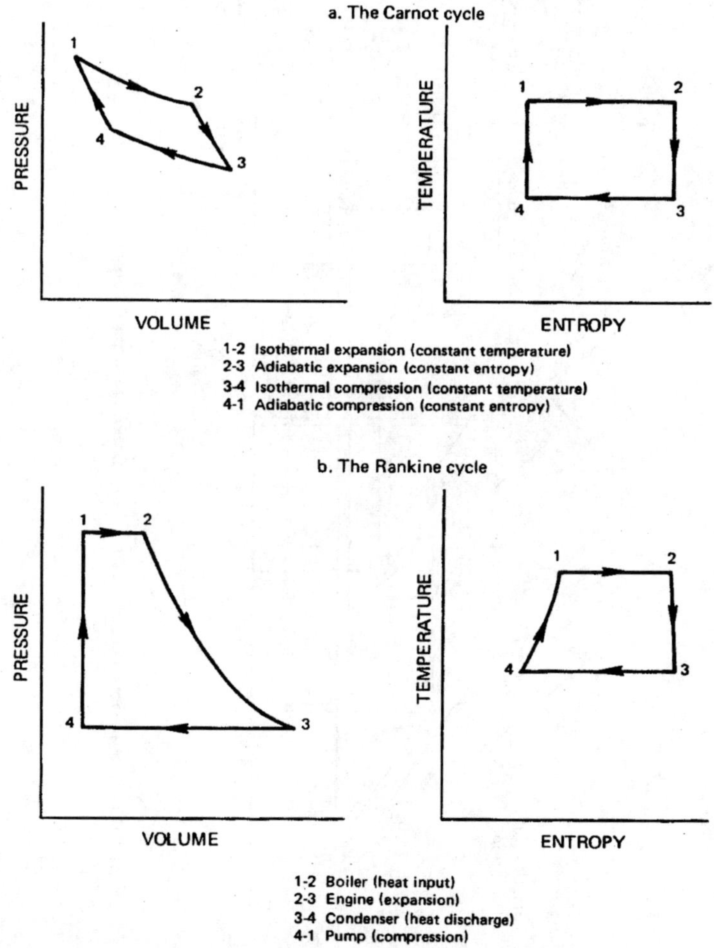

FIGURE 59 The Carnot and Rankine cycles.

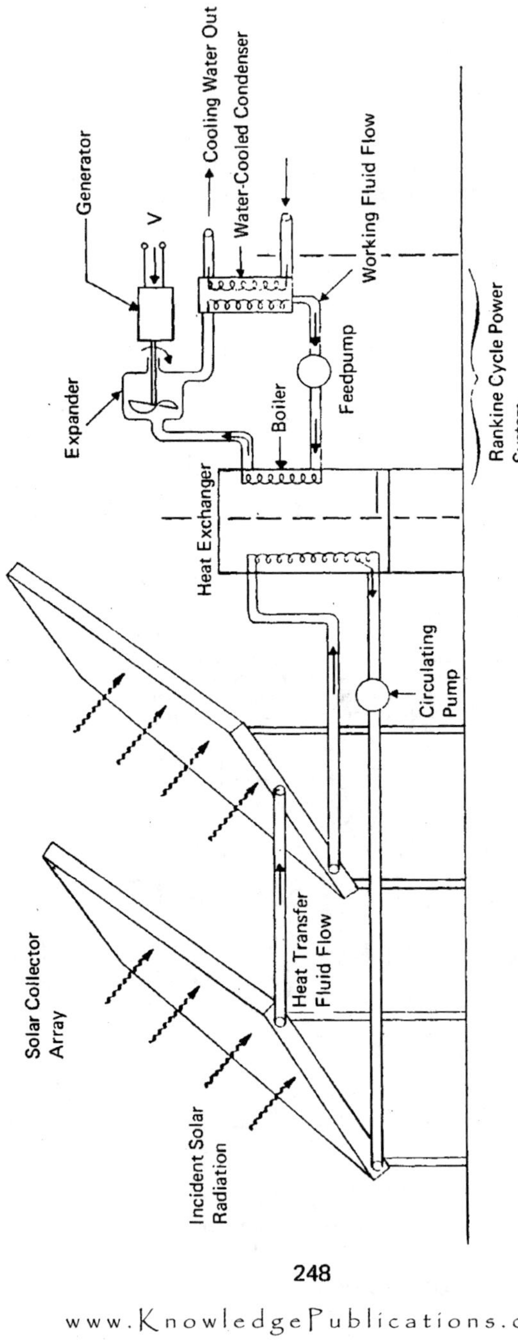

FIGURE 60 Solar-powered electric power plant. [Courtesy Arthur D. Little, Inc.]

Organic Rankine-cycle engines are particularly well suited for solar power application for a number of reasons:

- They have high thermal efficiency* even when operating with the low to moderate temperatures (180-400°F [80-200°C]) achievable with flat-plate collectors or collectors using low levels of focusing.
- The cost of components is low because of the use of common materials of construction (mostly carbon steel) and relatively uncomplicated mechanical components. This is potentially very important to developing countries because of the possibility of using local materials and labor.
- Organic Rankine-cycle engines have high reliability as a result of their sealed construction that protects them from the harmful effects of the surrounding environment (dust, moisture, etc.).
- Since the organic working fluids have very low freezing points, there are no problems associated with freezing.
- Finally, they are adaptable for use over a wide power range from 1-kW pumping systems to multimegawatt power stations.

System Description and Operation

The engine consists of four major components:

- expander
- böiler
- feedpump
- condenser.

During operation, a heat-transfer fluid (typically pressurized water) flows through the collector array and is heated to a temperature in the 200-400°F (100-200°C) range depending on solar-collector configuration, solar flux, and engine operating conditions. (This would entail a system capable of operating at about 235 psig [16.5 kg/cm^2 gauge].)

This hot fluid is then used to vaporize the working fluid of the engine in a heat exchanger; a number of common refrigeration fluids are appropriate for use in the engine loop. The hot, high-pressure, working fluid is then used to drive the expander of the Rankine-cycle engine. For higher-power-output applications (>100 kW) the expander will be a turbine. Lower-power systems can use positive displacement configurations such as reciprocating or vane-type expanders. After leaving the expander, the working-fluid vapor is

*60-70 percent of the ideal cycle efficiency, given the heat input and rejection temperatures.

condensed and the liquid is then pumped back into the solar-collector heat exchanger, completing the cycle.

System Performance Estimates

An estimate of overall system efficiency depends on analysis of the efficiency of the major components.

ENGINE EFFICIENCY

The efficiency of the engine depends on both the ideal efficiency of the cycle under consideration (assuming 100-percent expander efficiency, etc.) and the actual efficiency of the major system components, such as the expander and feedpump. Tests by a number of firms for a range of component configurations indicate that with existing technology the following component efficiencies are obtainable:

- expander–70-85 percent
- feedpump–70-93 percent.

Engine performance calculated on the basis of an expander efficiency of 80 percent and a feedpump efficiency of 80 percent indicates that it is possible to achieve about 65 percent of the ideal Carnot efficiency when the engine is operated in its appropriate temperature range. The corresponding engine efficiencies, as a function of both heat-input and heat-rejection temperatures, are shown in Figure 61. As this figure indicates, heat-engine efficiency increases with increasing heat-input temperature and with decreasing heat-rejection temperature. The 120°F (~50°C) heat rejection temperature is considered to be typical of air cooling, 100°F (~40°C) of evaporative cooling, and 80°F (~30°C) of water cooling. The effect of condenser temperature is quite significant–particularly for systems operating with lower (200°F [~95°C]) heat-input temperatures–thus providing a strong incentive to use water cooling if water is available (such as for water-pumping applications).

ANNUAL COLLECTOR EFFICIENCY

Both focusing and non-focusing collectors can be used as the heat source for organic Rankine-cycle engines. Flat-plate (non-focusing) collectors are used here to illustrate the effect of thermal collector characteristics on overall system performance for the following reasons:

- Flat-plate collectors have the advantage of being capable of effective operation mounted in a fixed position. The collectors and their mounting structure are, therefore, structurally simple—an advantage for use in remote areas.

- Flat-plate collectors can utilize diffuse as well as direct radiation, making such a system appropriate over a wide range of geographical areas and varying climatic conditions.
- The technology of flat-plate collectors is advancing as a result of industry and government programs to develop solar heating and cooling systems (particularly in the United States).
- Organic Rankine-cycle engines can effectively use the low to moderate temperatures achievable with flat-plate collectors.

For estimating the performance of the collector/engine combination it is most realistic to use the *average annual* collector efficiency as a measure of collector performance rather than the efficiency at midday (i.e., high solar flux conditions) often referred to in the literature.

Calculated curves[2] for the annual collector efficiency in an arid region at 32.5° north latitude for a variety of solar collectors tilted in a fixed position (32.5° relative to horizontal) show that the collection efficiency can vary widely, depending on the collector configuration (Figure 62). All the curves assume selective coatings on the absorber plate. (A collector with a flat black

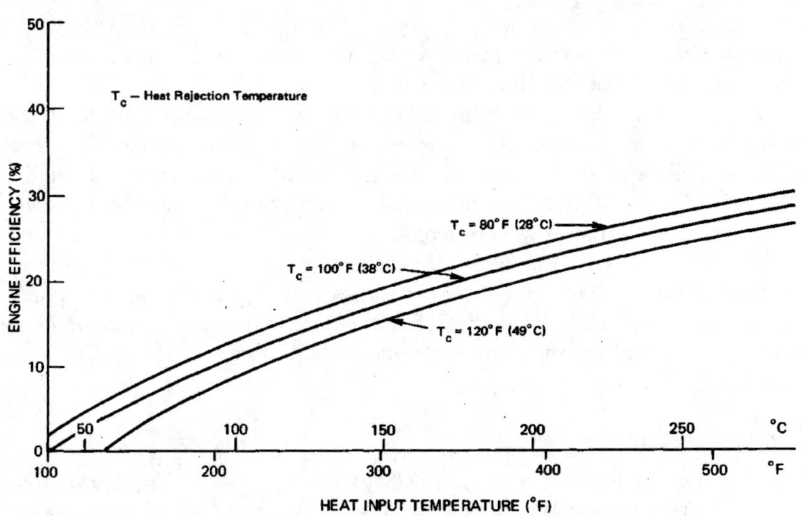

FIGURE 61 Heat engine efficiency as a function of temperature for an engine operating at 65 percent of Carnot efficiency. [Courtesy Arthur D. Little, Inc.]

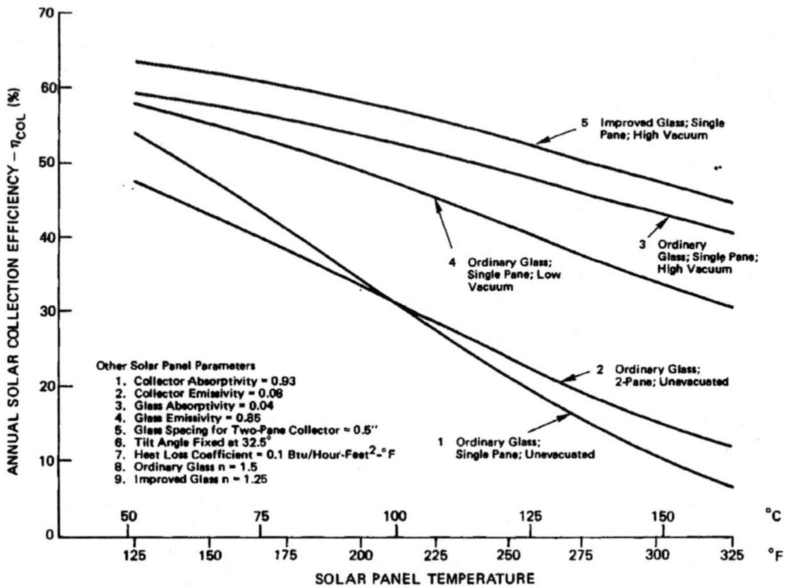

FIGURE 62 Annual solar collector performance as a function of collector temperature (for Yuma, Arizona). [Source: Bartoszek, *et al.,* Ref. 2]

coating and on two cover plates would have an efficiency level lower by approximately 10 percent than that indicated by Curve 1.)

As indicated, a large improvement in collector performance is achieved by using some form of convection suppression between the absorber plate and the cover plate(s) to eliminate heat transfer losses by convection. Figure 62 indicates the use of partial vacuum; similar results, however, may be obtained by using cellular structures (honeycombs).

Curve 5 indicates the improvement that would be achieved by using anti-reflective coatings or surface treatments on the glazing to reduce transmission losses. The benefits of anti-reflective coating would be even more pronounced for the lower-performance collectors typified by Curves 1 and 2.

SYSTEM EFFICIENCY

The overall system efficiency, which to a great extent determines the collector area requirements, is defined as:

$$\eta_s = \eta_e \times \eta_c$$

where:

η_s = overall system thermal efficiency
η_e = engine thermal efficiency
η_c = collector efficiency

Since increased collector temperature results in increased engine efficiency and decreased collector efficiency, there is an optimum operating temperature range for each engine/collector configuration. The variation in annual system efficiency is shown in Figure 63 for the collectors of Figure 62. The engine efficiency used in calculating these curves corresponds to the curve of Figure 61 with a condenser operating at 100°F (38°C).

As indicated, annual average efficiency levels are about 3.5 percent for collectors without some form of vacuum suppression and 6.5 percent to 9 percent if collectors using vacuum suppression are utilized. It should be noted that the peak efficiency of these systems under high flux conditions at solar

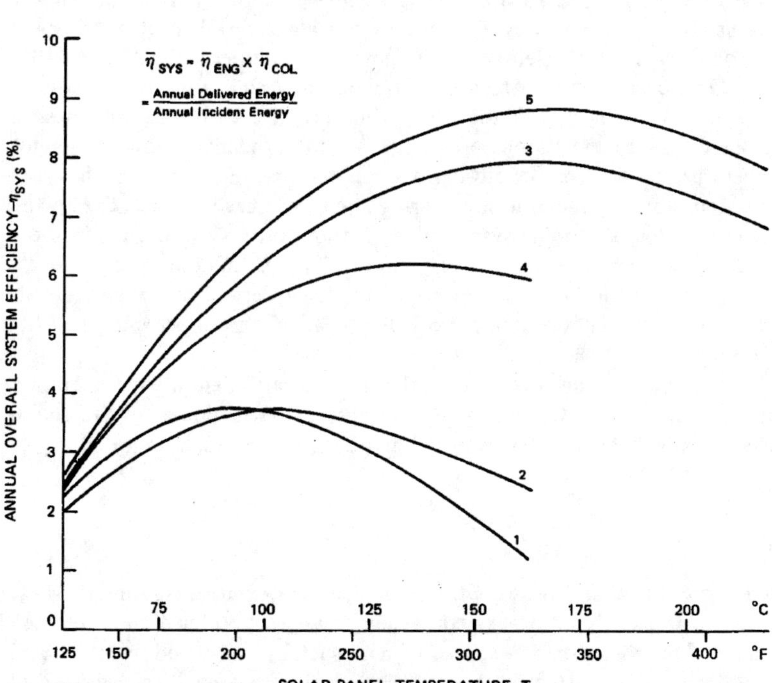

FIGURE 63 Annual overall system efficiency variation with collection temperature. [Source: Bartoszek, *et al.*, Ref. 2]

noon would be significantly (~50 percent) higher than those indicated in Figure 63. (These numbers are approximate; for a given design and application, it is possible to calculate more exactly what output can be expected.)

Comparison with Other Cycles

The heat generated in the solar collector array can, in principle, be used to operate any form of heat engine, including:

- Stirling cycle
- Rankine cycle (using water)
- Rankine cycle (using organic fluids)
- Ericsson cycle
- Closed cycle Brayton
- Hybrid cycle.

All of these engines have been suggested for use in solar thermal power systems. However, it is very difficult to operate gas-cycle engines (Stirling, Ericsson, Brayton) efficiently at the low temperatures (200-350°F [100-180°C]) associated with arrays of flat-plate solar collectors using low-level concentration, even though the ideal efficiency of the Stirling and Ericsson engines would be the Carnot efficiency. At these relatively low heat-input temperatures, the compression work in the gas cycle approaches the expansion work so that the net work output becomes very sensitive to the efficiency of the compression and expansion processes. Even with low operating temperature differences, the efficiency of the Rankine-cycle engines remains high (60-70 percent of the ideal Carnot efficiency). This is due to the fact that the pump work is a relatively small percentage of the expansion work.

If collector arrangements are used that can operate at elevated temperature levels (1,000°F [550°C] or higher) the engines based on gas cycles could, however, be very attractive and would merit serious consideration.

STIRLING ENGINES

Unlike the Rankine engines, Stirling engines are external-combustion heat engines that usually utilize air or other gases as working fluid. They are capable of operating on any source of heat such as sun, wood, coal, or field waste. Invented in 1816 by Robert Stirling, they were competitive with steam engines of the day, particularly with regard to fuel efficiency and safety.

(Figure 64 shows the basic Stirling cycle.) In the early 1900s thousands were sold as coal-burning water-pumping machines that produced from 50 to 500 watts of delivered power at about 2 percent overall thermal efficiency.[3] These pumping engines used air at atmospheric pressure for a working fluid, and a cast-iron hot end. As a result, they were very large and heavy for their power. However, they were exceptionally quiet, durable, and easy to operate and maintain; they were displaced only by the advent of widely distributed electric power in the 1920s.[4]

Toward the end of the Second World War there was launched an intensive development effort that has resulted in today's powerful, efficient, and quiet Stirling engines (Figures 65 and 66) intended for automotive use. In order for these large crank-type engines to come into commercial use, a number of very difficult design problems must still be solved.

As a result, and because of the nearly universal availability and very low cost of internal-combustion engines, the traditional crank-type Stirling engine remains in the development labs instead of on the road or in the field. No such engines producing more than a few watts of power are at present available on the market.

Free-Piston Stirling Engines

The practical difficulties of producing crank-type Stirling engines have led to the development of a great variety of free-piston linear-motion Stirling engines that, while they operate on exactly the same cycle as the crank machines (Figure 67), avoid many of their problems. Furthermore, their design makes the construction of small engines more easily achieved. Power is removed from these engines by various linear-motion pumps or alternators.[5] It should be emphasized that *all of these engines are under development* and are not yet commercially available in production quantities that would reflect a realistic manufacturing cost. These engines are under development in the United States under the sponsorship of the Energy Research and Development Administration (space power plant), the National Institutes of Health (artificial heart), and the American Gas Association (gas-fired air conditioner). Private groups are also involved in developing some of these devices.

Figures 68 through 71 show several means of power extraction that are useful with linear motion engines. All of these have been built in prototype quantity and operated successfully. The inertia pump (Figure 68) is being developed in the United States for commercial use in a gas-fired air-conditioner system in which the free-piston Stirling engine (FPSE) drives an inertia pump that pumps Freon-12 around a conventional cooling cycle. When lower temperatures for food freezing are required, an even simpler machine can be used. In this, an FPSE heat engine drives an FPSE heat pump;

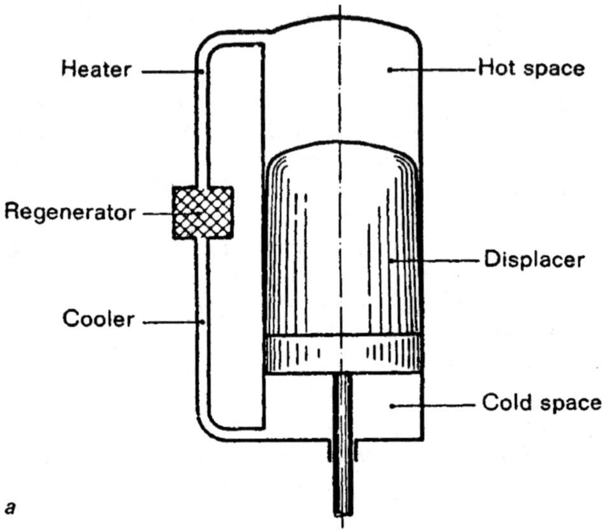

a

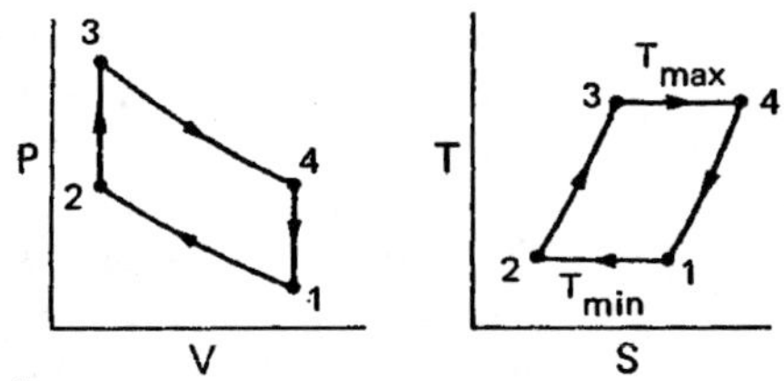

b

1-2 Isothermal compression at the lower temperature.
2-3 Heat input at constant volume, raising gas to upper temperature, increases pressure still further.
3-4 Isothermal expansion at upper temperature.
4-1 Gas cooled to lower temperature at constant volume.

FIGURE 64 The Stirling Cycle. [Courtesy Philips Research Laboratories, Eindhoven]

a. Principle of the displacer system.
b. The thermodynamic cycle.
c. Diagram of the cycle.

c

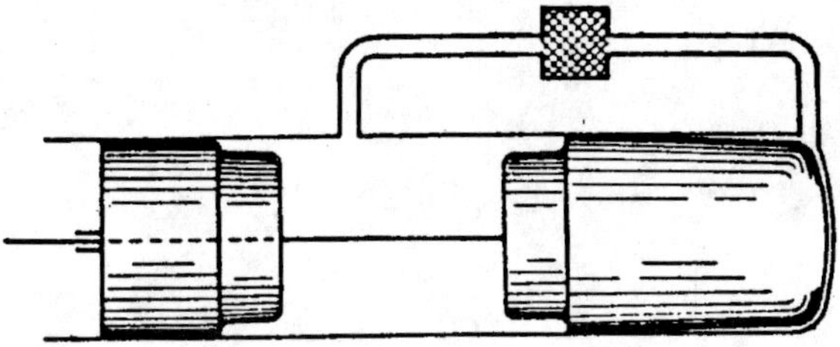

1

Piston at bottom dead centre. Displacer at top dead centre. All gas in cold space.

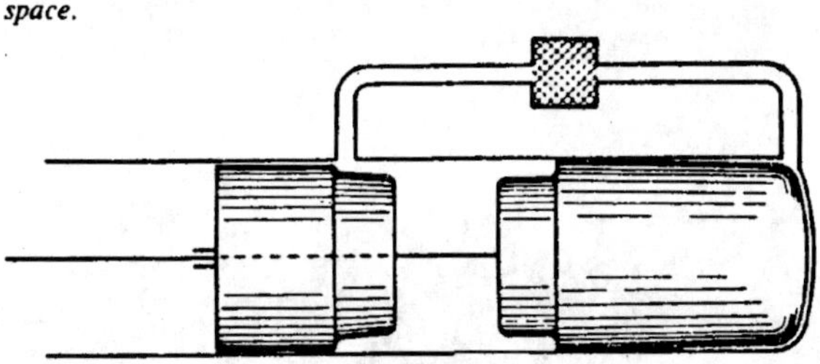

2

Displacer remaining at top dead centre. Piston has compressed gas at lower temperature.

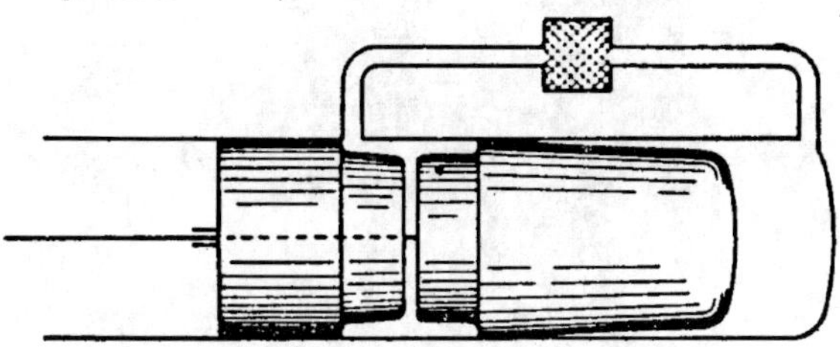

3

Piston remaining at top dead centre. Displacer has shifted gas through cooler, regenerator and heater into hot space.

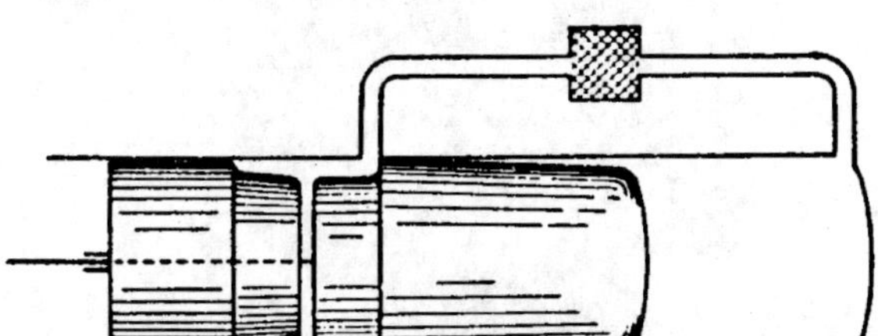

4

Hot gas expanded. Displacer and piston have reached bottom dead centre together. With piston stationary, displacer now forces gas through heater, regenerator and cooler into cold space, thus re-attaining situation **1**.

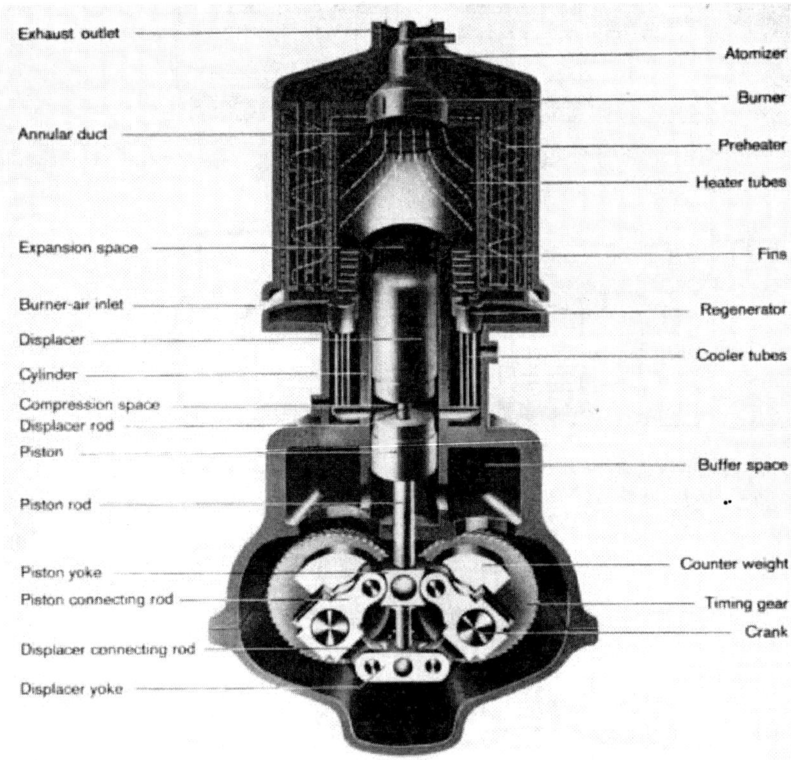

FIGURE 65 The Philips rhombic engine. [From: R. J. Meijer: 1970. *Philips Technical Review* 31:168-85]

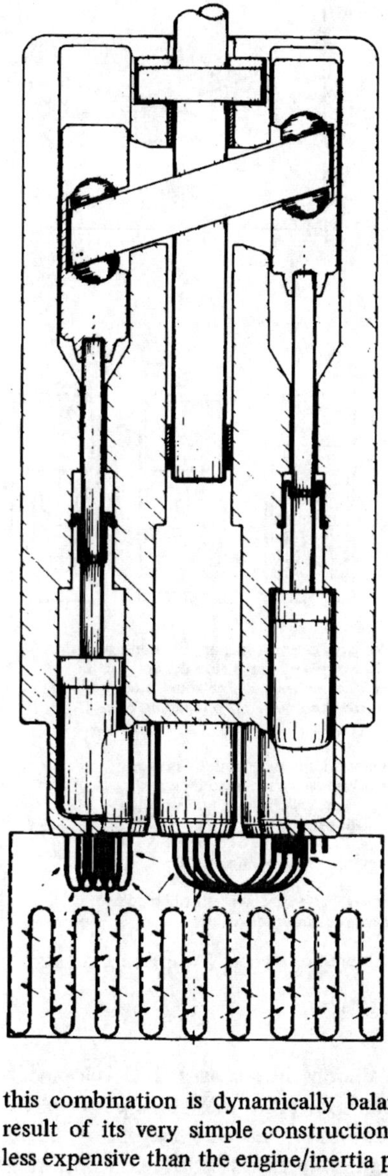

FIGURE 66 Swashplate Stirling engine. [From: R. J. Meijer. 1970. *Philips Technical Review* 31:168-85]

this combination is dynamically balanced and hermetically sealed, and as a result of its very simple construction, would probably be more reliable and less expensive than the engine/inertia pump combination.

A variant of the FPSE is the free-cylinder engine, illustrated in Figure 69, that operates from solar energy concentrated by a Fresnel lens.

A free-piston/linear-alternator engine is under development in the United

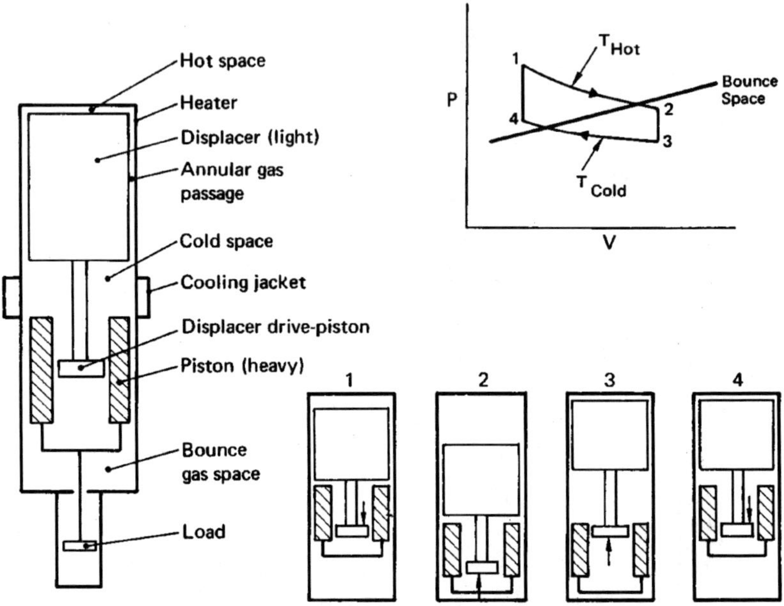

1-2 Working gas in hot space, working-gas pressure higher than bounce-gas pressure. Expanding gas drives displacer and piston down. With a lower mass than the piston, the displacer accelerates more rapidly and soon contacts the piston and both move together, compressing bounce gas.

2-3 Bounce-gas pressure greater than working-gas pressure. Pressure differential acting on displacer drive-piston, while bounce gas is still being compressed by piston, moves displacer up, shuttling working gas into cold space. Working-gas pressure drops rapidly.

3-4 Bounce-gas pressure drives piston up, compressing cold working gas.

4-1 Rise in working-gas pressure drives displacer down, shuttling cold gas into hot space. Working-space pressure rises rapidly.

FIGURE 67 Free-piston Stirling engine—principle of operation. [Adapted from Beale, *et al.,* Ref. 8]

States in a project supported by the Energy Research and Development Administration, intended to result in a very long-life 1-kW space power plant heated by a radio isotope. The linear-alternator machine may be made to operate on solar power, using either high temperature with a concentrating collector, or low temperature such as is available from a flat-plate collector. The use of a flat-plate collector to operate an FPSE diaphragm pump is illustrated in Figure 70. If the flat plate is used, condensing working fluids rather than an ideal gas might be more appropriate, since they produce much more work per cycle.[6]

Diaphragm Engine

A particularly attractive variation for electric-power generation is shown in Figure 71.[7] This machine uses flexing diaphragms and springs, without any sliding fits or seals. The displacer is driven by the force transmitted from the cylinder to the displacer through the displacer support-spring, as the displacer oscillates at a resonant frequency determined by its mass and support-spring stiffness. Since the motion of the cylinder is 180° out of phase with the

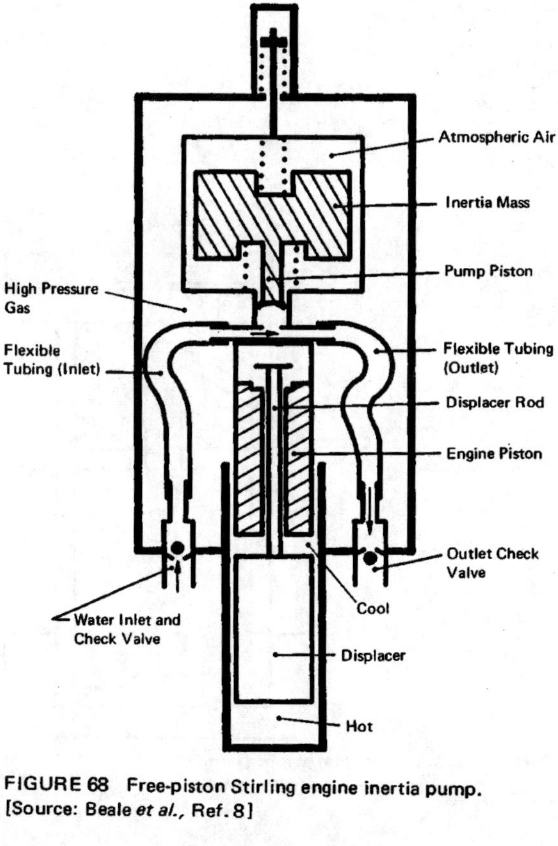

FIGURE 68 Free-piston Stirling engine inertia pump. [Source: Beale *et al.*, Ref. 8]

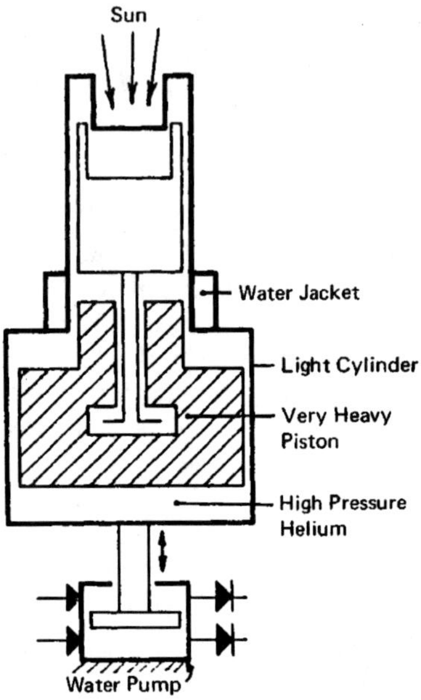

FIGURE 69 Free-cylinder Stirling engine. Power is taken from the cylinder, which moves in reaction to the motion of the heavy piston. [Source: Beale *et al.*, Ref. 8]

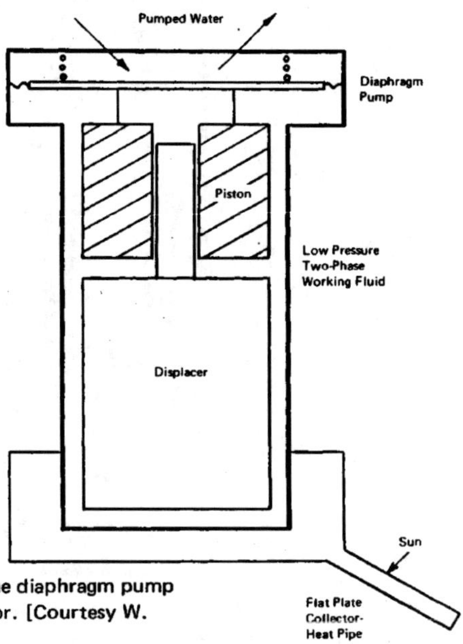

FIGURE 70 Free-piston Stirling engine diaphragm pump operated from a flat-plate solar collector. [Courtesy W. Beale]

diaphragm, a net energy input to the displacer is achieved. The present experimental models run at 110 Hz and produce about 25 W. Larger sizes are possible. Propane-heated and radioisotope-heated versions are operating. A propane-heated field-trial prototype is providing the electrical power for the UK National Data Buoy, and a similar machine powering a large marine light has been demonstrated commercially at the International Association of Lighthouse Authorities Exhibition at Ottawa in August 1975. This machine is being actively developed at the Atomic Energy Research Establishment, Harwell, England, and is now being offered commercially by two UK firms.[8]

Fluidyne Engine

One of the more recent variants of the free-piston Stirling engine is the "Fluidyne" pump being developed at Harwell.[12] This engine, illustrated in Figure 72, uses columns of fluid in tubes to displace the working gas (air) between the hot and cold regions, and also for extracting output power. This lends itself to direct integration in water-pumping systems.

At present, the following results have been achieved with some of the experimental Stirling engines of the type discussed here; better performance is expected with development.

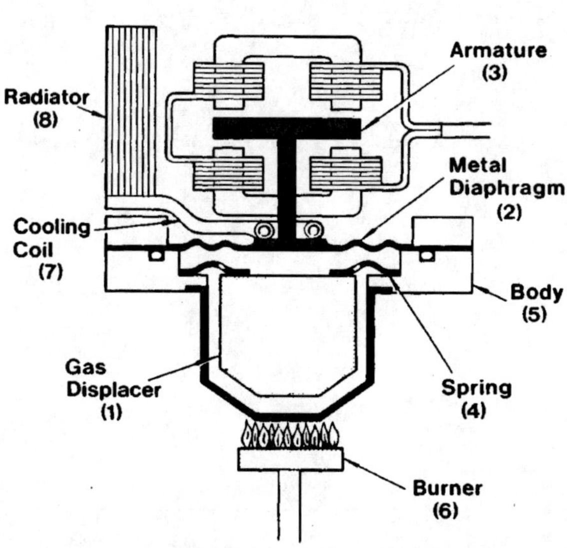

FIGURE 71 Harwell diaphragm engine. [Source: Harwell Bulletin No. 74/26]

Type of Load and Working Gas	Heat Source	Hot-End Temperature degrees C	Power Out	Thermal Efficiency	Reference
inertia Freon compressor (helium)	electric	650	2 kW	30%	9
inertia water pump (air)	electric	550	3 W	3%	9
linear alternator (helium)	electric	600	34 W	12%	9
linear alternator (air)	electric	500	3.6 W	4%	9
free cylinder water pump (helium)	sun (Fresnel lens)	450	6 W	3.5%	9
free cylinder water pump (helium)	electric	600	70 W	14%	9
Harwell diaphragm engine alternator (helium)	electric	594	37.5 W	16.9%	10
Harwell diaphragm engine alternator (helium)	propane	450	31.8 W	10%	11
Harwell diaphragm engine alternator (helium)	Sr^{90}	280	10.7 W	7.8%	10

A demonstration FPSE as shown in Figure 68 has operated for 2,600 hours without wear or performance degradation.[9] Critical components of the Harwell diaphragm engines have been operated for 2½ years without failure.[6] Two of these engines, one propane-heated and one radioisotope-heated, have each operated for more than 7,000 hours without wear or performance degradation.[8]

THE STATUS OF CURRENT TECHNOLOGY

There is as yet no significant commercial manufacture of Stirling engines. It is possible, however, to construct a 150-W output, free-piston Stirling engine

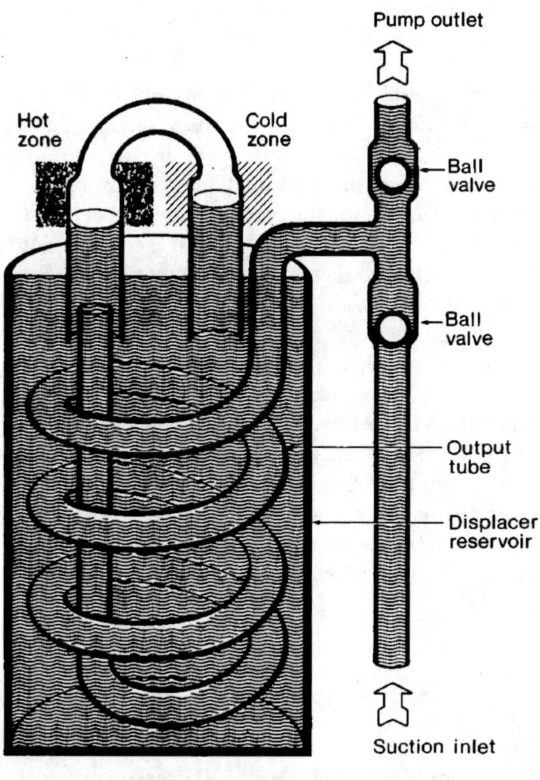

FIGURE 72 The Harwell "Fluidyne" pump. In operation, air is displaced between the hot and cold zones and the resulting changes in pressure force water up and through the pump, at the same time maintaining the rhythmic movement of the air. [Courtesy AERE Harwell]

weighing about 20 kg with all accessories, including power-output device (pump or alternator) and cooling system, with an overall efficiency of about 5 percent operating on solid fuels—or as high as 20 percent operating on bright sun with a Fresnel lens[9]—and costing approximately $100 as made in the United States. The performances and costs are approximations from present experience, and the costs may be lower if large numbers of engines are produced. The cost of the collector and tracker is not included, and is likely to be quite high (say $200 for a 100-W engine).

This machine could be expected to have a life of at least 4,000 hours and probably much more. The most likely failure modes would be corrosion of the hot end, and/or leakage of pressurized working fluid. It would be possible to repair it only at a central shop, but repair would be simple and cheap, probably requiring only replacement of the hot end and recharging. Internal moving components would operate on a gas film and should have no appreciable wear. Internal springs would be stressed only briefly during startup.

Whether the FPSE could be manufactured locally would depend on details of design. The only difficult operation is the very close fitting required on piston/cylinder and displacer rod. These might be designed for a hand-lapped fit, resulting in non-interchangeable matching pairs. Material throughout could be cast iron or steel if a low-temperature hot end is tolerated. Possibly a more desirable solution would be to use a relatively small amount of stainless steel for the hot end. This would greatly increase durability and might be less costly in the long run.

The diaphragm-type machine avoids the need for any sliding surfaces, so there is no wear and no close fits are required. Nevertheless, the present manufacturing costs of these machines in Great Britain are more than 100 times greater than the $100 estimate given for the 150-W FPSE. Furthermore, the diaphragm machines are much heavier for a given output than the FPSE machines. As with some thermoelectric and photovoltaic systems available commercially, the main present application of the diaphragm-type machine would be in situations where the ability to provide a few tens of watts of electric power, for a year or more without attention or refueling, is of great value.

RESEARCH AND DEVELOPMENT

Given the state of development of the free-piston Stirling engine, one would expect it to be commercially available in 5-10 years, if the demand warrants further commercial interest.

Further development of the diaphragm engine might also result in a device that would be available in 5-10 years. It is not only intrinsically extremely durable if properly designed and constructed, but also potentially very inexpensive in power levels below 100 W; the major cost is the alternator itself and the power diaphragm, which must be free of any flaws. It could be very largely assembled from pressings. Thus, if the potential market justified a substantial investment in design and tooling, there seems no reason why the cost should not be reduced to that of a small gasoline engine, if manufactured in equally large numbers.

The presently known technology would be adequate for power plants on the order of 100 W. For kilowatt levels of power, some work is needed to find designs of hot ends, solid-fuel burners, lenses, tracking mechanisms, and pumps appropriate for local manufacture in developing countries.

As a result of the active development programs being carried on for automotive, air-conditioning, artificial-heart, and space applications, rapid improvements in efficiency and durability of the Stirling engine can be expected, along with a reduction in costs.

USES OF THE STIRLING ENGINE IN DEVELOPING COUNTRIES

In view of the current state of development, there is little likelihood that Stirling engines will be used in developing countries to any great extent in the next 5 years. (The one possible exception is use of the large, slow, 1908-style engines discussed earlier, that could certainly be manufactured today in many developing countries.) If interest and demand in the industrialized countries result in manufacturing economies for the machines that exist in prototype form, some applications in developing countries may be possible in 5-10 years.

Stirling engines can be made to operate with either high-temperature sun-tracking collectors or with lower-temperature fixed collectors. The free-cylinder water pumps and the diaphragm linear alternator seem the most likely candidates for solar power.

Any of the Stirling engines mentioned could be designed to operate on solid fuel with power/volume ratio and efficiency dependent on the design of the heat exchangers. Solid-fuel stove/engine combinations can be devised to serve as heating or cooking units while providing electric power (Figure 73). These stoves could burn straw, dung, coal or any locally available fuel.

The free-cylinder engine with induction pump would have all its high-mortality components easily available, and its precision components completely protected from any possible contamination.

Another possibility would be to use an electric motor to drive a pump, and

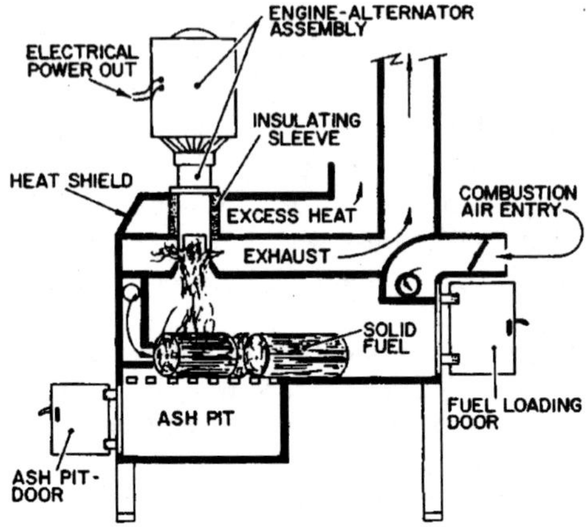

FIGURE 73 Combination wood stove/solid-fuel Stirling engine. [Courtesy W. Beale.]

use a Stirling engine, windmill, or other power source to maintain charge in a battery. This is a complex arrangement, but it has the advantage that many power sources can be used in parallel, and many uses can be made of the stored electric energy.

A third water-pumping alternative would be to use the 1908-style, large, slow, atmospheric engines. A modern design of these would have performance not less than the early 500-W 1-percent achievement. A possible advantage of this large and simple machine is that it requires lower tolerances and is more easily adapted to local foundry and machining capabilities than the more precisely finished high-pressure engines. It does, however, require far more metal for the same power. Another advantage of this rather primitive but appealing engine is its educational value. Its workings are obvious to the interested observer, which is not the case with the free-piston machines.

REFERENCES

1. Prigmore, Daryl R., and Barber, Robert E. 1974. A prototype solar powered, Rankine cycle system providing residential air conditioning and electricity. Paper No. 749026. *Ninth Intersociety Energy Conversion Engineering Conference Proceedings*. 26-30 August, San Francisco, California. pp. 326-333. New York: American Society of Mechanical Engineers.
2. Bartoszec, J. T.; Swanson, J. L.; and Teagan, W. P. 1975. A solar-powered water pumping system for the Yuma (Arizona) desalination plant. Paper No. 759114. *Record of the Tenth Intersociety Energy Conversion Engineering Conference. 18-22 August, University of Delaware, Newark,* pp. 765-771. New York: Institute of Electrical and Electronic Engineers.

3. Rider-Ericsson Engine Company. *Hot air pumping engines catalog–1908.* Lansing, Michigan: Robert Huxtable Company, Box No. 1107, Zip Code 48904.
4. Walker, Graham. 1973. *Stirling-cycle machines.* New York: Oxford University Press.
5. Beale, W. 1969. Free piston Stirling engine–some model tests and simulations. Paper 690230. Presented at the Society of Automotive Engineers Automotive Engineering Congress, 13-17 January, Detroit, Michigan.
6. Walker, Graham. 1974. Personal communication. University of Calgary, Mechanical Engineering Department, Alberta, Canada.
7. Cooke-Yarborough, E. H.; Franklin, E.; Geisow, J.; Howlett, R.; and West, D. D. 1974. Thermomechanical generator: an efficient means of converting heat to electricity at low power level. *Institute of Electrical Engineers, London, Proceedings* 121(7):749-751.

   Also Atomic Energy Research Establishment (of United Kingdom Atomic Energy Authority), Harwell, England (referred to after this as AERE), Report AERE-R7645.
8. Harrison, J. D. L. 1975. Personal communication. AERE Harwell, Energy Technology Support Unit.
9. Beale, W.; Holmes, W.; Lewis, S.; and Cheng, E. 1973. Free piston Stirling engines–a progress report. Paper 730647. Presented at the Society of Automotive Engineers Automotive Engineering Conference, 18-22 June, Chicago, Illinois.
10. Cooke-Yarborough, E. H., and Yeats, F. W. 1975. Efficient thermo-mechanical generation of electricity from the heat of radioisotopes. *Record of the Tenth Intersociety Energy Conversion Engineering Conference, 18-22 August, University of Delaware, Newark*, pp. 1003-1011. New York: Institute of Electrical and Electronic Engineers.

    Also Harwell Report AERE-R8036.
11. *Ocean '74: IEEE International Conference on Engineering in the Ocean Environment Record.* Vol. 1. 21-23 August, Halifax, Nova Scotia, pp. 226-231. New York: Institute of Electrical and Electronic Engineers.

    Also Harwell Report AERE-R7753.
12. West, C. D. 1974. Fluidyne heat engine. Paper No. 5. *Record of the International Solar Energy Society (UK Section) Conference on Solar Energy Utilization*, 9 July, Brighton Polytechnic, England, pp. 54-79. London: International Solar Energy Society, UK Section.

    Also Harwell Report AERE-R6775.

SUGGESTED READING

Martini, W. R. 1972. Developments in Stirling Engines. ASME Paper 72-WA/ENER-9. New York: American Society of Mechanical Engineers.

Zarinchang, Jafar. 1972. *The Stirling engine: a bibliography.* London: Intermediate Technology Development Group.

Walker, Graham. 1973. The Stirling engine. *Scientific American* 229(2):80-87. An excellent general summary of Stirling engine development. For a technical summary and comprehensive bibliography, see Ref. 4.

APPENDIX 5

Currently Manufactured Windmill Pumps and Generators

[This list is not necessarily complete but it has been compiled from the best information available to NAS staff. The information has been supplied by the developers; thus, neither the panel nor the NAS/NRC can assume responsibility for its accuracy.]

Manufacturer	Equipment
Aermotor Division of Braden Industries, Inc. Industrial Park P. O. Box 1364 Conway, Arkansas 72032 U.S.A. [Note: Manufacture of the fanwheel is done by licensee in Argentina, for units sold by Aermotor. Windmill heads are also manufactured by South African licensee.]	"Aermotor" water pumping windmill, multi-blade (18) fanwheel coupled to reciprocating cylinder pump.

Wheel Diam. (feet)	Stroke (inches)	Cost (windmill alone) (F.O.B. branch warehouse in U.S.)
6	5½ & 3	$ 465.
8	8 & 6	690.
10	10 & 7½	1165.
12	12 & 9	1945.
14	14 & 10	3125.
16	16 & 12	4235.

Stub towers, for mounting Aermotor windmill on owner-supplied tower, are available in heights from 3 feet to 14 feet, at prices ranging from $44 to $965, depending on the size of the fanwheel. Standard four-post towers are also available from 21 feet to 47 feet high, and costing from $535 to $3175, depending on the size of the fanwheel. (Costs F.O.B. branch warehouse in U.S.)

AERMOTOR PUMPING CAPACITY

Diameter of Cylinder (Inches)	Capacity per Hour, Gallons		Total Elevation in Feet — SIZE OF AERMOTOR					
	6 Ft	8-16 Ft	6 Ft	8 Ft	10 Ft	12 Ft	14 Ft	16 Ft
1¾	105	150	130	185	280	420	600	1,000
1⅞	125	180	120	175	260	390	560	920
2	130	190	95	140	215	320	460	750
2¼	180	260	77	112	170	250	360	590
2½	225	325	65	94	140	210	300	490
2¾	265	385	56	80	120	180	260	425
3	320	470	47	68	100	155	220	360
3¼		550			88	130	185	305
3½	440	640	35	50	76	115	160	265
3¾	—	730	—	—	65	98	143	230
4	570	830	27	39	58	86	125	200
4¼		940	—	—	51	76	110	180
4½	725	1,050	21	30	46	68	98	160
4¾	—	1,170	—	—	—	61	88	140
5	900	1,300	17	25	37	55	80	130
5¾	—	1,700	—	—	—	40	60	100
6	—	1,875	—	17	25	38	55	85
7	—	2,550	—	—	19	28	41	65
8	—	3,300	—	—	14	22	31	50

Capacities shown in the above table are approximate, based on the mill set on the long stroke, operating in a 15 to 20 mile-an-hour wind. The short stroke increases elevation by one-third and reduces pumping capacity one-fourth

Manufacturer	Equipment
Aerowatt 37, Rue Chanzy 75011 Paris France [Information supplied by U.S. representative: Automatic Power, Inc. P.O. Box 18738 Houston, Texas 77023]	"Aerowatt" wind-driven electric generator [battery charger]. Two-blade variable-pitch propeller, three-phase AC generator with 12- or 24-volt DC output from solid-state controller [rectifier/regulator]. Electrical brake.

Model No.	Propeller Diam. (feet)	Propeller Diam. (meters)	Power Generated at 16 mph [7.15 m/s] (kW)	Average produced in one year with average annual wind velocity of 16 mph [7.2 m/s] Power (kW)	Energy (kW-hrs)
24FP7	3.3	1.0	28	14	126
150FP7	6.7	2.04	130	72	628
300FP7	10.7	3.16	350	184	1600
1100FP7	16.7	5.09	1125 volt-amps	450	3900
4100FP7	30.7	9.34	4100 volt-amps	1500	13,100

Cost (F.O.B. Houston, Texas)

Windmill Generator*		Controller	
27FP7	$2,380.	24FP12 (12-volt)	$ 462.
150FP7	3,865.	150FP12 (12-volt)	1,070.
300FP7	5,680.	150FP24 (24-volt)	1,070.
1100FP7	9,740.	300FP24 (24-volt)	1,120.
4100FP7	22,360.		

***For standard machine with urethane-painted machined hardwood blades. Does not include tower or batteries.**

Manufacturer	Equipment
P. Andrag and Sons (Pty.) Ltd. P.O. Box 364 Belleville, Cape Province Republic of South Africa	Manufactures 8- and 10-foot fanwheels under license from Aermotor (U.S.). See entry under Aermotor for details.

Dempster Industries
P.O. Box 848
Beatrice, Nebraska 68310
U.S.A.

Water-pumping windmill, multi-blade fanwheel coupled to reciprocating cylinder pump.

Wheel Diam. (feet)	List Price*
6	465.
8	670.
10	1150.
12	1900.
14	2970.

*F.O.B. shipping-point

Typical pumping capacities, 18-mile-per-hour wind, 2-inch cylinder pump:

Wheel Diam. (feet)	Total Elevation (feet)	U.S. gallons per hour
6	95	130
8	135	195

Dunlite Electrical Co. Pty. Ltd.
21-27 Frome Street
Adelaide, S. A.
Australia

Wind-driven electric battery charger. Three-blade, 13-foot diameter variable-pitch propeller (automatically controlled by centrifugal governor) supplied in two basic models:

- 1000 watts, 12 volts DC
- 2000 watts, 115 volts DC

Manufacturer	Equipment
Electro GmbH Winterthur St. Gallerstrasse 27 CH-8400 Winterthur Switzerland	Wind-driven AC generators, for use as battery charger or as direct source of single- or three-phase AC power. Variable-pitch 2- or 3-blade propeller (except as noted) with manually adjusted tail fin.

Model	Rated Windspeed[a] (mph)	Rated Windspeed[a] (m/s)	Output Watts	Output Volts[b]		Diam. (meters)	No. of Blades	Cost (F.O.B. Factory Swiss Fr.) Generator[c]	Control Panel	Voltage Regulator
W50	40	18	50	DC	6/12/24	0.45 x 1m high[d]		2400.	310.	350.
W250	40	18	250	DC	12/24	0.66 x 1.3m high[d]		3100.	330.	380.
WV05	20	9	600	DC	25/36	2.5	2	3350.	330.	400.
WV15G	23	10	1200	DC	24/36/48/65/110	3.0	2	4100.	400.	150. [e]
WV25G	24	11	2200	DC	36/48/65/110	3.6	2	5150.	500.	150. [e]
WV35G	24	11	4000	DC	48/65/110	4.4	3	6850.	580.	150. [e]
WVG50G	26	12	5/6000	DC	65/110	5.0	3	7900.	680.	150. [e]
WV15W	23	10	1200	AC	110, 1-phase (30-70Hz)	3.0	2	4750.	900.	included
WV25D	25	11	2000	AC	110, 3-phase (40-70Hz)	3.6	2	5850.	1200.	included
WV35D	23	10	3500	AC	110/220, 3 phase (35-60Hz)	4.4	3	7600.	1550.	included
WVG50D	23	10	5000 f	AC	110/220, 3 phase (50-80Hz)	5.0	3	8700.	1550.	included

Automatic control for generators WV15-WVG50:

274-212	DC 24/36 for attended operation	S Fr. 1650.	F.O.B. Factory
266-213	DC 24/36 for unattended operation	S Fr. 2200.	F.O.B. Factory

Notes:
a. Speed at which wind plant develops maximum rated output.
b. WV05 and WV15G available with optional 12-volt output for 5% surcharge; WV25G available with optional 24-volt output for 5% surcharge.
c. Cost does not include tower, control panel, or regulator (except as noted).
d. Vertical-axis turbine.

The Heller-Aller Co.
Perry and Oakwood Streets
Napoleon, Ohio 43545
U.S.A.

"Baker" water pumping windmill, multi-blade fanwheel coupled to reciprocating cylinder pump.

Wheel Diam. (feet)	No. Fan Blades	Cost (windmill alone) (F.O.B. Napoleon, Ohio)
6	20	$ 401.20
8	36	522.90
10	30	712.70
12	32	1145.00

Pumping Capacities of Back-Geared "Baker" Windmills

(in 15-mile-per-hour wind)

Total Elevation in Feet	6 FOOT BAKER		8 FOOT BAKER		10 FOOT BAKER		12 FOOT BAKER	
	Diameter of Cylinder Inches	U. S. Gallons Per Hour	Diameter of Cylinder Inches	U. S. Gallons Per Hour	Diameter of Cylinder Inches	U. S. Gallons Per Hour	Diameter of Cylinder Inches	U. S. Gallons Per Hour
25	3	350	3½	900	4	1250	6	2400
35	2½	240	3	720	3½	925	5	1625
50	2¼	200	2½	450	3	700	4½	1425
75	2	160	2¼	350	2½	475	4	1125
100	2	150	2	250	2½	460	3	600
125	1⅝	120	1⅞	240	2	280	2½	525
150	....	....	1¾	220	2	280	2½	525
200	....	....	....	....	1⅞	260	2	325
250	....	....	....	....	1¾	215	2	325
300	....	....	....	....	....	....	1¾	200

The above capacities are approximate. By the total elevation in feet we do not mean the depth of the well, but the distance to the cylinder. Do not use pipe smaller than that for which cylinders are fitted.

While we recommend the above table larger cylinders may in many circumstances be used with satisfaction.

Manufacturer	Equipment
Maschinenfabrik Ludwig Bening 2847 Barnstorf Postfach 171 Federal Republic of Germany	"Lubing" wind-driven generator [battery charger], 24-volt output, 3-blade variable-pitch (centrifugal governor) propeller with blades of fiberglass-reinforced plastic.

Model		Diameter (meters)	Cost (F.O.B. Factory) (DM)	
M 022-3	G024-400	2.2	$5224.	Basic model; includes generator, rotor, rotor vanes, gearing, slip ring assembly, rectifier, voltage regulator, ammeters voltmeter
M 022-3-6	G024-400	2.2	7058.	Ditto, plus 6-meter tower, guy wires and fittings, foundation mounting bolts, hand winch, foundation plans.
M 022-3-9	G024-400	2.2	7448.	Ditto, but with 9-meter tower
M 022-3-12	G024-400	2.2	7858.	Ditto, but with 12-meter tower

"Lubing" wind-power pumps, 3-, 4-, and 6-blade rotors coupled to reciprocating brass pump, capable of maximum lifts of 23 to 52.5 feet (7 to 16 meters). Rotor blades aerodynamically shaped of molded plastic. 6-blade rotor equipped with centrifugal speed control on three blades, remaining three fixed. Prices not given.

Also available is a variety of pumps coupled to model M 022-3-6 (3-blade rotor mounted on 6-meter tower) for approximately DM 6500, F.O.B. factory, depending on the pump selected.

Winco Division
Dyna Technology, Inc.
P.O. Box 3263
Sioux City, Iowa 51102
U.S.A.

"Wincharger" Model No. 1222H wind-driven electric battery charger

- 6-foot diameter 2-blade propeller (wood) with centrifugally operated air-brake governor set to cut in at wind speeds above 23 miles per hour (11.8 meters per second)
- 200-watt, 12-volt DC generator
  starts charging in winds of 7 miles per hour (3.6 meters per second)
  maximum charging current 14 amperes at 23 miles per hour (11.8 meters per second)
  maximum voltage 15 volts
- 10-foot 4-leg tower (angle iron)
- control box complete with ammeter

Cost, complete–$445. F.O.B. factory.

APPENDIX 6

Commercial Developers of Wind Machines

[This list is not necessarily complete but it has been compiled from the best information available to NAS staff. The information has been supplied by the developers; thus, neither the panel nor the NAS/NRC can assume responsibility for its accuracy.]

1. American Energy Alternatives, Inc.
   P.O. Box 905
   Boulder, Colorado 80302
   U.S.A.

Prototype wind-driven generator based on "bicycle-wheel" turbine (see Figure 31) 8 feet in diameter, direct-drive alternator with rectifier supplying 12, 24, 48, etc. volts DC. Plans call for following models:

output	Rated Windspeed (mph)	Projected Cost (F.O.B. Boulder)
1.5	?	$2100-2200
2.5	33 (at 5000′ alt.)	$2650
5	42 (at 5000′ alt.)	$2900-3000

2. American Wind Turbine Co.
   1016 East Airport Road
   Stillwater, Oklahoma 74074
   U.S.A.

"Bicycle-wheel" turbine (see Figure 31) coupled to direct-drive, high-speed alternator, with solid-state field-modulated frequency conversion to supply constant frequency AC. Prototype system built, with plans for turbines 15 to 30 feet in diameter, including pumping systems using electric motor pumps. Plans include "tooling package" for manufacture by developing countries.

3. Windworks
   Box 329, Route 3
   Mukwonago, Wisconsin 53149
   U.S.A.

Nonprofit company, designs wind-driven generator systems, including synchronous inverters and towers, and provides design information on a consulting basis.

4. Noah Energy Systems GmbH
   Wippenhohner Str. 32
   D-5202 Hennef 1
   Switzerland

Prototype wind-driven generator providing 30 kW (variable frequency) at wind speeds of 10 meters/second and 90 kW maximum (in gusts). Rotor 12 meters in diameter, 6 fixed airfoil blades. Reportedly developed for application in developing countries, with local servicing. Projected cost is Fr. 75,000, FOB factory.

5. Jacobs Wind Electric Company, Inc.
   Box 722, Route 11
   Fort Myers, Florida 33901
   U.S.A.

Formerly manufactured the Jacobs Wind Electric Plant, 1930-1960. Currently designing a new wind-driven electric plant—no details available.

6. Grumman Aerospace Corp.
   Bethpage, New York 11714
   U.S.A.

Commercial development of sail-wind windmill developed at Princeton University. Grumman's version has three blades made of aluminum spars and Dacron sails, 12.5 feet in diameter, coupled to automobile alternator producing about 1 kW in 20-mph winds. Windmill also intended to be sold for pumping and as a heat generator by coupling to a gear pump with restricted supply of water. Prices not available. Grumman also is developing a controlled-pitch 3-blade 25-foot diameter 15-kW wind-driven generator with extruded aluminum blades and standard ("off the shelf") generator. Machine is expected to produce 35,000 kWh annually for average wind speeds of 14 mph. Price not available.

7. Dornier-System GmbH
   Postfach 1360
   7990 Friederichshafen
   Federal Republic of Germany

Prototype Darrieus windmill-generator, 10-kW capacity. Rotor fabrication designed for manufacture in developing countries.

APPENDIX 7

Windmill Towers for Construction on the Farm

(Reprinted from the catalog of Smith & Thayer Co., Boston, Massachusetts.)

For information on light-weight, low-cost tower of modern design, contact:

Windworks
Box 329
Route 3
Mukwonago, Wisconsin 53149
U.S.A.

DIRECTIONS

FOR BUILDING

36-FOOT DERRICK

FOR 8½ OR 10-FOOT

BACK GEARED IRON TURBINE WIND ENGINE.

LIST OF MATERIALS FOR 36-FOOT DERRICK FOR 8½ AND 10-FOOT ENGINES.

4 Pieces, 4x4, 18 feet long, for Main Post.
4 Pieces, 4x4, 20 " " " " "
1 Piece, 2x6, 12 feet long, for Lookout on which to build Platform.
2 Pieces, 2x4, 16 feet long, for Ladder.
2 Pieces, 2x4, 12 feet long, for Platform.
2 Pieces, 1x4, 12 feet long, for Ladder.
24 Pieces, 1x6, 16 feet long, for Braces, Collars and Girts.
4 Anchor Posts, 8 feet long, should be of solid stuff not less than 6x6.
4 Cross Anchors.
5 Pounds of 20d nails for Platform.
15 Pounds 10d nails for Braces.
8 4½x½ Bolts and Washers for Splicing Main Post.
8 10½x½ Bolts and Washers for bolting Main Posts to Anchor Posts.

The cost of above material will not exceed $12, and the cost of labor, building derrick and erecting mill will be from $8 to $12, as it requires the labor of two men two days. We do not furnish derricks, except on special agreement, as they are inconvenient and expensive to ship, and they can be made on the ground cheaper. They are usually furnished by the purchaser, or by the agent erecting the mill. Any ordinary carpenter can, from our specifications, give an estimate of its cost, and by following our description erect one without trouble.

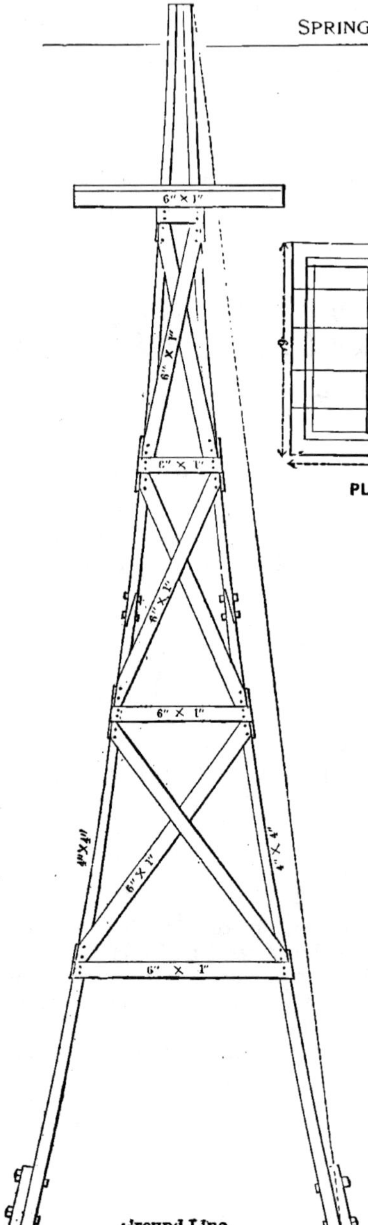

Diagram 3 Sides

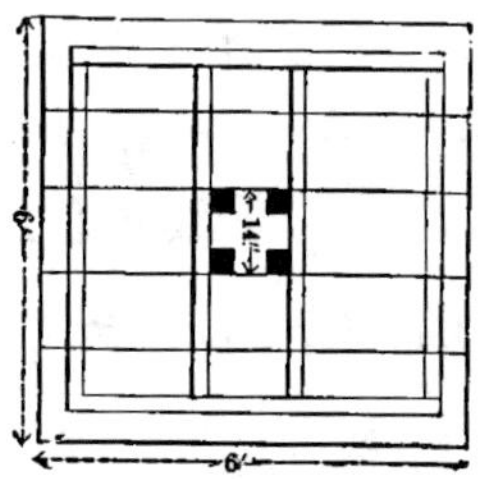

PLATFORM.

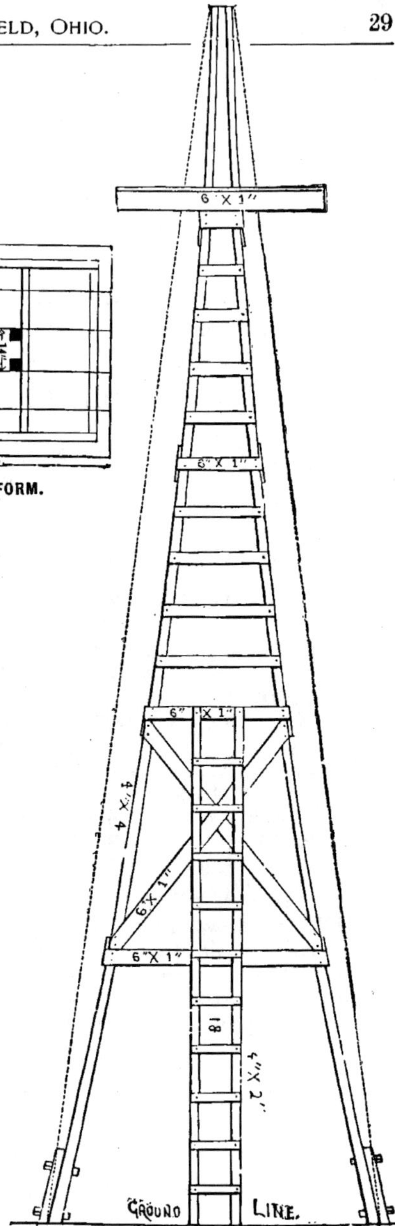

Diagram Ladder Side

DIRECTIONS FOR

36-Foot Derrick for 8½ and 10-Foot Engines.

Select the four posts, 4x4, for top of derrick, and cut out the inside corners of each post 2 inches at the top, tapering to a point or nothing in 4 feet. Splice the 4x4 posts togethor with a 2 foot splice, in the manner indicated in the cut, saw the ends square and see that the four posts are exactly the same length; lay them side by side and space off for the girts, making square across the four posts at once (as shown by illustration on page 31.) The top edge of the first girt should be 68 inches from top of derrick. The top edge of lower girt should be 90 inches from the lower end of posts; divide the distance between the upper and the lower girt equally for two other girts. Lay two posts side by side with their ends exactly even, insert a piece 1 inch thick, 2 inches wide, and 4 inches long, between the top ends, which will make the posts measure 9 inches across at the top, from outside to outside. Spread the posts so that they will measure 14 inches across from outside to outside, 62 inches from the top. Nail on the top collar, using the first one for a pattern by which to cut the remaining three, two of which should be 2 inches longer than the first to allow for lap. Observe the same rule in cutting all the girts. Nail on the first girt 9 inches from the top and second 68 inches from the top as marked. Go to the bottom of the posts and spread the ends 9 feet apart. (The rule is 1 foot spread for every 4 feet in height.) Nail a temporary stay on both to hold them there, then draw a chalk line from the bottom of one post to the top and around to the bottom of the other post. Measure from the line to the outside edge of each post and move the posts back and forth until the distance from the line to the outside edge of each post is the same at each girt; this will give you the correct spring of the posts (These directions should be strictly observed.) Nail on lower girt 90 inches from bottom, and remaining 2 at equal distances, as marked. Nail on diagonal braces across from each girt to the one above each way, and see that inside corner of each brace is even with inside of post, and trim outside corner even with outside edge of post; use first brace for a pattern to cut remaining 7 in that section by allowing 2 inches for lap on 4, as in the girts. Nail on lookout as sill for platform (2x6x6), let its lower edge rest on the upper edge of first girt. When this side of derrick is complete turn it upside down with girts and braces on ground, lay the remaining posts on top of the first and build that side of the derrick exactly like the first, being careful to nail the girts on exactly opposite the first. Insert the 1-inch strips between the upper and lower posts and nail on the two remaining collars at the top and the first girts, then block up under the lower posts until the posts have the same

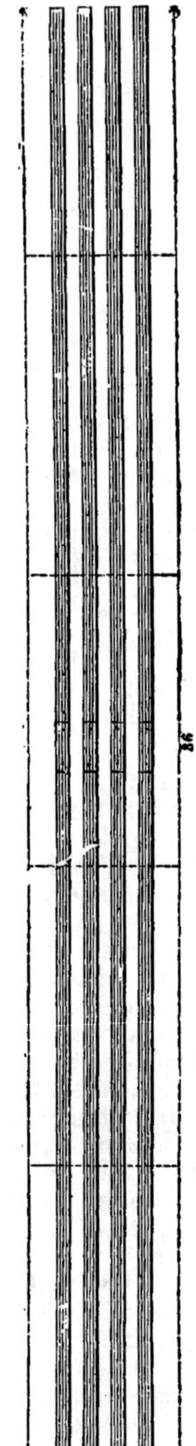

spring at each girt as they have the other way. Nail on the remaining girts on two sides, and braces on side and complete ladder side as shown in cut. Raise the derrick with pike poles. A rope fastened near the top will greatly facilitate the work of raising when the derrick is about half-way up. Drop a plumb line from the center of the derrick at the top, and when the derrick stands perfectly plumb bolt to the anchor posts, which should be not less than 6 inches square and 4 feet in the ground, with a cross anchor near the bottom of each anchor post. Complete platform as shown in cut on page 29.

DIRECTIONS FOR

36 foot Derrick for 12 foot Engine.

Select the four posts, 4x6, for top of derrick, and cut off the inside of each post at the top, tapering to a point or nothing in sixteen feet, making posts at top four inches square. Cut out the inside corners of each post 2 inches at the top, tapering to a point or nothing in 5 feet. Splice the 4x6 posts together with a 2 foot splice,

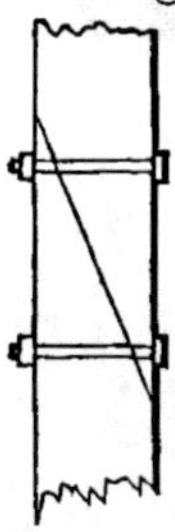

in the manner indicated in the cut, saw the ends square and see that the posts are exactly the same length; lay them side by side and space off for the girts, marking square across the four posts at once (as shown by illustration on page 37.) The top edge of the first girt should be 80 inches from top of derrick. The top edge of lower girt should be 96 inches from the lower end of posts; divide the distance between the upper and the lower girt equally for two other girts. Lay two posts side by side with their ends exactly even, insert a piece 2 inches thick, 2 inches wide, and 4 inches long, between the top ends, which will make the posts measure 10 inches across at the top from outside to outside. Spread the posts so that they will measure 16 inches across from outside to outside, 72 inches from the top. Nail on the top collar, using the first one for a pattern by which to cut the remaining three, two of which should be 2 inches longer than the first to allow for lap. Observe the same rule in cutting all the girts. Nail on the collar 9 inches from the top and first girt 80 inches from the top as marked, Go the bottom of the posts and spread the ends 9 feet apart. (The rule is 1 foot spread for every 4 feet in height). Nail a temporary stay on both to hold them there, then draw a chalk line from the bottom of one post to the top and around to the bottom of the other post. Measure from the line to the outside edge of each post and move the posts back and forth until the distance from the line to the outside edge of each post is the same at each girt; this will give you the correct spring of the posts. (These directions should be strictly observed.) Nail on lower girt 96 inches from bottom, and remaining 2 at equal distances, as marked. Nail on diagonal braces across from each girt to the one above each way, and see that inside corner of each brace is even with inside of post, and trim outside corner even with outside edge of post; use first brace for a pattern to cut remaining 7 in that section by allowing 2 inches for lap on 4, as in the girts. Nail on lookout as sill for platform (2x6x6,) let its lower edge rest on the upper edge of first girt. When this side of derrick is complete turn it upside down with girts and braces on ground, lay the remaining posts on top of the first and build that side of the derrick exactly like the first, being careful to nail the girts on exactly opposite the first. Insert the two-inch strips between the upper and lower posts and nail on the two remaining collars at the top and the first girts then block up under the lower posts until the posts have the same

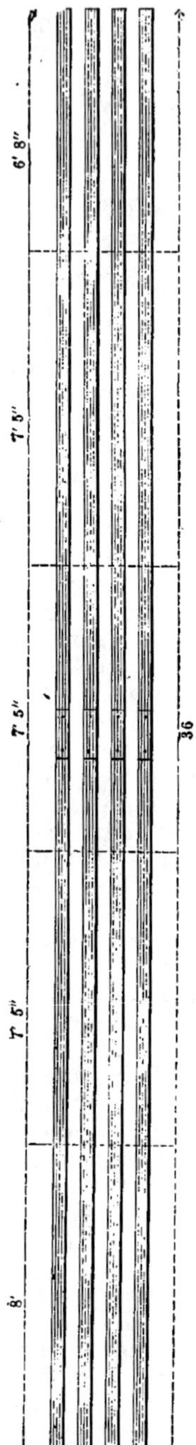

spring at each girt as they have the other way. Nail on the remaining girts on two sides, and braces on one side and complete ladder side as shown in cut. Raise the derrick with pike poles. A rope fastened near the top will greatly facilitate the work of raising when the derrick is about half way up. Drop a plump line from the center of the derrick at the top, and when the derrick stands perfectly plumb bolt to the anchor posts, which should be not less than 6x8 inches square and 5 feet in the ground, with a cross anchor near the bottom of each anchor post. Complete platform as shown in cut on page 29.

QUADRANTS FOR TRANSMITTING POWER.

FROM WIND ENGINES.

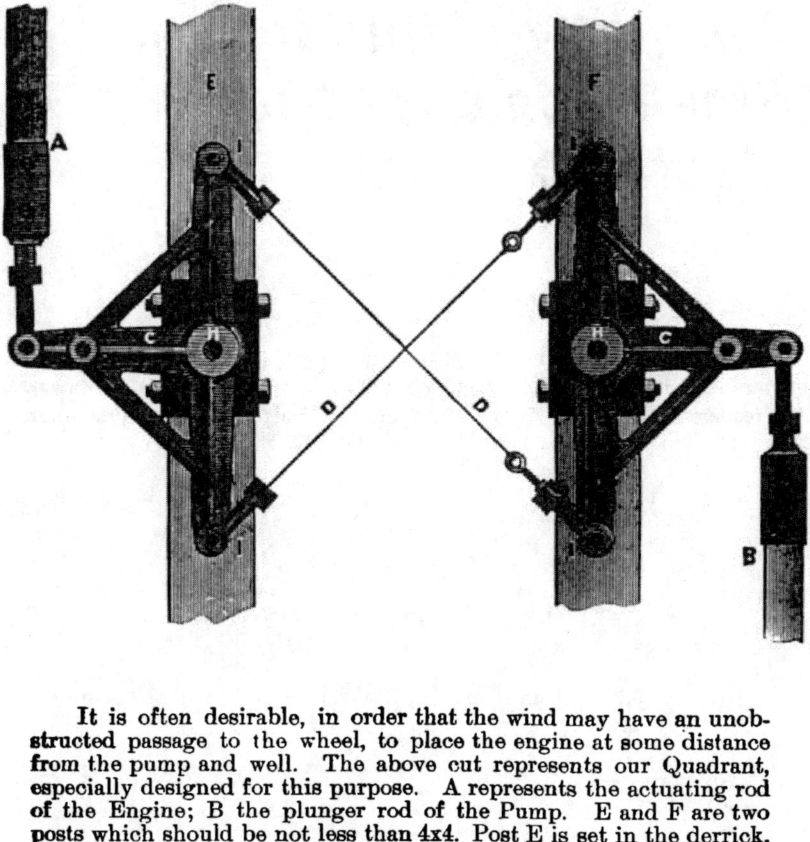

It is often desirable, in order that the wind may have an unobstructed passage to the wheel, to place the engine at some distance from the pump and well. The above cut represents our Quadrant, especially designed for this purpose. A represents the actuating rod of the Engine; B the plunger rod of the Pump. E and F are two posts which should be not less than 4x4. Post E is set in the derrick, its center 9 inches from the center of the actuating rod A, the top, braced to the derrick, and bottom of post firmly imbedded in the ground. Post F is securely fastened to the platform of well, its center 9 inches from center of plunger rod of Pump B. The two Quadrants C are placed on stud H, on casting bolted to posts E and F, and are connected together by wires D D, crossing each other. By this arrangement we obtain the same motion on the pump as if it was placed directly under the derrick, the down stroke of actuating rod of Engine giving a down stroke of plunger rod of Pump with very little loss of power.

APPENDIX 8

Currently Manufactured Small-Scale Hydropower Machinery

[Information included in this appendix has been supplied by the manufacturers. It is not presented as complete but is meant to serve as a convenience to readers who might wish to request further details from the manufacturer.]

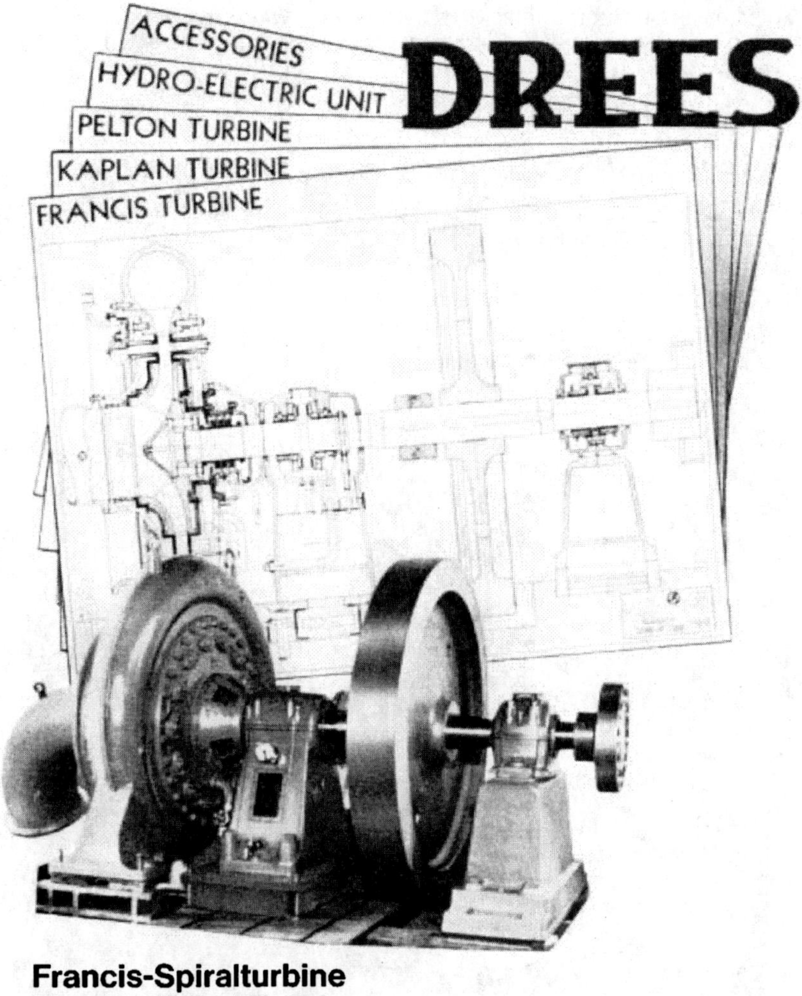

Francis-Spiralturbine
Francis Spiral Turbine
Turbine a Bâche Spirale Francis
Turbina Espiral Francis

DREES & CO GMBH · 476 WERL i. WESTF.
MASCHINENFABRIK UND EISENGIESSEREI · ABT.: WASSERTURBINENBAU
Schützenstraße 36-38 - Telefon 02922 5071/5072 - Telegramme: Dreesco Werl
Telex: 8421404 dres d

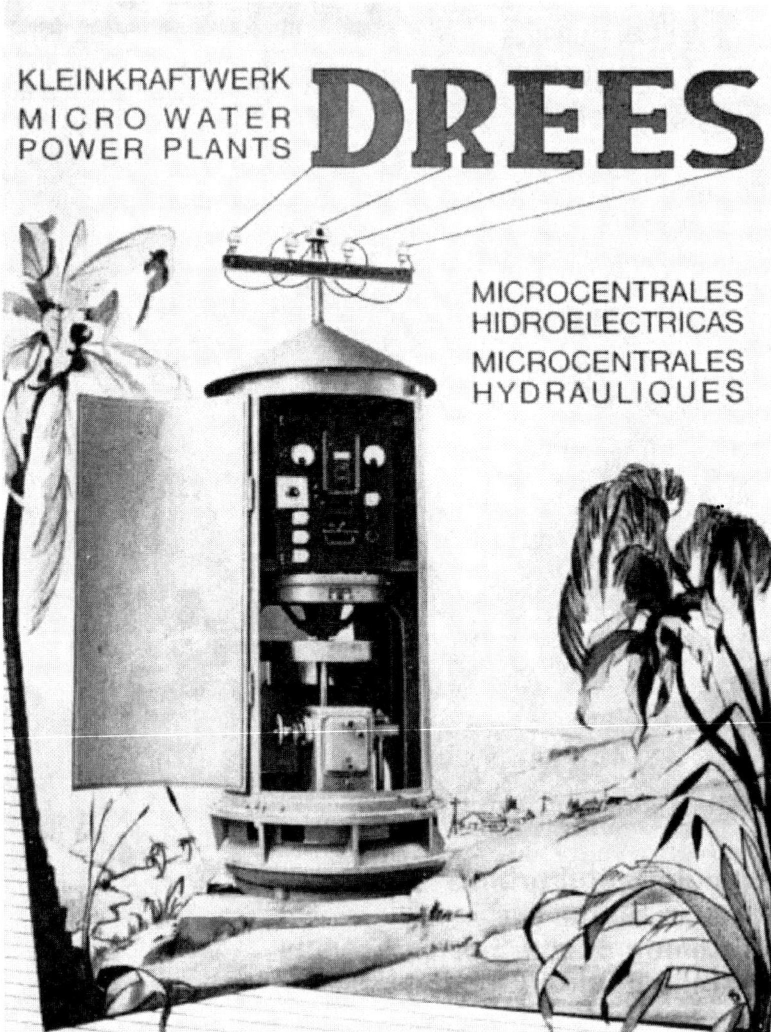

Turgo Impulse Wheel—suitable for heads of water between 50 and 400 feet (15 and 120 meters) with outputs up to 500 B.H.P.

Examples of turbine capacities and prices, supplied by the manufacturer:

kW	Head (ft)	Flow (ft^3/min)	Type	Price
10	50	376	Francis	£6,300
10	100	106.5	Turgo	£4,500
25	50	512	Francis	£7,800
10	200	45.5	Pelton	£4,200

Small spiral cased Francis turbines—suitable for water heads of 25 to 200 feet (7.5 to 60 meters) with outputs up to 100 or 150 horsepower.

THE JAMES LEFFEL & CO.
SPRINGFIELD, OHIO, U. S. A.

Rating Tables of Hoppes Hydro-Electric Units

ELECTRICAL CAPACITY	Head in Feet	Style	Water in Cubic Feet Per Minute
½ KILOWATT OR 500 WATTS	* 8	HL	104
	* 9	HJ	92
	* 10	HJ	82
	* 11	F	74
	* 12	F	68
1 KILOWATT OR 1000 WATTS	* 8	IL	190
	* 9	HL	175
	* 10	HL	155
	11	HL	140
	12	HJ	127
	13	HJ	118
	14	HJ	110
	15	HJ	105
	16	HJ	100
	17	HJ	94
	18	HJ	90
	19	F	84
	20	F	80
	21	F	76
	22	F	74
	23	F	72
	24	F	70
	25	F	68
2 KILOWATTS OR 2000 WATTS	* 8	JP	330
	* 9	JP	290
	* 10	JP	260
	* 11	HL	245
	* 12	HL	225
	* 13	HL	215
	14	HL	190
	15	HL	178
	16	HL	166
	17	HL	156
	18	HJ	153
	19	HJ	148
	20	HJ	140
	21	HJ	133
	22	HJ	127
	23	HJ	120
	24	F	116
	25	F	110
3 KILOWATTS OR 3000 WATTS	8	LR	470
	9	JP	415
	10	JP	370
	11	JP	340
	12	JP	310
	13	JP	280
	14	JP	260
	15	IL	250
	16	IL	240
	17	HL	225
	18	HL	210
	19	HL	200
	20	HL	190
	21	HL	180
	22	HL	170
	23	HL	165
	24	HJ	162
	25	HJ	158

ELECTRICAL CAPACITY	Head in Feet	Style	Water in Cubic Feet Per Minute
5 KILOWATTS OR 5000 WATTS	8	OT	760
	9	OT	600
	10	LR	590
	11	LR	535
	12	LR	490
	13	JP	470
	14	JP	435
	15	JP	400
	16	JP	365
	17	JP	340
	18	JP	330
	19	JP	320
	20	JP	315
	21	HL	300
	22	HL	290
	23	HL	285
	24	HL	275
	25	HL	260
7½ KILOWATTS OR 7500 WATTS	11	OT	800
	12	OT	740
	13	LR	680
	14	LR	630
	15	LR	590
	16	JP	550
	17	JP	515
	18	JP	490
	19	JP	480
	20	JP	450
	21	JP	430
	22	JP	410
	23	JP	400
	24	JP	390
	25	JP	380
10 KILOWATTS OR 10000 WATTS	12	OT	980
	13	OT	900
	14	OT	840
	15	OT	780
	16	LR	715
	17	LR	670
	18	LR	650
	19	JP	610
	20	JP	580
	21	JP	550
	22	JP	525
	23	JP	500
	24	JP	490
	25	JP	480

Head in feet referred to above is illustrated on page 294.
Style refers to various sizes of Hoppes Hydro-Electric Units. See dimensions on 294.
Quantity of water the unit will use at full rated capacity is listed above in cubic feet per minute.
*Units marked with * furnished in direct current only.

Standard rating for alternating current units is 3 phase 60 cycle and either 120 or 240 or 480 volts.
These Hoppes Hydro-Electric Units may also be furnished for 50 cycle current.
When you write us, please give full particulars on your electrical requirements.

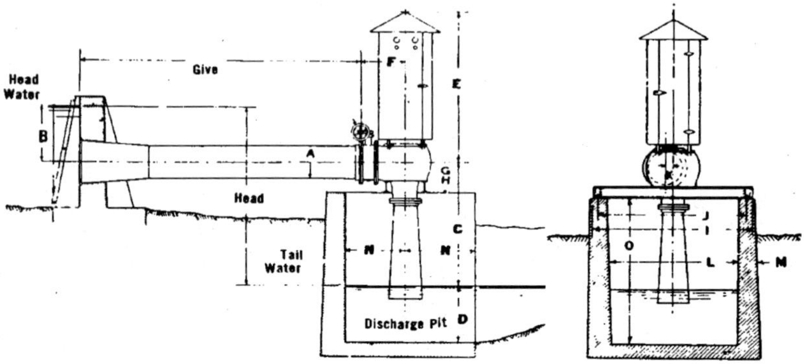

Style	Flow (cfm)	Price
HJ	82	$ 3,850.00
F	68	3,550.00
HL	155	5,200.00
HJ	105	4,525.00
F	68	4,150.00
JP	260	5,300.00
HL	178	4,850.00
JP	370	6,750.00
IL	250	5,950.00
HJ	158	5,200.00
LR	590	9,950.00
JP	400	8,950.00
HL	260	5,400.00
OT	800	11,950.00
LR	590	9,375.00
OT	980	12,550.00
LR	715	10,250.00
JP	610	7,300.00
JP	480	7,300.00

GENERAL DIMENSIONS
Style Sizes

	F	HJ	HL IL	JP	LR	OT
A	6″	10″	12″	16″	18″	20″
B	18″	24″	24″	36″	36″	36″
C	60″	60″	60″	60″	60″	60″
D	24″	30″	30″	36″	42″	48″
E	58″	72″	72″	88″	90″	90″
F	20½″	18″	18″	22″	23″	27″
G	5½″	8⅝″	9⅝″	11⅛″	11⅞″	14¼″
H	5″	5″	5″	6″	6″	7″
I	72″	72″	72″	84″	84″	96″
J	66″	66″	66″	78″	78″	90″
K	3⅛″	4¾″	4″	4″	4″	4½″
L	60″	60″	60″	72″	72″	84″
M	8″	8″	8″	9″	10″	10″
N	30″	30″	30″	36″	42″	48″
O	73½″	73⅜″	75⅜″	78⅞″	84⅛″	86¾″

OSSBERGER-TURBINENFABRIK

WEISSENBURG/BAYERN

Germany - Allemagne - Alemania

Schliessfach - P.O. Box - 8 P - Apartado 32 - Telefon 23 62

Schema einer Patent-Ossberger-Wasserturbine

Layout of a Patent Ossberger Water Turbine

Schéma d'une Turbine Hydraulique Brevetée Ossberger

Esquema de una Turbina Hidráulica Patente Ossberger

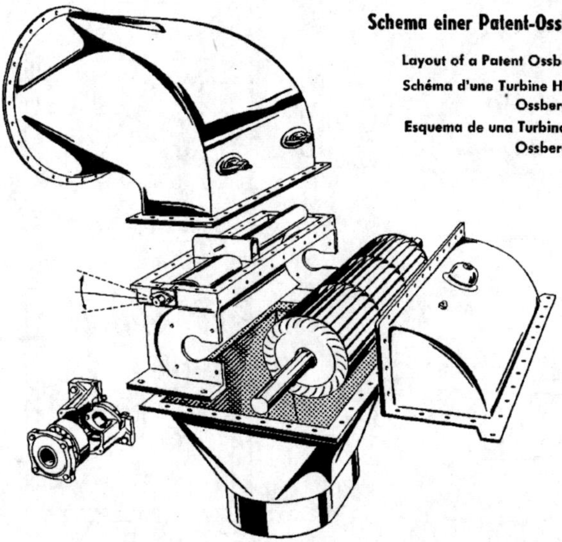

Charakteristiken:

- 2-Zellen-Bauweise
- Ganzstahlbauweise
- Wenig drehende Teile
- Leitapparat bestehend aus einer Drehschaufel im Dreipunktsystem
- Wasserzu- und Abführung durch Rohrleitungen
- Nur 5 Schmierstellen
- Unempfindlich gegen verschmutztes Triebwasser
- Garantie für günstigsten Wirkungsgrad ab $^1/_6$ Q
- Garantie für geringen Verschleiß und große Lebensdauer
- Garantie für unkomplizierten Einbau an ungefährdetem Standort, anspruchslose Wartung und Pflege

Salient features:

- 2-cell construction
- All steel construction
- Few rotating parts
- Guide equipment consists of one rotary blade with a 3 point contact
- Water supply and discharge through piping
- Only 5 lubricating points
- Unaffected by polluted propellant water
- Guarantees highest efficiency from $^1/_6$ Q upwards
- Guarentees low wear and long life
- Guarantees trouble-free installation on a safe site, undemanding in maintenance or servicing

Caractéristiques:

- Construction à double compartiment
- Construction tout acier
- Peu de pièces tournantes
- Directrice se composant d'une palette rotative en un système à trois points
- Amenée et évacuation des eaux par conduites
- Cinq points de graissage seulement
- Insensible aux eaux polluées ou chargées de feuilles
- Garantie d'un degré d'efficacité le plus favorable à partir de $^1/_6$ Q
- Garantie d'une usure minimum et d'une longévité maximum
- Simplicité d'installation garantie. Entretien et soins quasi nuls

Caracteristicas:

- Tipo de dos celdas
- Modelo totalmente de acero
- Menos piezas giratorias
- El órgano de conducción consta de una paleta directriz con el sistema de tres puntos
- Conducciones de entrada y salida de agua por tuberías
- Solamente cinco puntos de engrase
- Insensible para aguas motrices ensuciadas
- Garantía para rendimientos favorables a partir de $^1/_6$ Q
- Garantía de un desgaste mínimo y gran duración
- Garantía para sencilla instalación en lugares fuera de peligro, modesta vigilancia y conservación

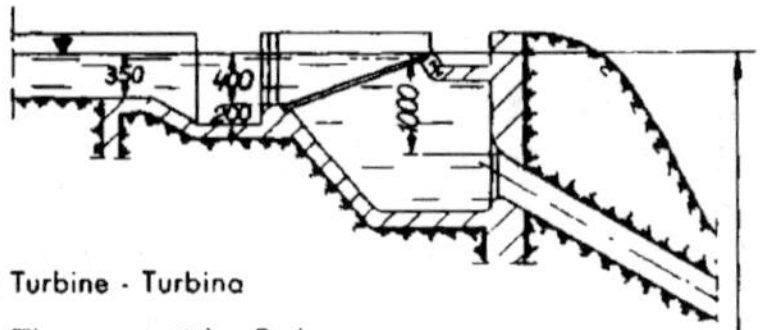

Einbaubeispiel

Mounting sketch - Projet d'installation
Ejemplo para instalación

1. Turbine - Turbina
2. Übergangsstück - Reducer - Raccord de réduction - Adaptador
3. Krümmer - Bend - Courbure - Codo
4. Druckrohrleitung - Penstock - Conduite forcée - Tuberia de presión
5. Trägerrahmen - Fundament frame - Cadre de fondation - Bastidor
6. Riemenscheibe - Pulley - Poulie - Polea
7. Außenlager - Bearing - Palier - Chumacera
8. Feinrechen - Screen - Grille - Rejilla
9. Spülschütze - Sluice gate - Vanne - Compuerta de descarga

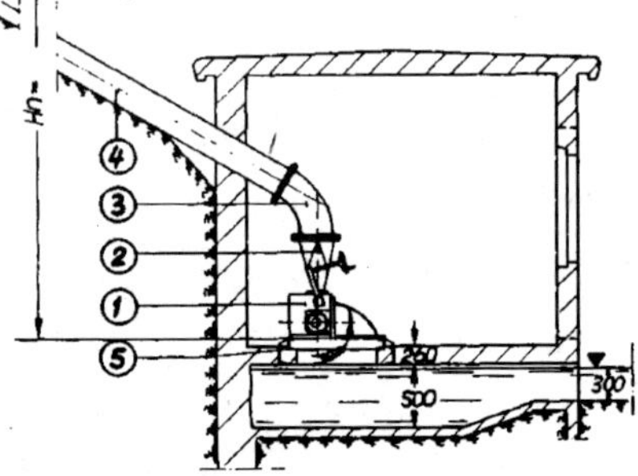

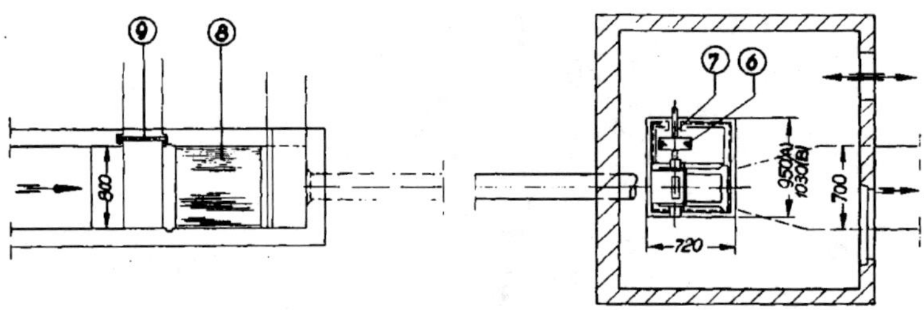

PATENT-OSSBERGER-WASSERTURBINE TYPE: **UNIVERSAL A**

H m	Gefälle · Head · Chute · Caida							Maße · Measures Dimensions · Medidas 1030 x 750 x 950 mm	Preis Price Prix Precio
	4	5	6	7	8	9	10	Gewicht · Weight · Poids Pesos 370 kg	
Q Litr./sec.	Wassermenge · Supply · Débit · Caudal								
	49	55	60	66	70	75	80		
N PS/HP/CV/HP	Leistung · Output · Puissance · Potencia								
	2,1	2,9	3.8	4,9	6,0	7,2	8,5		
n U/min. · r.p.m. - t/min.	Drehzahl · Speed · Vitesse · Velocidad								
	380	430	470	510	550	580	615		
mm	Rohrleitung · Penstock · Conduite forcée · Tuberia								
	250	250	250	250	250	250	250		

PATENT-OSSBERGER-WASSERTURBINE TYPE: **UNIVERSAL B**

H m	10	15	20	25	30	950 · 750 · 950 mm	340 kg	Preis Price Prix Precio
Q Litr./sec.	22	27	31	35	37,5			
N PS · HP · CV ·	2,3	4,3	6,6	9,3	12			
n U/min. r.p.m, t/min.	615	755	870	980	1070			
mm	200	200	200	200	200			

Im Lieferumfang enthalten - Parts of delivery - Le prix de livraison comprend - El precio contiene: Pos. 1, 2, 3, 5, 6, 7, 8

RIFE RAM AND PUMP WORKS

RIFE HYDRAULIC ENGINE MANUFACTURING CO.
Box 367, Millburn, New Jersey 07041

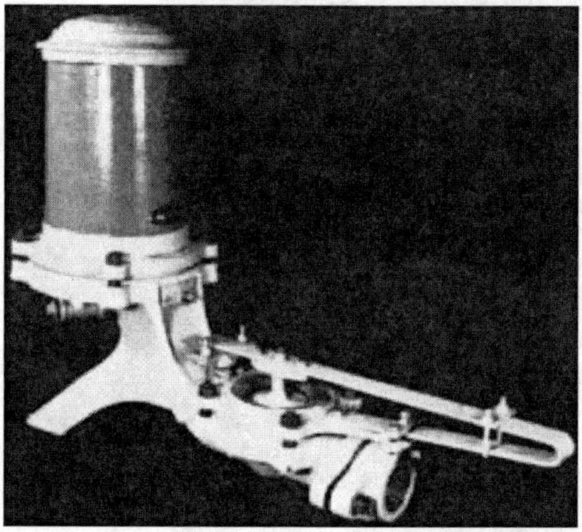

RIFE "Series B" Ram

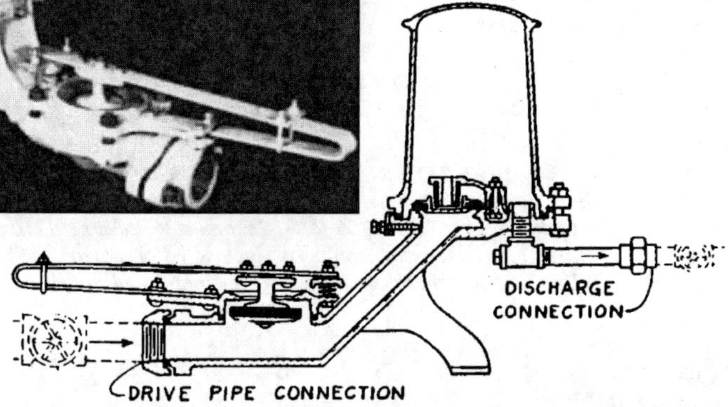

RIFE "Everlasting" Standard Ram

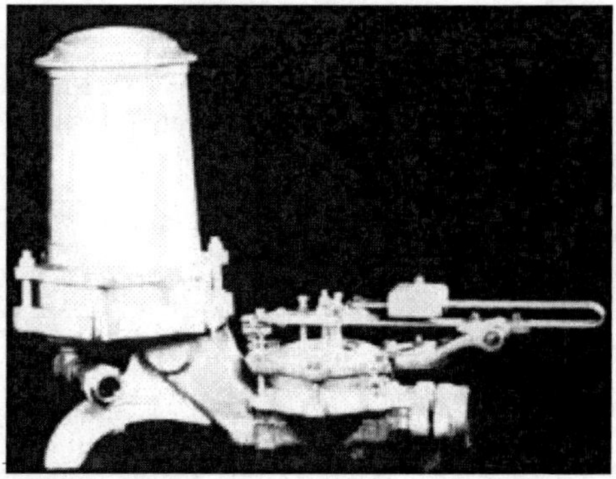

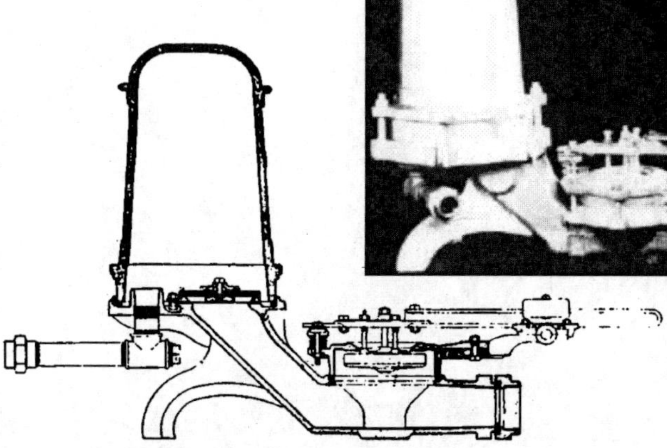

SPECIFICATIONS OF RIFE HYDRAULIC RAMS
New High Capacity Models

RAM NO.	INTAKE (drive) PIPE SIZE	DISCHARGE (delivery) PIPE SIZE	INTAKE CAPACITY (Gals. per min. used) Min. Normal Max.	MINIMUM VERTICAL FALL required in feet	CRATED WEIGHT (SINGLE-ACTING) in pounds (approx.)
RIFE "NEW MODEL" SERIES B RAMS (4 Bolt Design)					
Maximum Vertical Fall 15 ft.–Maximum Vertical Lift 150 ft.					
Unit includes iron gate strainer for intake end of drivepipe					
10B	1-1/4"	3/4"	1½ – 6 – 7	2	94
15B	1-1/2"	3/4"	6 – 10 – 13	2	102
20B	2"	1"	8 – 18 – 20	2½	150
25B	2-1/2"	1"	12 – 28 – 35	2½	202
30B	3"	1-1/4"	20 – 40 – 55	3	263
RIFE "EVERLASTING" STANDARD RAMS					
A more rugged development of the previous Series "A" 6 Bolt Design					
Maximum Vertical Fall 25 ft. – Maximum Vertical Lift 250 ft.					
All steel and iron parts zinc coated or plated					
Unit includes special steel strainer for intake end of drivepipe					
10S	1-1/4"	3/4"	3 – 7 – 10	3	160
15S	1-1/2"	3/4"	5 – 11 – 15	3	171
20S	2"	1"	10 – 20 – 25	3½	249
25S	2-1/2"	1"	15 – 30 – 45	3½	273
30S	3"	1-1/4"	25 – 45 – 70	4	339
40S*	4"	2"	35 – 90 – 125	4	565
60S*	6"	3"	75 – 225 – 350	4	1325
80S	8"	4"	Repair parts only		

***#40S and #60S equipped with steel air chamber. #60S includes special welded pipe nipple to assemble inclined drivepipe to level ram.**

APPENDIX 9

The Helical Rotary Screw Expander and Its Potential for Geothermal Application

The helical screw expander basically consists of a pair of counter-rotating tapered screws, as shown in the accompanying drawing. These screws have been utilized as pumping devices for many years and have proved to be long-lived and reliable. The original pump design, developed in Sweden in the 1930s, is known as the Lysholm screw. This device gradually came to be used as a gas compressor in commercial installations until, by the 1950s, its use was rather extensive. Experimental use of this device–in the reverse application as gas expander–as a prime mover with geothermal sources began in 1971.[1]

The construction, and hence the operation, of the machine are illustrated in the drawing. The channels formed by the meshing rotating screws provide a path for the liquid, which permits continuous flashing of steam as the pressure decreases, and all of the fluid–liquid plus vapor–is carried through the prime mover throughout the entire expansion process. As described by MacKay and Sprankle, "The geothermal fluid flows through the internal nozzle control valve and at high velocity enters the high-pressure pocket formed by the meshed rotors, the rotor case bore surfaces, and the case end face, designated by A in the two figures. As the rotors turn, the pocket elongates, splits into a V, and moves away from the inlet port to form the region designated by B. With continued rotation, the V lengthens, expanding successively to C, D, and E as the point of meshing of the screws appears to retreat axially from the expanding fluid. The expanded fluid at low pressure is then discharged into the exhaust port."[1]

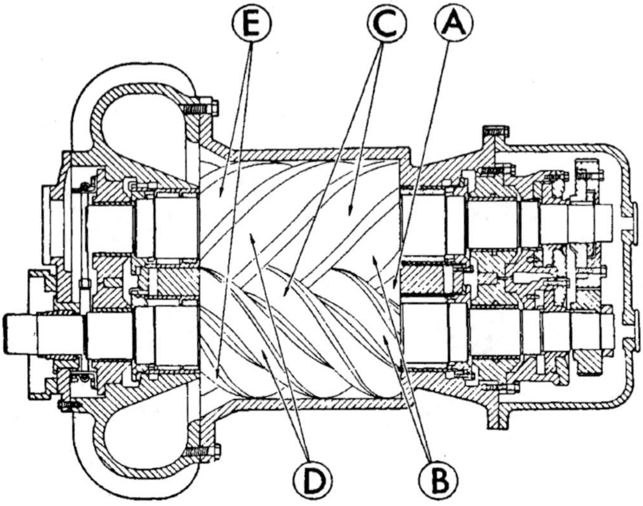

PLAN SECTION VIEW

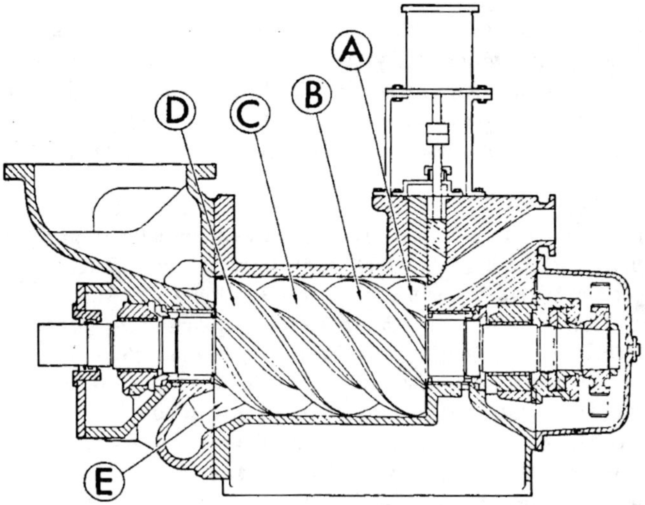

SIDE SECTION VIEW

The potential advantages of the helical rotary screw expander lie in possible economies of operation when certain mineral-rich brines are used to generate electric power. Mineral deposition on the screw surfaces may be beneficial by eliminating the need for close machining tolerances. As deposits build up, clearances disappear and better seals are formed between the counter-rotating screws and between the screws and the case. This continual lapping action, as scale grinds against scale, also serves to prevent total encrustation and may provide a "self-healing" mechanism if the case or the rotors should be scarred, by permitting mineral deposits to fill the depression.

Tests show that the device is tolerant of a wide range of geothermal fluids. It is capable of running on geothermal brines that have the ability to deposit silica and carbonate scale in conventional systems, and also seems to be very tolerant of geothermal fluids containing clay, silt, and sand without significant interference with its operation as a prime mover. Tests indicate that it is also tolerant of a wide range of enthalpy of the geothermal fluid. The low-enthalpy fluids produce less electrical power than do the high-enthalpy ones, but in tests the system has shown itself capable of operating with a remarkable degree of effectiveness over significant enthalpy ranges.

This expander can be manufactured by any of the helical rotary screw compressor manufacturers. With international capability for manufacturing, the lead time for producing small electrical power generators of this type would be relatively short, and standardization would permit more easy maintenance. As with conventional equipment, these power generating units could be readily mobile[1] and their portability would make for flexibility in utilization.

If further development and field use of the helical rotary screw expander should bear out the test results, geothermal sources not exploitable with current techniques might become economically exploitable. Its thermodynamic efficiency is somewhat lower than that of conventional gas turbines, but energy efficiency will be less important than flexibility and reliability in rural applications. Projected costs for complete generator systems are of the order of $100 to $200 per kilowatt. These capital costs, inherent simplicity, and potential reliability in a technologically adverse environment make it of considerable potential value for rural applications in developing countries.

REFERENCE

1. McKay, R. A., and Sprankle, R. S. 1974. Helical rotary screw expander power system. In *Proceedings of the Conference on Research for the Development of Geothermal Energy Resources, Pasadena, California, 23-25 September*, pp. 301-307. Pasadena: California Institute of Technology, Jet Propulsion Laboratory.

ADVISORY COMMITTEE ON TECHNOLOGY INNOVATION

MEMBERS

GEORGE BUGLIARELLO, President, Polytechnic Institute of New York, Brooklyn, New York, *Chairman*

E. R. PARISER, Senior Research Scientist, Department of Nutrition and Food Science, Massachusetts Institute of Technology, Cambridge, Massachusetts

LEWIS PERINBAM, Vice President Special Programs, Canadian International Development Agency, Ottawa, Canada

CHARLES A. ROSEN, Staff Scientist, Stanford Research Institute, Menlo Park, California

VIRGINIA WALBOT, Department of Biology, Washington University, St. Louis, Missouri

BOARD ON SCIENCE AND TECHNOLOGY FOR INTERNATIONAL DEVELOPMENT

DAVID PIMENTEL, professor, Department of Entomology and Section of Ecology and Systematics, Cornell University, Ithaca, New York, *Chairman*

CARL DJERASSI, Department of Chemistry, Stanford University, Stanford, California, *Chairman*, 1973-1975

MEMBERS

GEORGE S. HAMMOND, Foreign Secretary, National Academy of Sciences

RUTH ADAMS, American Academy of Arts and Sciences, Boston, Massachusetts

EDWARD S. AYENSU, Director, Department of Botany, Smithsonian Institution, Washington, D.C.

JOHN D. BALDESCHWIELER, Division of Chemistry and Chemical Engineering, California Institute of Technology, Pasadena, California

IVAN L. BENNETT, JR., Dean, School of Medicine, New York University Medical Center, New York, New York

DWIGHT S. BROTHERS, International Agricultural Development Service, The Rockefeller Foundation, New York, New York

GEORGE BUGLIARELLO, President, Polytechnic Institute of New York, Brooklyn, New York

CHARLES DENNISON, Former Vice President, International Minerals and Chemicals Corporation, New York, New York

Board on Science and Technology for International Development

Commission on International Relations

National Academy of Sciences–National Research Council

2101 Constitution Avenue, Washington, D.C. 20418, USA

Advisory Studies and Special Reports

Reports published by the Board on Science and Technology for International Development are sponsored in most instances by the U.S. Agency for International Development and are intended for free distribution primarily to readers in developing countries. A limited number of copies is available for distribution on a courtesy basis to readers in the United States and other industrialized countries who have institutional affiliation with government, education, or research and who have professional interest in the subject areas treated by the reports.

Single copies of published reports listed below are available free while the supplies last. Requests should be made on your organization's letterhead. Other interested readers may buy the reports listed here from the National Technical Information Service (NTIS) whose address appears below.

5. **The Role of U.S. Engineering Schools in Development Assistance.** 1976. 30 pp. Examines opportunities and constraints facing U.S. engineering schools in mobilizing their resources to aid developing countries. NTIS Accession No. PB 262-055. $4.50

7. **U.S. International Firms and R,D & E in Developing Countries.** 1973. 92 pp. Discusses aims and interests of international firms and developing-country hosts and suggests that differences could be mitigated by sustained efforts by the firms to strengthen local R,D & E capabilities. NTIS Accession No. PB 222-787. $6.00

8. **Ferrocement: Applications in Developing Countries.** 1973. 89 pp. Assesses state of the art and cites applications of particular interest to developing countries—boat-building, construction, food and water storage facilities, etc. NTIS Accession No. PB 220-825. $6.50

14. **More Water for Arid Lands: Promising Technologies and Research Opportunities.** 1974. 153 pp. Outlines little-known but promising technolgies to supply and conserve water in arid areas. NTIS Accession No. PB 239-472. $8.00 (French-language edition is available from Office of Science and Technology, Development Support Bureau, Agency for International Development, Washington, D.C. 20523 or through NTIS, Accession No. PB 274-612. $8.00.)

16. **Underexploited Tropical Plants with Promising Economic Value.** 1975. 187 pp. Describes 36 little-known tropical plants that, with research, could become important cash and food crops in the future. Includes cereals, roots and tubers, vegetables, fruits, oilseeds, forage plants, and others. NTIS Accession No. PB 251-656. $9.00

17. **The Winged Bean: A High Protein Crop for the Tropics.** 1975. 43 pp. Describes a neglected tropical legume from Southeast Asia and Papua New Guinea that appears to have promise for combatting malnutrition worldwide. NTIS Accession No. PB 243-442. $4.50.

18. **Energy for Rural Development: Renewable Resources and Alternative Technologies for Developing Countries.** 1976. 305 pp.
(French-language edition is available from Office of Energy, Development Support Bureau, Agency for International Development, Washington, D.C. 20523.)

19. **Methane Generation from Human, Animal, and Agricultural Wastes.** 1977. 131 pp. Discusses means by which natural process of anerobic fermentation can be controlled by man for his benefit, and how the methane generated can be used as a fuel. NTIS Accession No. PB 276-469. $7.25.

21. **Making Aquatic Weeds Useful. Some Perspectives for Developing Countries.** 1976. 175 pp. Describes ways to exploit aquatic weeds for grazing, and by harvesting and processing for use as compost, animal feed, pulp, paper, and fuel. Also describes utilization for sewage and industrial wastewater treatment. Examines certain plants with potential for aquaculture. NTIS Accession No. PB 265-161. $9.00.

22. **Guayule: An Alternative Source of Natural Rubber.** 1977. 80 pp. Describes a little-known bush that grows wild in deserts of North America and produces a rubber virtually identical with that from the rubber tree. Recommends funding for guayule development. NTIS Accession No. PB 264-170. $6.00

23. **Resource Sensing from Space: Prospects for Developing Countries.** 1977. 203 pp. An examination of current and prospective applications of interest to the LDCs, certain implications for long-term governance of a remote sensing system, and desirable technical cooperation initiatives to diffuse user capabilities. NTIS Accession No. PB 264-171. $9.25.

25. **Tropical Leguemes: Resources for the Future.** 1979. 331 pp. Describes plants of the family Leguminosae, including root crops, pulses, fruits, forages, timber and wood products, ornamentals, and others. NTIS Accession No. PB 298-423. $12.00.

26. Leucaena: Promising Forage and Tree Crop for the Tropics. 1977. 118 pp. Describes *Leucaena leucocephala*, a little known Mexican plant with vigorously growing, bushy types that produce nutritious forage and organic fertilizer as well as tree types that produce timber, firewood, and pulp and paper. The plant is also useful for revegetating hillslopes and providing firebreaks, shade, and city beautification. NTIS Accession No. PB 268-124. $7.25.

28. Microbial Processes: Promising Technologies for Developing Countries. 1979. 198 pp. Discusses the potential importance of microbiology in developing countries in food and feed, plant nutrition, pest control, fuel and energy, waste treatment and utilization, and health.

29. Postharvest Food Losses in Developing Countries. 1978. 202 pp. Assesses potential and limitations of food loss reduction efforts; summarizes existing work and information about losses of major food crops and fish; discusses economic and social factors involved; identifies major areas of need; and suggests policy and program options for developing countries and technical assistance agencies. NTIS Accession No. PB 290-421. $9.25.

30. U.S. Science and Technology for Development: Contributions to the UN Conference. 1978. 226 pp. Serves the U.S. Department of State as a major background document for the U.S. national paper, 1979 United Nations Conference on Science and Technology for Development. Includes an overview section plus five substantive sections as follows: 1) industrialization; 2) health, nutrition, and population; 3) food, climate, soil, and water; 4) energy, natural resources and environment; and 5) urbanization, transportation, and communication.

Related Publications

Other reports (prepared in cooperation with BOSTID) available from the above address are:

An International Centre for Manatee Research. 1975. 34 pp. Describes the use of the manatee, a large, almost extinct, marine mammal, to clear aquatic weeds from canals. Proposes a research laboratory to develop manatee reproduction and husbandry. Published by the National Science Research Council of Guyana. NTIS Accession No. PB 240-244. $4.50.

Natural Products for Sri Lanka's Future. 1975. 53 pp. Report of a 1975 workshop with the National Science Council of Sri Lanka. Identifies neglected and unconventional plant products that can significantly contribute to Sri Lanka's economic development. Published by National Science Council of Sri Lanka. NTIS Accession No. PB 251-520. $5.25.

Workshop on Solar Energy for the Villages of Tanzania. 1978. 167 pp. Report of a workshop with the Tanzania National Scientific Research Council, Dar es Salaam, Tanzania. Reviews state-of-the-art of small-scale solar energy devices, and suggests short- and long-range projects using them in villages. Published by Tanzania National Scientific Research Council. NTIS Accession No. PB 282-941. $9.00.

Out-of-Print Publications

The following out-of-print BOSTID reports are available only from the National Technical Information Service unless otherwise noted. To order, send report title, NTIS Accession Number, and amount indicated. (Note: Prices are current for April 1979 and are subject to change without notice.) Pay by NTIS Deposit Account, check, money order, or American Express account. U.S. orders without prepayment are billed within 15 days; a $5.00 charge is added. Prices for foreign buyers are double the prices indicated below, and payment in full must be enclosed. Send order to:

National Technical Information Service
Springfield, Virginia 22161, USA

1. East Pakistan Land and Water Development as Related to Agriculture. January 1971. 67 pp. Reviews World Bank proposed action program in land and water management. NTIS Accession No. PB 203-328. $5.25.

2. The International Development Institute. July 1971. 57 pp. Endorses concept of new science-based technical assistance agency as successor to AID; examines its character, purposes, and functions. NTIS Accession No. PB 203-331. $5.25.

3. Solar Energy in Developing Countries: Perspectives and Prospects. March 1972. 49 pp. Assesses state of art, identifies promising areas for R & D, and proposes multipurpose regional energy research institute for developing world. NTIS Accession No. PB 208-550. $5.25.

4. Scientific and Technical Information for Developing Countries. April 1972. 80 pp. Examines problem of developing world's access to scientific and technical information sources, provides rationale for assistance in this field, and suggests programs for strengthening information infrastructure and promoting information transfer. NTIS Accession No. PB 210-107. $6.00.

6. Research Management and Technical Entrepreneurship: A U.S. Role in Improving Skills in Developing Countries. 1973. 40 pp. Recommends initiation of a systematic program and indicates priority elements. NTIS Accession No. PB 225-129. $4.50.

9. Mosquito Control: Some Perspectives for Developing Countries. 1973. 63 pp. Examines biological control alternatives to conventional pesticides; evaluates state of knowledge and research potential of several approaches NTIS Accession No. PB 224-749. $6.00.

10. Food Science in Developing Countries: A Selection of Unsolved Problems. 1974. 81 pp. Describes 42 unsolved technical problems with background information, possible approaches to a solution, and information sources. NTIS Accession No. PB 235-410. $6.00.

11. Aquatic Weed Management: Some Perspectives for Guyana. 1973. 44 pp. Report of workshop with the National Science Research Council of Guyana describes new methods of aquatic weed control suitable for tropical developing countries. NTIS Accession No. PB 228-660. $5.25.

12. Roofing in Developing Countries: Research for New Technologies. 1974. 74 pp. Emphasizes the need for research on low cost roofs, particularly using materials available in developing countries. NTIS Accession No. PB 234-503. $6.00.

13. Meeting the Challenge of Industrialization: A Feasibility Study for an International Industrialization Institute. 1973. 133 pp. Advances concept of an independent, interdisciplinary research institute to illuminate new policy options confronting all nations. NTIS Accession No. PB 228-348. $7.25.

15. International Development Programs of the Office of the Foreign Secretary, by Harrison Brown and Theresa Tellez. 1973. 68 pp. History and analysis, 1963-1973; lists staff/participants and publications. NTIS Accession No. PB 230-543. $5.25.

20. Systems Analysis and Operations Research: A Tool for Policy and Program Planning for Developing Countries. 1976. 98 pp. Examines utility and limitations of SA/OR methodology for developing country application and means for acquiring indigenous capabilities. NTIS Accession No. PB 251-639. $6.50.

24. Appropriate Technologies for Developing Countries. 1977. 140 pp. Examines fundamental issues and inter-relationships among economic, political and social factors relating to choice of technologies in developing countries. Discusses criteria of appropriateness and suggests policies for improving technical decisions. NTIS Accession No. PB 80-124-118. $11.00

Other out-of-print reports (prepared in cooperation with BOSTID) available from the National Technical Information Service are:

Products from Jojoba: A Promising New Crop for Arid Lands. 1975. 30 pp. Describes the chemistry of the oil obtained from the North American desert shrub *Simmondsia chinensis*. NTIS Accession No. PB 253-126. $4.50.

Aquatic Weed Management: Some prospects for the Sudan and the Nile Basin. 1975. 57 pp. Report of a 1975 workshop with the Sudanese National Council for Research. Suggests modern and innovative methods for managing the water hyacinth. Published by National Council for Research–Agricultural Research Council of Sudan. NTIS Accession No. PB 259-990. $5.25.

Ferrocement, a Versatile Construction Material: Its Increasing Use in Asia. 1976. 106 pp. Report of a 1974 workshop with the Asian Institute of Technology, Bangkok, Thailand. Surveys applications of ferrocement technology in Asia and the Pacific Islands. Includes construction of grain silos, water tanks, roofs, and boats. Published by Asian Institute of Technology. NTIS Accession No. PB 261-818. $6.50.

International Consultation on Ipil-Ipil Research. 1978. 1972 pp. Report of a 1976 conference sponsored with the Philippine Council for Agriculture and Resources Research, Los Baños, Laguna, Philippines. Contains background papers and workshop session summary reports on ipil-ipil (*Leucaena* spp.). (Companion volume to report no. 26 above.) NTIS Accession No. PB 280-161. $8.00.

Reports in Preparation (working titles)

BOSTID will fill requests for single copies of reports in preparation upon publication as outlined at the beginning of this section.

27. Firewood Crops: Shrub and Tree Species for Energy Production.
31. Food, Fuel, and Fertilizer from Organic Wastes.
32. The Water Buffalo: Its Potential for Development.
33. The Potential for Alcohol Fuels in Developing Countries.
34. Revegetating the Range: Selected Research and Development Opportunities.
35. Aerial Seeding of Forests.

SUPPLEMENT

Energy for Rural Development

Renewable Resources and Alternative Technologies for Developing Countries

Advisory Committee on Technology Innovation
Board on Science and Technology for
International Development
Commission on International Relations
National Research Council

NATIONAL ACADEMY PRESS
Washington, D.C. 1981

NOTICE: The project that is the subject of this report was approved by the Governing Board of the National Research Council, whose members are drawn from the Councils of the National Academy of Sciences, the National Academy of Engineering, and the Institute of Medicine. The members of the committee responsible for the report were chosen for their special competences and with regard for appropriate balance.

This report has been reviewed by a group other than the authors according to procedures approved by a Report Review Committee consisting of members of the National Academy of Sciences, the National Academy of Engineering, and the Institute of Medicine.

The National Research Council was established by the National Academy of Sciences in 1916 to associate the broad community of science and technology with the Academy's purposes of furthering knowledge and of advising the federal government. The Council operates in accordance with general policies determined by the Academy under the authority of its congressional charter of 1863, which establishes the Academy as a private, nonprofit, self-governing membership corporation. The Council has become the principal operating agency of both the National Academy of Sciences and the National Academy of Engineering in the conduct of their services to the government, the public, and the scientific and engineering communities. It is administered jointly by both Academies and the Institute of Medicine. The National Academy of Engineering and the Institute of Medicine were established in 1964 and 1970 respectively, under the charter of the National Academy of Sciences.

This report has been prepared by the Advisory Committee on Technology Innovation, Board on Science and Technology for International Development, Commission on International Relations, National Academy of Sciences–National Research Council, for the Office of Science and Technology, Bureau for Development Support, Agency for International Development, Washington, D.C., under Contract No. AID ta-C-1433.

Library of Congress Catalog Number: 81–81065

CONTRIBUTORS TO THIS VOLUME

WILLIAM L. HUGHES, Engineering Energy Laboratory, and Clark A. Dunn Professor of Engineering, Oklahoma State University, Stillwater, Oklahoma 74078, USA, *Editor*

CHARLES A. BAREBO, Rodale Resources, Inc., 576 North Street, Emmaus, Pennsylvania 18049, USA

WILLIAM BEALE, Sunpower, Inc., 6 Byard Street, Athens, Ohio 45701, USA

GEORGE C. CHANG, Penn College of Engineering, Cleveland State University, Cleveland, Ohio 44115, USA

WILLIAM DELP, Independent Power Developers, Inc., Route 3, Box 174H, Sand Point, Idaho 83864, USA

DARRELL M. DODGE, Energy Systems Group, Rocky Flats Plant, Rockwell International, P.O. Box 464, Golden, Colorado 80401, USA

JACK J. FRITZ, Office of Energy, Bureau for Development Support, U.S. Agency for International Development, Washington, D.C. 20523, USA

DERRICK P. GRIMMER, Environmental Appropriate Technology, Inc., P.O. Box 278, Fremont, Missouri 63941, USA

RICHARD HARWOOD, Organic Gardening & Farming Research Center, Rodale Press, Inc., 33 East Minor Street, Emmaus, Pennsylvania 18049, USA

STEVE HOGAN, Solar Energy Research Institute, 1617 Cole Boulevard, Golden, Colorado 80401, USA

REBECCA C. MCDONALD, Earth Resources Laboratory, National Space Technologies Laboratories, National Aeronautics and Space Administration, NSTL Station, Mississippi 39529, USA

PAUL A. NELSON, Chemical and Electrical Systems, Energy and Environmental Technology, U.S. Department of Energy, Argonne National Laboratory, 9700 South Cass Avenue, Argonne, Illinois 60439, USA

GENE NIGHTINGALE, Institutional Issues, Wind Systems Program, Small Wind Electrical Conversion Systems, Energy Systems Group, Rocky Flats Plant, Rockwell International, P.O. Box 464, Golden, Colorado 80401, USA

JOSEPH R. NORTON, Head Emeritus, School of General Engineering, Oklahoma State University, Stillwater, Oklahoma 74078, USA

DANIEL O'NEIL, Office of the Director, Engineering Experiment Station, Georgia Institute of Technology, Atlanta, Georgia 30332, USA

KEN PYKKONEN, Small Wind Electrical Conversion Systems, Energy Systems Group, Rocky Flats Plant, Rockwell International, P.O. Box 464, Golden, Colorado 80401, USA

R. G. RAMAKUMAR, Department of Electrical Engineering, Oklahoma State University, Stillwater, Oklahoma 74078, USA

ROBERT W. REX, Republic Geothermal, Inc., 11823 East Slauson Avenue, Santa Fe Springs, California 90670, USA

ROLF T. SKRINDE, Olympic Associates Co., 1214 John Street, Seattle, Washington 98109, USA

JOEL V. STAFFORD, Energy Systems Group, Rocky Flats Plant, Rockwell International, P.O. Box 464, Golden, Colorado 80401, USA

KURT H. VAUS, Olympic Associates Co., 1214 John Street, Seattle, Washington 98109, USA

B. C. WOLVERTON, Earth Resources Laboratory, National Space Technologies Laboratories, National Aeronautics and Space Administration, NSTL Station, Mississippi 39529, USA

JOHN G. HURLEY, Staff Coordinator, Board on Science and Technology for International Development, Commission on International Relations, National Research Council

SABRA BISSETTE, Staff Editor, Board on Science and Technology for International Development, Commission on International Relations, National Research Council

Preface

In 1976, the National Academy of Sciences published *Energy for Rural Development: Renewable Resources and Alternative Technologies for Developing Countries*, a report prepared by an ad hoc panel of the Advisory Committee on Technology Innovation (ACTI). ACTI was formed in 1971 by the Board on Science and Technology for International Development (BOSTID). The 1976 report was an outgrowth of ACTI's ongoing interest in renewable energy sources for rural areas.

The original book (published in both English and French) presented information on a variety of subjects, including direct uses of solar energy (heating, cooling, distillation, crop drying, photovoltaics), indirect uses of solar energy (wind power, hydropower, photosynthesis, biomass), geothermal energy, and energy storage. Both a nontechnical section and a technical section on each of these subjects were included. The report was not intended to be a detailed "how-to" manual on energy devices; rather, it was intended to report on the state of the art, tell who was doing what, and direct the reader where to go for detailed information. Assessments were made of the short-, intermediate-, and long-term applicability of each technology for developing countries.

In the 5 years that have passed since publication of *Energy for Rural Development*, significant changes have occurred in the world energy picture. The world's heavy dependence on oil and rapidly rising energy costs have affected the developing countries in particular. It is critical, therefore, that up-to-date information on renewable energy resources and alternative technologies, preferably those available locally, be provided.

Although there have been few remarkable new discoveries in the past 5 years, steady progress has been made in research and development on renewable energy resources and alternative technologies. This supplement includes information on new technologies developed during this period and on advances made in technologies described in the original volume. Like that volume, this report serves merely to direct the reader where to go for more information, and is not intended to be a "how-to" manual or detailed catalog. Almost all of the subjects in the original book are discussed again—some only briefly. There are new discussions of the use of pedal power and conditioning of electric power.

Because the information presented here builds on material provided in the 1976 report, the original panel was not reconvened to prepare this supplement. Instead, William L. Hughes, chairman of the original panel, called upon various specialists to assess developments over the past 5 years in renewable resources and alternative technologies and served as editor for the material that they provided.

A number of scientists, listed in the front of this volume, made valuable contributions to this report. Their efforts and support are greatly appreciated. In addition, the comments and criticisms of those who reviewed the manuscript proved particularly useful. They include: Charles E. Backus, Arizona State University; J. L. Birchfield, Georgia Institute of Technology; James C. Bresee, North Carolina Energy Institute; William R. Breslin, Volunteers in Technical Assistance; Norman L. Brown, Agency for International Development; David E. Cooper, Southern California Edison Co.; Erich Farber, University of Florida at Gainesville; Jack J. Fritz, Agency for International Development; Elmer L. Gaden, Jr., University of Virginia; Stephen A. Goulden, Cetus Corporation; Carl J. Heise, Lorton, Virginia; Ronald Judkoff, Solar Energy Research Institute; Thomas A. Lawand, Brace Research Institute; George Löf, Colorado State University; M. M. McGinty, Eagle-Picher Industries, Inc.; John C. Orth, The Montana Energy and MHD Research and Development Institute, Inc.; David Pimentel, Cornell University; David W. Rabenhorst, Johns Hopkins University; Roger Revelle, University of California at San Diego; Warren Rice, Arizona State University; Albert D. Sheppard, Georgia Institute of Technology; Neil N. Turner, General Electric Co.; and C. C. Warnick, Idaho Water Resources Research Institute.

LIST OF ABBREVIATIONS

a	acre
AC	alternating current
amp	ampere
amp-hr	ampere-hour
Btu	British thermal unit
°C	degrees Celsius
cfm	cubic feet per minute
cm	centimeter
DC	direct current
°F	degrees Fahrenheit
ft	foot
ft^3	cubic foot
ft-lb	foot-pound
gal	gallon (U.S.)
GWp	peak gigawatt
ha	hectare
hp	horsepower
hr	hour
Hz	Hertz
J	joule
in.	inch
°K	degrees Kelvin
kg	kilogram
kJ	kilojoule
km	kilometer
kmph	kilometers per hour
kVA	kilovolt ampere
kW	kilowatt
kWe	kilowatt-electric
kWp	peak kilowatt
l	liter
lb	pound
m	meter
m^3	cubic meter
μ	micron
mi	mile
MJ	megajoule
mm	millimeter
mph	miles per hour
m/sec	meters per second
MW	megawatt
MWp	peak megawatt
ppm	parts per million
°R	degrees Rankine
rpm	revolutions per minute
sec	second
W	watt
Wp	peak watt
Wh	watt-hour
yr	year

Contents

Introduction

This supplement to *Energy for Rural Development: Renewable Resources and Alternative Technologies for Developing Countries* is concerned with small, low-cost, locally operated energy systems. For that reason, it does not include alternatives such as tidal energy, large-scale wave energy, and ocean thermal gradient systems, which may well become important in the future, but will probably require large-scale capitalization. This does not imply, however, that large capital projects are necessarily inappropriate for developing countries. Where potential loads, cost-benefit ratios, ecological considerations, and financing are favorable, such projects should be considered.

With some exceptions, most of the technologies discussed in the original volume and this supplement are not ready for extensive application in developing countries. Either the cost is too high relative to current conventional energy sources or the current technology is not sufficiently reliable. Yet the energy technologies discussed here are based on well-established scientific principles, seem the most susceptible to significant improvement through purely technological innovations, and can probably soon be applied. They represent, therefore, highly promising possibilities for the developing countries in the future.

Applying any energy technology to developing countries entails a number of quality and cost considerations.

In past years there has been a tendency to compare energy outputs in terms of kilowatt-hours (or Btus) produced. Thus a fire burning 25 kg (55 lb) of wood per hour is equivalent to a 100-kW electric generator by such comparisons, even though the former could never perform the functions of the latter.

It seems desirable, therefore, to make a distinction between "low-grade" and "high-grade" energy. Low-grade energy is, generally, low-temperature heat. Some examples are relatively unconfined thermal energy, such as heat from open fires or flat plate solar collectors, or exhaust heat from steam electric power plants. Liquid and gaseous fuels represent high-grade energy sources.

Anything that requires mechanical or electrical power usually requires a high-grade energy source. Although some developing countries have abundant biomass that could provide adequate total energy, changing low-grade energy to high-grade energy to fuel vehicles and perform other useful work is generally expensive, difficult, and inefficient. For uses that require only low-grade energy, such as space heating, crop drying, sterilization, and pasteurization, direct use of available low-grade energy is usually more efficient.

Thus the overall energy problem facing developing nations—and many developed nations as well—is threefold:

1 To use low-grade energy as effectively as possible
2 To find better ways to convert low-grade energy to high-grade energy
3 To use both natural and synthesized high-grade energy as effectively as possible

Regarding costs, quite often the per unit (e.g., kilowatt-hour) cost of an alternate energy system is greater than the cost from a large, capital-intensive system for the same kind of mechanical or electrical energy. What is often forgotten, however, is that the rural villager in a developing country seldom has a choice between two such systems. Hypothetically, his options for producing electricity might include a wind generator, a solar photovoltaic system, or a small generator requiring fuel. In addition to choosing among these options, he must decide whether he in fact needs electricity to improve his life, or whether he can obtain an equal or better improvement at less risk by spending his time, effort, and money in other ways.

Costs must also be considered at the national level. Most developing countries, by necessity, have strict controls on the use of scarce foreign exchange for the purchase of imported energy, goods, and services. This, in turn, means that energy systems used in developing countries should be manufactured locally to the extent consistent with reasonable economy. Energy-use planning in any developing country should, therefore, determine what can and cannot be produced locally.

Since many of the technologies in this book will involve operational problems, there is a significant need for pilot programs to identify the day-to-day difficulties and provide workable (and publishable) solutions.

ORGANIZATION OF THIS BOOK

The organization of this supplement is similar to that of the original volume. Part I, devoted to direct uses of solar energy, discusses the current state of the art in photovoltaics and passive and active uses of solar energy in building designs. These discussions do not, of course, cover the range of possibilities of solar energy, but like all the other sections of this book are intended to supplement the original publication.

Part II is concerned with indirect uses of solar energy, including wind energy, energy from falling water, and energy from biomass. The wind energy section reviews the application of small wind power systems, in particular Rockwell International Corporation's wind program, sponsored by the U.S. Department of Energy, which has been testing and operating a wide variety of small wind systems for some years. The section on energy from falling water covers the well-established technology of small-scale hydroelectric systems (which have much applicability now wherever sites are available) and the isothermal hydraulic compression of air, a technology touched on only briefly in the original book. The section on energy from biomass is divided into four parts:

- The anaerobic digestion of human, animal, and agricultural wastes
- The use of aquatic plants as a method of sewage processing and subsequent production of biogas and fertilizer by anaerobic digestion
- The production of alcohols for liquid fuels
- The pyrolytic reduction of biomass to combustible gases, liquids, and char

Part III is concerned with other energy technologies. New developments in the field of geothermal energy are presented. A section concerned with energy storage discusses advanced batteries applicable to wind and photovoltaic systems, and presents a realistic evaluation of "super flywheels," devices that store energy in mechanical form. Completely new material is presented on the possibilities of using pedal power more productively for many jobs by employing special devices that utilize mechanical leverage. This is followed by an up-date on Stirling engines, external combustion engines that could use a variety of solid or liquid fuels, as well as solar energy inputs. Finally, a short discussion is presented on the conversion (or more properly, conditioning) of electrical energy from renewable sources to conventionally usable electricity.

Appendixes to this report include a list of manufacturers of energy-related devices, centers of research and development undertaking work on such devices, and a list of suggested reading.

The original volume of *Energy for Rural Development* included a nontechnical section for readers unfamiliar with technologies it described. Because this supplement does not include such a section, readers may wish to refer to the earlier book for definitions and descriptions of technologies discussed in this volume. To facilitate this, an index to both books is included at the end of this report.

PART I
DIRECT USES OF SOLAR ENERGY

Photovoltaics

Encouraging advances have been made in photovoltaic devices and systems applications during the past 5 years; pricing and efficiency goals have been met and, in many cases, exceeded. In the United States, costs per peak watt dropped from $22/Wp in 1976 to as low as $7/Wp in 1980. The industry is beginning to mature, and a glance at the number of companies currently making either cells or systems reflects the industry's enhanced status (see Appendix A).

Worldwide, research and development activities directed toward photovoltaic devices (solar cells) have increased dramatically. Further, manufacturers of these devices—an industry restricted for the most part to industrialized countries—realize that the production volumes needed to continue the trend of declining manufacturing costs must result from growing demand. With an increase in production volumes and improvement of manufacturing techniques, actual production of solar cells could be transferred eventually to locales in developing countries with adequate potential as manufacturing sites. Initially, however, most developing countries will have to continue to import these devices.

Applications of photovoltaic systems in developed and developing countries have increased in complexity from the initial powering of educational television sets in remote locales, to the powering of water pumps for potable water, to current projects in Mali, Upper Volta, Saudi Arabia, and the southwestern United States to satisfy the other mechanical and electrical power needs of isolated villages and towns. These latter projects will provide baseline data on both the technical aspects (i.e., reliability, maintenance, lifetime, etc.) and, equally important, the socioeconomic impacts of decentralized rural electrification by photovoltaic systems.

Manufacturing techniques for terrestrial photovoltaic systems have been greatly improved during the last 5 years, resulting in price reductions. Several governments have established programs to meet reduced price goals for silicon-based solar cell modules. The U.S. goals for photovoltaic modules call for a 1986 price of $.70/Wp (in 1980 U.S. dollars) at a production level of 500 MWp. Present quotations of $7-$10/Wp, with an estimated 1980 production of 2 MWp, fall in line with the U.S. goals monitored by the U.S. Department of Energy (DOE). Actual DOE goals call for prices for photovoltaic modules (in quantities of tens of kilowatts) to decrease to $2.80/Wp by 1982, $.70/Wp by 1986, and $.15-$.50/Wp by 1990 (all in constant 1980 U.S. dollars). A yearly budget of well over $150 million per year is being projected over the next 8 years to meet these goals (U.S. Department of Energy 1979).

In France, the Commissariat à l'Energie Solaire (COMES) has established price goals of $1.60-$2.50/Wp by 1985, and COMES and the Centre National de la Recherche Scientifique are each spending $4-$5 million on photovoltaic research. In the Federal Republic of Germany, the prospect of developing export potential stimulated the Ministry of Research and Technology to establish an 8-year, $76-million program to perfect a photovoltaic system utilizing polycrystalline silicon. The program goal is to develop arrays cheaper than $2.40/Wp by 1985. Both Italy and Great Britain have smaller, established research programs. The European Community plans to spend over $23 million in the next 4 years for photovoltaic system development. A major goal is to build several generating plants of 300-400 kW each. Japan established a multimillion dollar project in 1974 to develop a strong manufacturing base for photovoltaic devices by 1985. India has established a price goal of $2.80/Wp by 1985, a price considered to be the threshold for economic water pumping in rural India. Efforts and economic goals of the Soviet Union have not been publicized. An excellent review of many programs under way outside the United States is given in an article by Palz (1978). In addition to these programs, private industry is spending tens of millions of dollars on research and development. (See Appendix B for a list of photovoltaic research programs under way.)

RECENT TECHNOLOGICAL ADVANCES

Single-Crystal Silicon Photovoltaic Arrays

The largest single program attempting cost reductions on single-crystal silicon photovoltaic arrays is the DOE-sponsored Low-Cost Solar Array Project administered by the Jet Propulsion Laboratory of the California Institute of Technology. This program is designed to achieve the DOE goals of $2.80/Wp by 1982 and $.70/Wp by 1986. The program is divided into four tasks (production of polycrystalline silicon material, formation of poly-

TABLE 1 The 1986 price allocation guidelines (1980 U.S. dollars) for the Low-Cost Solar Array Project

	Goals	
Silicon (polycrystalline)	\$14/kg	\$.035–.117/Wp[a]
Sheet alternatives:		
–Czochralski ingot and slicing	\$28.90/m² of wafer	\$.204/Wp
–Heat Exchange Method (HEM) ingot and slicing	\$35/m² of wafer	\$.246/Wp
–Edge-defined film growth (EFG)	\$24.20/m² of wafer	\$.214/Wp
–Web dendritic	\$39.30/m² of wafer	\$.297/Wp
–Silicon on ceramic (SOC)	\$20.30/m² of wafer	\$.196/Wp
Cell fabrication	\$21/m² of cells	\$.141–.192/Wp[a]
Encapsulation materials	\$14/m² of module	\$.098–.139/Wp[a]
Module assembly	\$14/m²	\$.098–.139/Wp[a]
Module totals		\$.70/Wp

[a]This range is caused by the use of different sheet technologies.

crystalline silicon into large, single-crystal sheets, encapsulation, and development of production processes and equipment). Some associated support activities are engineering, operations, and project analysis and integration.

The goals of the overall Low-Cost Solar Array Project are summarized in Table 1. The goals are aggressive, but many manufacturers are confident that the 1986 DOE goals are attainable.

In France, the Centre de Recherches Nucléaires and the Laboratoires d'Electronique et de Physique Appliquée are looking at improved or low-cost diffusion techniques for the manufacture of single-crystal silicon. Most of the remaining industrialized countries are monitoring the single-crystal programs and preparing for eventual entry into the marketplace.

Multigrained Silicon and Amorphous Silicon

Two other outgrowths of silicon solar cell technology have emerged: the use of multigrained silicon and amorphous silicon. The multigrained (or polycrystalline silicon) cells have been demonstrated to have high efficiency (11 percent) and are being mass-produced by two companies. Heliotonic GmbH, a subsidiary of Wacker-Chemitronic GmbH (FRG), is producing a cast polycrystalline silicon in brick form. With government support, Wacker has established a program with AEG-Telefunken in which Wacker supplies the polycrystalline silicon bricks and Telefunken produces the modules. The cells will cost less than the Czochralski-grown, single-crystal cells, and the process is more amenable to mass production with higher production rates. Semix,

Inc. (a subsidiary of the U.S.-based Solarex Corporation), which is manufacturing polycrystalline silicon in the same form (a cast brick), has set up a production facility, and reportedly is making a 15 percent-efficient cell at a cost of \$4-\$6/Wp. Although efficiencies that high have only been documented in the laboratory, the price range is reasonable compared to single-crystal cell prices.

Development of thin-film polycrystalline silicon cells is still in the research stage. DOE, through the Solar Energy Research Institute (SERI) in Golden, Colorado, is sponsoring 17 contracts at a total cost of \$2.9 million. More research needs to be done even before cost estimates can be made for thin-film polycrystalline silicon cells.

Cells based on amorphous silicon-hydrogen alloys are an even more infant technology than the thin-film silicon, yet hold great promise to provide a very low-cost cell. A principal advantage is that such cells would absorb light many times more effectively than single-crystal silicon. DOE, through SERI, is devoting almost \$3 million in 17 contracts to attaining an efficiency of 7 percent for present amorphous cells.

Japanese laboratories are actively developing amorphous silicon cells, and efficiencies of 4.3 percent have been reported by Sanyo. Great Britain is also actively pursuing the amorphous silicon fields at both Dundee University (which has reported a laboratory cell efficiency of almost 5 percent) and the University of Sheffield.

Cadmium Sulfide/Cuprous Sulfide

Materials other than silicon can be used for photovoltaic devices. Cadmium sulfide (CdS) in conjunction with cuprous sulfide (Cu_2S) has been used for over 15 years for photovoltaic conversion. Theoretical efficiencies of 16 percent have been calculated, but only a 9.1 percent conversion efficiency has been reported, with practical limits felt to be just over 10 percent.

The importance of encapsulation was realized when the deleterious effects of humidity on bare cells were discovered. This discovery led to research on the stability of cells. SES, Inc., a subsidiary of Shell Oil Co. (USA), has established a production facility and has been field testing new cell designs for the past 2 years. They anticipate making the modules commercially available and cost-competitive with the silicon modules by the early 1980s. DOE hopes to demonstrate a thin-film cell by 1990 that can be produced for \$.15-\$.40/Wp. Both SES, Inc. and the Institute of Energy Conversion (IEC) of the University of Delaware find these figures attainable, but IEC feels that a price of \$.50/Wp should be possible by 1984.

Studies by Westinghouse (USA) indicate that 1986 prices for the CdS/Cu_2S cell will range from \$.75 to \$.86/Wp for an 8 percent-efficient module with a size of 2 ft (.61 m) × 8 ft (2.44 m). The University of Stuttgart in the Federal Republic of Germany is establishing a pilot production line to produce cells with efficiencies close to 7 percent. Activities are also under way in France, Belgium, Mexico, Great Britain, Italy, Iran, India, and Japan to further the development of CdS-based cells.

Screen Printing and Spray Deposition

Two of the most exciting new prospects in the fabrication of cell materials are screen printing and spray deposition. The screen-printed cell has recently emerged in Japan. Screen printing is a very inexpensive process that is amenable to mass production. Reported efficiencies of over 8 percent have prompted DOE to encourage research activities in the United States on this process as well as on the spray deposition technique, a low-cost technique in which the CdS is sprayed onto the substrate and heated to form the active layer.

Gallium Arsenide

Another material that has achieved commercial feasibility is gallium arsenide (GaAs). Like silicon, this material is single-crystal, but it is capable of higher conversion efficiencies than silicon and is much more costly to process. GaAs is being considered for applications where concentrated sunlight is used; however, concentrators might not be used extensively in developing countries because of the more complex system required to track the sun (as opposed to the stationary flat-plate modules).

Of greater significance to flat-plate technology is the research being conducted to develop either thin-film or amorphous GaAs. The thin-film (or polycrystalline) GaAs effort has achieved efficiencies of 6-9 percent. Cal Tech's Jet Propulsion Laboratory is also working in this field, and hopes to reach 15 percent efficiency at $.70/Wp. Amorphous GaAs work is being started at Harvard University, but results will probably not be available for several years. Work on GaAs is also being undertaken in France, Japan, Great Britain, and by the Commission of the European Communities, principally in concentrator applications.

Other Research

Research is being supported in such areas as electrochemical photovoltaic cells, and with cadmium telluride, indium phosphide, cuprous oxide, polyacetylene, zinc phosphide, cuprous selenide, and a variety of lesser-known materials.

Balance of System (BOS)

A field of increasing importance, and one that has been neglected for years, is "balance of system" of a photovoltaic installation. BOS represents all of the system except the photovoltaic modules and includes structures, power conditioners, and storage. Work is being done to develop new low-cost support structures, increase efficiencies of power converters, and lower the costs of storage, as well as to examine new storage techniques. As these programs advance, the prices of complete systems could drop and system reliability will be improved. Present lifetime expectations are 20 years for modules, 10-12 years for power converters, and 8-10 years for storage (if batteries are used). It is hoped that costs of complete systems without storage will reach approximately $2/Wp by 1986.

Information on photovoltaic devices has become increasingly available over the last several years, but information on the needs of people in developing countries has only recently begun to flow to manufacturers. (See Appendix C for listing of recent publications on photovoltaics.)

RECENT APPLICATIONS OF PHOTOVOLTAICS IN RURAL AREAS

Telecommunications

In 1975, over 4,000 photovoltaic-powered educational television sets (ETVs) were installed in Niger and the Ivory Coast (Weiss 1976). The success of this system in education in rural villages has been demonstrated, and expansion is anticipated. An Ivory Coast firm projecting the production of more than 13,000 photovoltaic-powered sets in the next decade (Sloop and Watt 1979). Weiss and Pak (1976) estimate a worldwide demand of 10 MW for ETVs by the early 1980s and conclude that photovoltaic-powered sets presently have an economic advantage over alternative systems.

Another remote telecommunications application, operated since 1977 by the Mexican government, utilizes 36 W of photovoltaic power, operates a rural telephone station, and is unique in that all components of the system were designed and produced in Mexican laboratories. It is anticipated that 70 more stations will be installed in rural areas and a solar cell pilot production line is being set up to produce 20 kW/yr (Perez, *et al.* 1978).

Irrigation

In Mead, Nebraska, a 25-kWp irrigation system, larger than most rural applications, has provided important base-line information on silicon module reliability and system design (Figure 1). This project, sponsored by DOE, has shown a module failure rate of 2 percent in 2 years (Forman 1979) and has provided important statistical information on such aspects as soiling and cleaning (Forman 1978). In the 2 years that it has been in operation, the system has provided power with less down time than the local utility. Fur-

FIGURE 1 A 25-kWp photovoltaic-powered irrigation system at Mead, Nebraska, sponsored by DOE.

ther U.S. applications of photovoltaic devices to irrigation systems have been identified and analyzed (see Twersky and Fischbach 1978).

A study pertinent to irrigation systems conducted by the U.S. National Aeronautics and Space Administration (NASA) at its Lewis Research Center states that under present cost goals, a utility grid extension of greater than 5-7 mi (8-11 km) makes the use of photovoltaic systems attractive (Rosenblum, *et al.* 1978b). The study also projects a 6-GWp market in foreign water pumping applications, and a significant market penetration in 1981 or 1982 as systems become economically feasible. But the study cautions that a 4-year lead time is necessary to evaluate and test planned systems, and that those systems will be paced by costs, government attitudes, and sources of funding.

Several other sources have also presented prospects for photovoltaic power in pumping applications in rural areas (see Agarwal 1978; Sangal, *et al.* 1978; Campana, *et al.* 1976).

The Republic of Mali has had extensive experience with photovoltaic systems (W. Mackie, SERI, Golden, Colorado, personal communication 1979). At least five systems have been installed, most with French assistance, and are used primarily for pumping water in rural areas (Figure 2). Figure 3 shows a 16-kWp system being installed at a hospital. This system will provide power for water pumping, refrigeration for pharmaceuticals, and lighting and cooling for the operating rooms.

Under a program financed by the U.S. Agency for International Development (AID) and administered by the Solar Energy Research Institute, four more water pumping stations will be installed in four regions. The 2-kW photovoltaic array-powered stations will also feature 20-30 m^3 (706-1,059 ft^3) of water storage, and are a part of a completely packaged pumping

FIGURE 2 Remote photovoltaic-powered water pump in Mali. (Photograph courtesy of W. Mackie, SERI, Golden, Colorado)

FIGURE 3 A 16-kWp photovoltaic array being installed at the Djiola Hospital in Mali. (Photograph courtesy of W. Mackie, SERI, Golden, Colorado)

system offered by Pompes Guinard of France. No electrical storage is planned for the systems, which will be installed by Peace Corps Volunteers.

Total Village Power Needs

Photovoltaic systems that meet the major electrical power needs of a village are also being implemented. It has been estimated that a 10-kWp system is necessary to meet the initial total demand of a typical village of 500 in Upper Volta or Ethiopia (Kern 1978; BDM Corporation 1978). These demands would include potable water supply, irrigation, ETVs, school/shop lighting, milling, refrigeration, and site-specific light cottage industry applications.

One study of the use of photovoltaic power systems in 19 developing countries concludes that those systems can presently operate at an energy cost of \$1.60-\$2.20/kWh (Hein and Siddiqi 1978). This cost will likely be reduced to \$.25-\$.50/kWh by 1983-85. The study also points out the advantages of the modularity of photovoltaic systems in overcoming the problem of initial high capital outlays for unused central station capacities.

Two further studies have examined the market potential and costs of village power systems. In the first, the authors estimate a potential of 10 GWp for foreign market applications in village power, with an average usage of 24 Wp per person (Rosenblum, *et al.* 1978a). The authors conclude that if DOE price goals are met, 20 GWp could be required by the year 2000. A second study analyzes the energy costs of photovoltaic systems versus diesel and utility extension (Rosenblum, *et al.* 1979). As shown in Figures 4 and 5, photovoltaic systems are presently competitive in applications

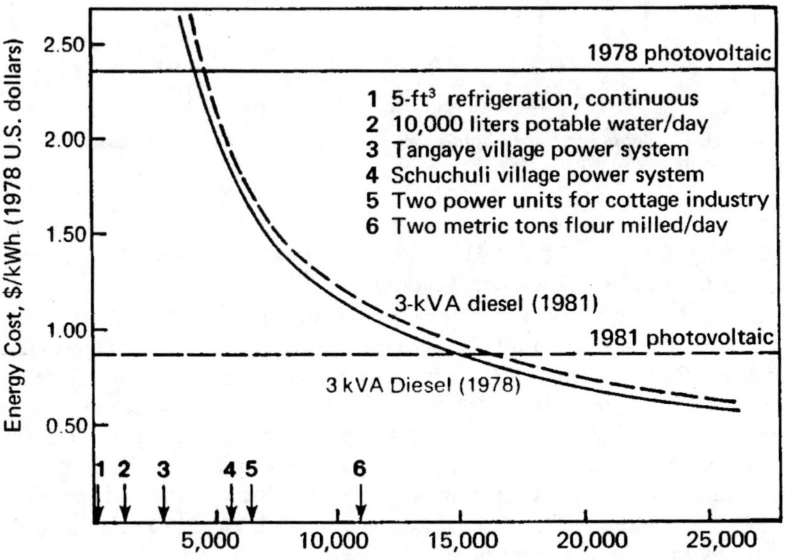

FIGURE 4 Photovoltaic and diesel energy cost comparisons. (Source: Rosenblum, *et al.* 1979)

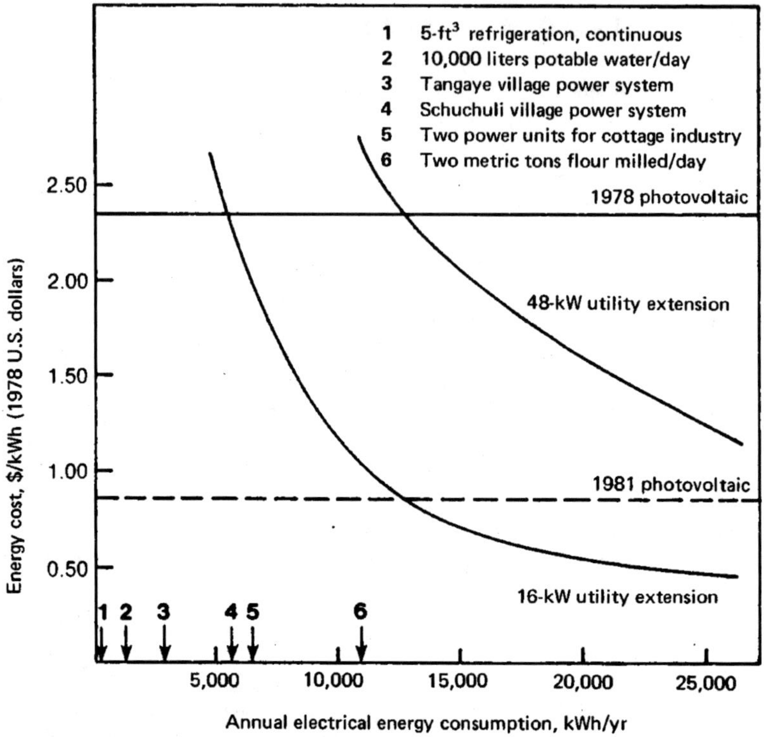

FIGURE 5 Comparisons of photovoltaic and utility extension energy costs. (Source: Rosenblum, *et al.* 1979)

requiring less than 5,000 kWh/yr (∿1 kW capacity), and by 1981 will be competitive for many larger applications.

NASA-Lewis Research Center has successfully implemented two photovoltaic stand-alone village power systems (Rosenblum, *et al.* 1979; Ratajczak and Bifano 1979). The first of these two systems was installed on March 1, 1979, in Tangaye, Upper Volta. This project, sponsored by AID, is designed to study the social and economic effects of providing an electric-powered water pump and grain mill, which relieves the village women of their daily 1- to 2-hour routine of hand flour milling and water lifting. Financial considerations limited the photovoltaic system size to a 1.8-kWp, 120-volt DC array, with 540 amp-hr of battery storage (Figure 6).

The second application is a 3.5-kWp, 120-volt DC photovoltaic array installed in the village of Schuchuli, which is inhabited by about 100 people and located on the Papago Indian reservation in southwestern Arizona. The

FIGURE 6 A view of the 1.8-kWp solar cell arrays (right center) in the village of Tangaye, Upper Volta, that power a water pump and grain mill. The pump supplies the water-dispensing tank (center). The grain mill, operated as a cooperative by the village, is located in the building behind the tank. This installation, funded by AID and the Government of Upper Volta and under the technical management of the NASA–Lewis Research Center, is intended to explore potential new energy. (Photograph courtesy of NASA–Lewis Research Center, Cleveland, Ohio)

power supply is used for water pumping, lighting, refrigeration, and services for village housekeeping.

The system, consisting of 24 photovoltaic panels, was inaugurated in December 1978. Direct-current systems were used to avoid the power losses, costs, and additional complexities of the DC-AC inverters, and battery storage of 2,380 amp-hr capacity was incorporated to provide power at night or during low insolation periods. Portions of the array may be disconnected automatically to prevent overcharging of the batteries, and automatic load shedding prevents excessive discharging of the batteries.

Saudi Arabia and the United States have undertaken a joint project to provide a 350-kWp system to supply the electrical needs of Al Jubaylah in Saudi Arabia. This system, with complete electrical control and battery storage, will have a diesel generator back-up to supply additional power needs. Although presently in the design stage, it is anticipated that the system will be completed by 1983, with technological transfer playing a significant role in the project.

REFERENCES

Agarwal, A. (1978) Solar energy and the Third World, *New Scientist* 77(1089):357-359.

BDM Corporation (1978) Photovoltaic Power Systems: Market Identification and Analysis. U.S. Report No. HCP/M2533-01/2. BDM Corporation, McLean, Virginia, USA.

Campana, A. D., A. Castiel, A. Perez, J. Roger, D. Dupuy, P. Lavit, and M. Lepert (1976) Realization and testing of a pumping system powered by solar cells, *Proceedings, Unesco/WMO Symposium on Solar Energy*. August 30-September 3, 1976, Geneva, Switzerland. WMO-No. 477. World Meteorological Organization, Geneva, Switzerland.

Forman, S. (1978) Endurance and Soil Accumulation Testing of Photovoltaic Modules at Various MIT/LL Test Sites. Lincoln Lab Document No. C00-4094-23. Massachusetts Institute of Technology, Lexington, Massachusetts, USA.

Forman, S. (1979) Photovoltaic module performance and degradation at various MIT/LL test sites, *Proceedings, 1979 International Congress of the International Solar Energy Society*. May 28-June 1, 1979, Atlanta, Georgia. ISES, Victoria, Australia.

Hein, G. and T. Siddiqi (1978) Utilization of Solar Energy in Developing Countries: Identifying Some Potential Markets. Report No. NASA TM-73964. NASA-Lewis Research Center, Cleveland, Ohio, USA.

Kern, E. (1978) Solar Energy Alternatives for the United States Embassy and for Rural Development Projects within the Republic of Upper Volta. Lincoln Lab Document No. C00-4094-5. Massachusetts Institute of Technology, Lexington, Massachusetts, USA.

Palz, W. (1978) Trends in photovoltaic development outside the United States, *Proceedings, Thirteenth Photovoltaic Specialists Conference.* June 5-8, 1978, Washington, D.C. Institute of Electrical and Electronics Engineers, New York, New York, USA.

Perez, E., J. del Valle Padilla, A. Urbano, E. Servin, and H. Romero (1978) Photovoltaic rural telephonic station realized under Mexican technology, *Proceedings, Thirteenth Photovoltaic Specialists Conference.* June 5-8, 1978, Washington, D.C. Institute of Electrical and Electronics Engineers, New York, New York, USA.

Ratajczak, A. F. and W. J. Bifano (1979) Description of Photovoltaic Village Power Systems in the United States and Africa. Report No. NASA TM-79149. NASA-Lewis Research Center, Cleveland, Ohio, USA.

Rosenblum, L., W. J. Bifano, W. A. Poley, and L. R. Scudder (1978a) Photovoltaic Village Power Application: Assessment of the Near-Term Market. Report No. NASA TM-73893. NASA-Lewis Research Center, Cleveland, Ohio, USA.

Rosenblum, L., W. J. Bifano, L. R. Scudder, W. A. Poley, and J. P. Cusick (1978b) Photovoltaic Waterpumping Applications: Assessment of the Near-Term Market. Report No. NASA TM-78847. NASA-Lewis Research Center, Cleveland, Ohio, USA.

Rosenblum, L., W. J. Bifano, G. Hein, and A. F. Ratajczak (1979) Photovoltaic Power Systems for Rural Areas of Developing Countries. Report No. NASA TM-79149. NASA-Lewis Research Center, Cleveland, Ohio, USA.

Sangal, S., Y. Gnanainder, and E. Bhaskar (1978) An experimental study of photovoltaic water pumping systems for rural India, *Proceedings, Thirteenth Photovoltaic Specialists Conference.* June 5-8, 1978, Washington, D.C. Institute of Electrical and Electronics Engineers, New York, New York, USA.

Sloop, J. and M. Watt (1979) Solar energy in Africa: survey of needs and activities; potential applications of small solar systems, *Proceedings, 1979 International Congress of the International Solar Energy Society*. May 28-June 1, 1979, Atlanta, Georgia. ISES, Victoria, Australia.

Twersky, M. and P. Fischbach (1978) Irrigation Systems for the Solar Photovoltaic Energy Program. Lincoln Lab Document No. C00-4094-21. Massachusetts Institute of Technology, Lexington, Massachusetts, USA.

U.S. Department of Energy (1979) National Photovoltaic Program. Multi-Year Program Plan. Report No. DOE/ET-0105-C. U. S. Department of Energy, Washington, D.C., USA.

Weiss, C. (1976) Solar photovoltaic cells in developing countries, *Proceedings, Twelfth Photovoltaic Specialists Conference*. November 15-18, 1976, Baton Rouge, Louisiana. Institute of Electrical and Electronics Engineers, New York, New York, USA.

Weiss, C. and S. Pak (1976) Developing Country Applications of Photovoltaic Cells. Paper presented at the National Solar Photovoltaic Program Review Meeting, January 20, 1976, Washington, D.C., USA.

Sun/Earth-Tempered Building Designs

The principal function of a well-designed building is to provide human comfort. An architecturally integrated design, combining good thermal characteristics and concern for human comfort and aesthetics, is the emerging proper house design in industrialized countries, although it has been the basis for construction of dwellings in many parts of the developing world for centuries.

When heating is needed to maintain human comfort, a *sun-tempered design* uses a building as solar collector and stores heat within the mass of the building. One form of this design, *passive solar*, maximizes a building's solar gains, minimizes thermal losses, and maintains human comfort without the use of mechanical energy. It includes such elements as solar energy collection and thermal storage, control, moderation, and distribution. An *active solar* system requires equipment such as solar collectors, heat storage containers, and electrical or other man-made energy to power pumps and fans for air control and distribution.

Sun-tempered designs differ principally in the placement of the thermal storage mass and retrieval of the stored heat by reradiation and natural convection. House designs that make use of solar energy range from strictly passive to strictly active, with many designs falling into the hybrid class that combines elements of active and passive systems.

Earth-tempered (or earth-sheltered) designs make use of the earth to moderate extremes in the ambient temperature (Figure 7). Several feet below the earth's surface the temperature is relatively constant year-round, and very constant diurnally. At this level, the temperature is also higher than the ambient air temperature in winter and cooler in summer. Covering or surrounding a building with earth provides an effective "flywheel effect," which maintains more uniform interior temperature on a diurnal basis. In addition,

FIGURE 7 Modern, earth-sheltered home near Stillwater, Oklahoma. (Photograph courtesy of Energy Engineering Laboratory, Oklahoma State University)

earth sheltering of a building reduces seepage of heated or cooled air through fissures in the structure.

In most developing countries, the existing, traditional rural architecture has usually met environmental constraints by making use of the materials, labor, and fuel available. For example, a raised hut, open on all sides for natural ventilation, with a large overhanging roof for shade, an open interior for air circulation, and built entirely of indigenous materials, is a building form found widely in the hot, humid climate of Southeast Asia.

Many of the building designs described here briefly are directed toward providing heat and thus may be more applicable in those developing countries with high altitudes or in the northern latitudes. Furthermore, since electricity is often not available in these areas, they are restricted to the use of strictly passive solar designs. Should photovoltaic cells or some other decentralized sources of electric power become economically available in developing nations, hybrid or active solar designs could come into eventual use for both heating and cooling, particularly in schools, public office buildings, etc. For that reason, they are included here.

SUN-TEMPERED DESIGNS

Features common to all passive solar sun-tempered building designs include good insulation; large glass windows (glazings) facing south (in the Northern Hemisphere); small north, east, and west windows; a large interior thermal mass; and various other thermal conservation techniques (e.g., double glazing, night-time insulation of windows, air-lock entrances). Some form of auxiliary heat is required; wood stoves have been popular with and aesthetically pleasing to owners of passive solar homes.

Direct Gain Design

This design uses large south-facing (north-facing in the Southern Hemisphere), double-glazed and/or night-insulated windows to collect the incoming solar radiation. To carry out this simple, potentially inexpensive design, a certain amount of wall and floor surface area must be constructed of materials appropriate for thermal storage, such as adobe, brick, concrete, or ceramic tile. The extent of the surface area is more critical than the thickness of the materials. The mass of the building prevents the interior from overheating during the day and overcooling at night. Building mass is most effective as thermal storage if it is directly exposed to the incoming solar by radiation. This necessitates use of bare stone floors and masonry walls for best thermal performance. Insulated drapes or shutters that can be opened when the sun is available and closed at night or during extended periods of cloudy weather are helpful in reducing heat losses through the large south window and in maintaining a radiantly uniform environment.

Mass Wall Design (Trombe Wall)

Trombe walls (named after French engineer Felix Trombe) are generally constructed of adobe, stone, or concrete, with vents top and bottom. For this design, a vertical, massive wall is constructed directly behind a south-facing window to absorb and store the incoming solar energy (see Figure 8). Typically an air gap of about 4 in. (10.16 cm) is allowed between the interior

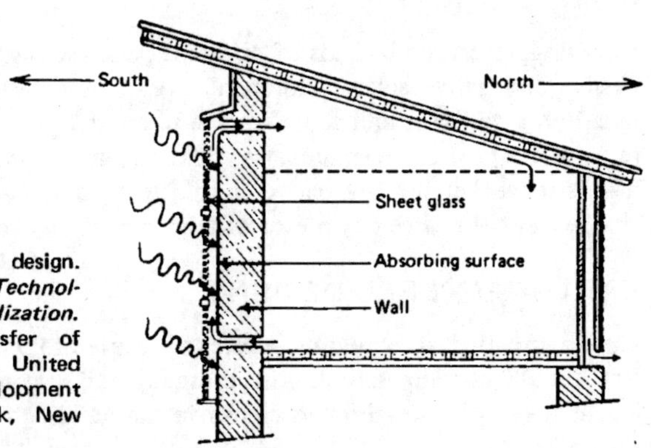

FIGURE 8 Mass wall design. (Source: UNIDO (1978) *Technology for Solar Energy Utilization.* Development and Transfer of Technology Series No. 5. United Nations Industrial Development Organization, New York, New York, USA, p. 23.)

surface of the window and the exterior surface of the wall, which is usually painted black for maximum absorption. The outer surface of the wall and the air gap become quite hot on sunny days, approximately 130°-160°F (55°-70°C), even in cold climates.

Heat is transferred into a building in two ways. If heat is required during the day, vents are placed high and low in the wall to transfer heat (approximately 30 percent) immediately by convection. The remaining heat is conducted through the wall. Depending on wall thickness, which is typically 8-16 in. (20-40 cm), the inner surface reaches its maximum temperature approximately 6-8 hours after maximum irradiation. Thus heat is available in the evening when heating loads on the building tend to be greatest. Mass walls are a popular retrofit design.

Mass walls can also be constructed of containers filled with water. Heat is transferred more rapidly through these walls than masonry walls. However, the high heat capacity of the water and the high storage effectiveness achieved as a result of the isothermal tendencies of the fluid permit energy to be stored at lower temperatures. This increases the efficiency of the water wall system somewhat beyond that of the mass wall system.

Greenhouse Designs

A solar greenhouse does not need large amounts of man-made heat for proper operation. It requires double glazing and/or night insulation, a storage mass, the proper glazing tilt (a compromise between heat and plant production in all but the months around the winter solstice), and an insulated upper roof and northern wall. For an attached solar greenhouse, the northern wall is replaced by an adjoining structure that receives excess heat from the greenhouse during the day. Solar greenhouse designs have become popular in recent years, particularly to those undertaking solar retrofit of structures.

Hybrid Designs

Such designs can be hybrids of different passive designs, or a combination of passive and active solar components. A common example of the latter is based on a collector and some storage in an architecturally integrated design (e.g., direct gain or greenhouse), with the control and distribution of reserve heat storage handled by active means (e.g., a rock storage bin thermally charged and discharged by means of thermostatically controlled air fans).

EARTH-TEMPERED DESIGNS

Earth tempering of a building reduces exposure of the building skin to swings in the ambient temperature, both diurnally and seasonally, and reduces heating or cooling losses through cracks in the building's thermal envelope (see Figure 7).

Earth-tempered buildings should have well-insulated roofs and upper outside walls in order to minimize heat transfer to the earth that is closest to the ambient air and hence shows the greatest temperature changes (seasonally and diurnally) and differences from the interior air temperature. To retain the earth, the walls and roof must be strong and massive. Stone, concrete, and heavy timbers are widely available materials that have been used successfully. Good drainage and waterproofing are necessary to prevent dampness and condensation problems.

Earth itself is a poor insulator, with a relatively high thermal conductivity. Because of its high heat capacity, earth does have a large thermal reactance to varying temperature differences across its boundaries; that is, earth tempering resists changes in temperature between inside and out.

Earth Roof Designs

An earth roof, usually combined with piling up of earth (berming) around the walls of a dwelling (generally done on the north, east, and west walls of the structure in the Northern Hemisphere), further reduces temperature swings. The thickness of the earth roof should be no more than about a meter (3.2 ft) for structural reasons, unlike earth berming where earth can be deposited several meters thick in a lateral direction. The use of (exterior) insulation under earth roof coverings is even more crucial than with earth-bermed designs.

Underground Designs

Underground designs are the extreme in earth-tempered schemes. The walls and roof are totally underground, often several meters or deeper, and light is admitted through skylights or clerestories. In designs using passive solar features, the south wall is exposed to the winter sun. Underground structures may be dug into the south-facing side of a hill or constructed in a flat area with total earth berming (no north, east, or west windows) as a final step. The method selected depends most critically on the depth of the water table.

Earth Cooling

Earth cooling is one of the benefits of earth-tempered designs, particularly in areas where high humidity does not permit effective cooling by evaporation or radiation from above-ground structures. Earth roofs are usually planted in grasses and small plants which help cool the building and stabilize the roof.

Another cooling method is use of air-cooling tunnels through which outside air is drawn from some cool spot (e.g., the shaded northern side of a hill) and vented into the northern or cooler side of the building. Warm air exits from roof vents, drawing in cooler air by thermosiphoning. Cooling by this method is currently being investigated, since the cooling reservoir is depleted more quickly in an earth tunnel than in underground designs. Because the volume of earth that can interact with the fresh air on a daily thermal discharge/recharge cycle is limited, earth tunnels require careful design, both for thermal transfer and for dehumidification. Earth tunnels should be particularly attractive for use with above-ground passive solar buildings in humid climates.

PART II
INDIRECT USES OF SOLAR ENERGY

Wind Energy

Cost projections for advanced wind machines under development show that the gap between wind energy costs and the cost of utility power is narrowing, but largely due to dramatic increases in the prices of oil, coal, and nuclear energy. Unfortunately, these increased energy costs will eventually result in higher material and production costs for wind systems.

Wind system manufacturers have shown increased interest in exporting their products. Aerowatt (France), Lubing (FRG), and Dunlite (Australia) systems are widely accepted throughout the world. However, in general, U.S. manufacturers have not yet begun to investigate foreign markets, and few of them have even the beginnings of a foreign distribution network in place. This situation may change dramatically within the next 5 years.

This chapter describes high-technology advances that, by reducing wind energy costs and increasing machine reliability, may make manufactured wind systems attractive to developing countries in the near future. Low-technology advances that are now applicable to developing countries are also described, including information on newer "home-built" systems that can be constructed using locally available materials and labor. In addition, this chapter includes a description of some of the accessories now available for wind machines, an overview of commercially available wind machines, and a description of the research programs under way in several countries.

Increased interest in wind energy in the United States has resulted in the publication of numerous helpful books, reports, and other publications. For developing countries, the most informative among these are *Wind Power for Farms, Homes and Small Industry* (Park and Schwind 1978) and *A Siting Handbook for Small Wind Energy Conversion Systems* (Wegley, *et al.* 1978) (see Appendix C for a complete listing).

HIGH-TECHNOLOGY ADVANCES

No important changes in the configurations of currently available wind systems have appeared since 1976, and recent advances have been made using technology that has been known for at least 40 years. (See Appendix A for a listing of manufacturers of wind energy systems and Appendix B for a listing of research institutions addressing wind energy systems.)

Changes that have occurred in the state of the art for wind turbine generators (WTGs) during the past 4 years, include (1) replacement of the upwind rotor with the downwind rotor as the most commonly used horizontal-axis configuration, (2) emergence of the vertical-axis rotor as a major competitor with the horizontal-axis machine, and (3) growing prevalence of alternating current-output WTGs. Other important developments are the increasing role of governments (particularly the United States) in supporting development of these systems and advances in mechanical-output wind machine technology.

Electrical-Output Wind Machines (Wind Turbine Generators)

Downwind Versus Upwind Rotors. Upwind machines have dominated the world market for 40 years and are, for the most part, still available. Increasingly, however, downwind rotors are being used in the design of wind machines. Such rotors can reduce system complexity and, by eliminating the tail and tail-furling mechanisms, could be less costly. Systems using downwind rotors include Enertech (Figure 9), Grumman, Kedco, Whirlwind, and Storm Master (Wind Power Systems, Inc.)—all developed within the last 4 years by U.S. manufacturers. In addition, all but one of the small horizontal-axis prototypes being developed under U.S. Government sponsorship have downwind rotors.

Despite the potential economic advantages of the downwind rotor, its superiority over tail-oriented systems has not been shown from an operational standpoint. System reliability, performance, and cost should be considered more important than rotor configuration when selecting or designing a system.

Darrieus and Other Vertical-Axis Rotors. The Darrieus vertical-axis rotor (Figure 10) has received considerable attention in the United States and Canada. First designed by J.G.J.M. Darrieus in 1927, this system has three advantages over horizontal-axis systems because it does not require: (1) a rotor-orientation system, since its entire rotor area is always exposed to the wind; (2) slip rings to pass energy to the base of the tower as with horizontal-axis systems; and (3) pitch control for its blades. Inherent in the third advantage are two disadvantages of this system. Because the orientation of the blade with respect to the axis remains stationary throughout each revolution, the angle at which the wind strikes the airfoil is not always optimum, which results in reduced performance as compared with ideal horizontal-axis systems. Further, because one or both of the blades are not in an optimum position at any given point in the rotor path, the Darrieus needs start-up assistance, provided either by an auxiliary motor or an auxiliary rotor that self-starts. Theoretically, the Darrieus rotor is inherently stronger than horizontal-äxis rotors because the curved blades are attached at each end of the central mast.

Sandia Laboratories (Albuquerque, New Mexico), under a contract with the U.S. Department of Energy (DOE), has developed a troposkein (turning rope) Darrieus blade to increase rotor life and reduce dynamic frequency coupling between the rotor and tower. Dominion Aluminum Fabricating, Ltd. (Canada) has been marketing 14-kW and 34-kW systems for several years. Alcoa Corporation (USA) has developed an inexpensive blade extrusion method for troposkein Darrieus rotors and is the major source of blades for machines produced by Tumac, Dynergy, and other U.S. manufacturers.

FIGURE 9 Enertech 1500 downwind turbine generator.

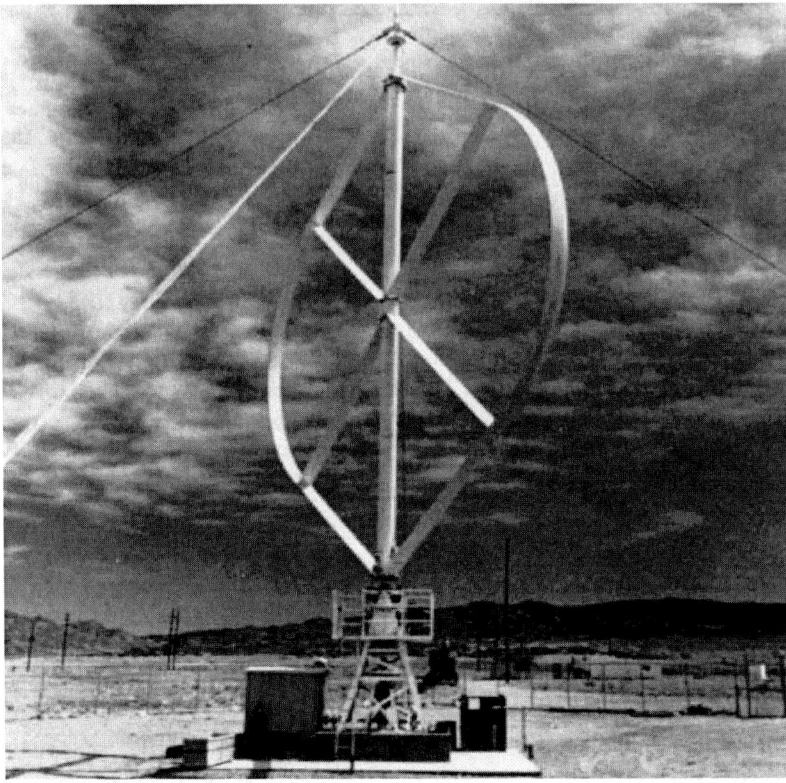

FIGURE 10 A 17-m (56-ft) Darrieus wind system located at Sandia Laboratories, Albuquerque, New Mexico.

Several U.S. companies (e.g., Pinson Energy Corporation and McDonnell Douglas Corporation) have developed vertical-axis "cyclogiros," which overcome the performance losses of the Darrieus by using blade pitch regimens that are scheduled throughout each rotor revolution (Figures 11 and 12). However, the increased performance of these machines over the Darrieus is somewhat offset by the complexity of their mechanical and electrical control systems.

In Great Britain, P.I. Enquiries, Ltd. is marketing the Variable Geometry Vertical-Axis Wind Turbine, which has "coned," not adjustable-pitch, blades for protection in high winds. Coning reduces the swept area of the rotor, which in turn reduces blade loading and angular velocity (revolutions per minute). One advantage of this system is that only one strut is used to support each blade instead of the two commonly used in cyclogiro systems. Coning is

FIGURE 11 Pinson C2E Cycloturbine (4 kW).

centrifugally activated at a prescribed rpm by springs attached to the blades by long tie-wires.

A simpler vertical-axis design—one that may be particularly useful to developing countries because of its structural simplicity and ease of manufacture—is the straight-bladed, pitchless, vertical-axis wind turbine. Its rotor blades are only attached at their bottom ends to the generator shaft; thus no central tower is required as in Darrieus systems. A major disadvantage of this machine is the loss of swept area compared with horizontal-axis blades of equal length, which results from the fact that the swept area of this rotor describes a cone, whereas the swept area of a horizontal-axis wind turbine describes a circle. This loss must be made up by eliminating slip rings and by using a blade pitch mechanism and/or very inexpensive blade materials. A prototype has been developed in the United States by the Oregon Offshore Public Power

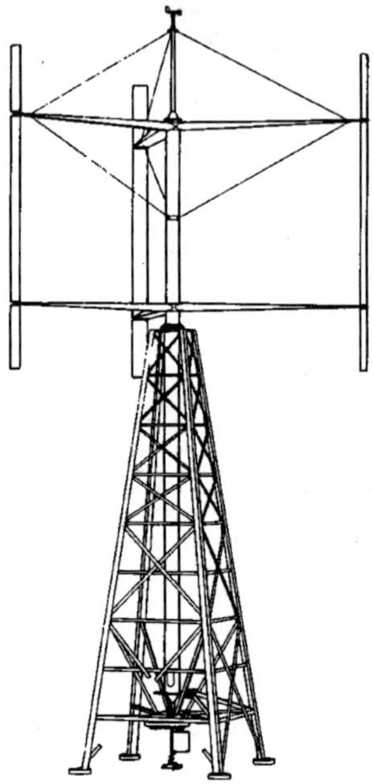

FIGURE 12 McDonnell Douglas 40-kW Giromill for electrical or mechanical applications (prototype).

Services*; however, little or no controlled testing of this design has been performed.

Vertical-axis machines currently available cost from $9,000 to $15,000 for a 2-4-kW unit compared with $3,000-$8,000 (1979 prices) for horizontal-axis machines of similar output.

Electrical-Output Systems. Of the more than 50 electrical-output models available in the United States during 1979, over half had generators that produced alternating current. Both synchronous and induction generators are used for these machines, but one drawback of the latter, especially for use in developing countries, is that they must be tied to an electrical grid to produce the 60-Hz (or 50-Hz) power commonly used by appliances and motors. Power from synchronous generators can be used directly, but they are more costly than induction generators.

The simplest system for AC-output utility interconnections (from the user's point of view) is now represented by the 3-blade downwind Enertech 1500, a 1.5-kW (rated) system that drives a 1.65-kW induction generator. The machine's integrated solid-state control system can be connected directly to a utility through a simple 3-pronged plug inserted in a standard electrical outlet.

*4000 N.E. Columbia Road, Portland, OR 97303, USA.

U.S. Government-Sponsored Development

In 1976, in an effort to stimulate private industry's production of wind machines of appropriate sizes and characteristics for high-potential applications, the U.S. Energy Research and Development Administration contracted with Rockwell International for the development of a national test site for small windmills at Rocky Flats, near Golden, Colorado. As part of the overall project's hardware development program, outside organizations were funded to design and develop small wind energy conversion systems (SWECS) in three power ranges–1-2 kW, 8 kW, and 40 kW (Table 2). Although the prototypes developed under this program are high-technology units with relatively sophisticated power and control systems, they could soon become applicable in developing countries.

TABLE 2 Characteristics of small wind energy conversion systems (SWECS) currently being developed in the United States (tentative).*

	Developer	kW output @ 20 mph	Diameter	Generator	Configuration/ features
High reliability (1–2 kW)	Enertech Corp.	2	5 m (16.4 ft)	Alternator	2-Blade downwind, horizontal axis, variable pitch
	North Wind Power Co., Inc.	2	5 m (16.4 ft)	Direct-drive alternator	3-Blade upwind, horizontal axis, cyclically fixed pitch
	Aerospace Systems Pinson	1	2.4 m × 4.6 m (8 ft × 15 ft)	Alternator	3-Blade, vertical axis, cyclically variable pitch
8 kW	Windworks, Inc.	8 kW	10 m (33 ft)	Direct-drive alternator	3-Blade downwind, horizontal axis, variable pitch
	United Technologies Research Center	9 kW	9.5 m (31 ft)	Induction	2-Blade downwind, horizontal axis, variable pitch
	Grumman Energy Systems	10 kW	10 m (33 ft)	Induction	3-Blade downwind, horizontal axis, variable pitch
40 kW	Kaman Aerospace Corp.	40 kW	19.5 m (64 ft)	Induction	2-Blade downwind, horizontal axis, variable pitch
	McDonnell Douglas Corp.	40 kW	12.8 m × 17.7 m (42 ft × 58 ft)	Induction	3-Blade, vertical axis, cyclically variable pitch

*At the conclusion of each contract for these devices, design information and production drawings will be available from the National Technical Information Service, 5285 Port Royal Road, Springfield, VA 22151, USA.

The three sizes of SWECS currently under development could fulfill widely different applications. The 8-kW wind systems, adequate for individual residences and small farms in the United States in moderate wind regimes (except for heating purposes), might well be useful for village use and irrigation pumping in developing countries. As most applications for 8-kW machines in the United States are expected to involve interconnections with existing electric power sources, these systems are being designed to provide 120/240-volt AC, 60-Hz output for tie-in with a utility power line (see Figures 13, 14, and 15). By contrast, the 1-2-kW systems (24-volt DC for charging batteries) are intended for noninterconnected remote application where servicing would be on an annual basis at best. These systems must satisfy stringent reliability requirements. Finally, the 40-kW systems are intended for irrigation applications as well as for small communities or factories. Two types of output are specified for this size range: (1) mechanical output for use with deep-well pumps, and (2) electrical output for both

FIGURE 13 An 8-kW prototype system developed by Windworks (USA) which incorporates a direct-drive, 10-kW alternator with samarium–cobalt magnets.

FIGURE 14 An 8-kW prototype system with a flex-beam rotor developed by United Technologies Research Center (USA).

tie-in with a utility and stand-alone applications that require 440-volt AC, 60-Hz output.

Mechanical-Output Wind Machines

Significant developments in mechanical-output systems include the previously mentioned, relatively large (40-kW) prototypes designed for electrical or mechanical output. Two mechanical-output wind machines not mentioned in the first edition of *Energy for Rural Development* are being manufactured by Sparco (Denmark) and Bowjon (USA). The Sparco Windmill is available with a diaphragm (Model D) or piston (Model P) pump. It is a very small machine with a diameter of only 1.27 m (4.1 ft), and its two cast-aluminum blades generate enough torque to pump up to 225 1 (59 gal) of water per hour for low-lift (up to 10-m) operations.

The Bowjon, a somewhat larger machine, uses an air compressor to generate the power required to run an "air lift pump" for deep submersed pumping. This windmill has a 2.4-m (8-ft) diameter rotor with four galvanized steel blades, and it is capable of pumping 908 1/hr (240 gal/hr) through a 2.54-cm (1-in.) water line when lift is 9.1 m (30 ft) and submergence is 18.3 m (60 ft). Submersions of up to about 60 m (200 ft) are possible using a 1.25-cm (.5-in.) water line (about 190 1/hr). One particularly attractive feature of this

FIGURE 15 An 8-kW prototype developed by Grumman Energy Systems (USA) incorporates three extruded aluminum blades in a coned downwind rotor.

windmill is that it can be located as far as .4 km (.25 mi) from the well or water source, which allows the user greater siting flexibility. The U.S. price for this system in 1979 was $700.

LOW-TECHNOLOGY ADVANCES

Low-technology wind systems can be easily built and maintained using local labor. Although these systems may be less efficient than more sophisticated systems, overall performance per unit cost may well prove superior due to the use of locally available labor and second-hand parts.

In addition to the home-built mechanical-output windmills described in the first edition of *Energy for Rural Development*, most mechanical-output systems manufactured by Heller-Aller (USA), Aermotor, (USA), Dempster (USA), Lubing (FRG), and Sparco (Denmark) qualify as low-technology systems because of their simple design. Furthermore, several new systems have emerged since 1976, including the Princeton sailwing (Figures 16 and 17) and a modification called "The Quixote."

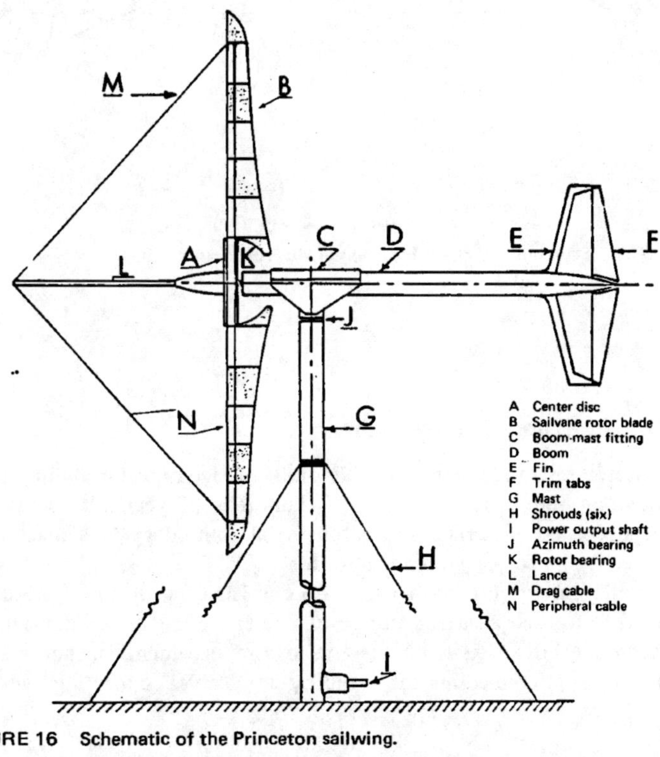

FIGURE 16 Schematic of the Princeton sailwing.

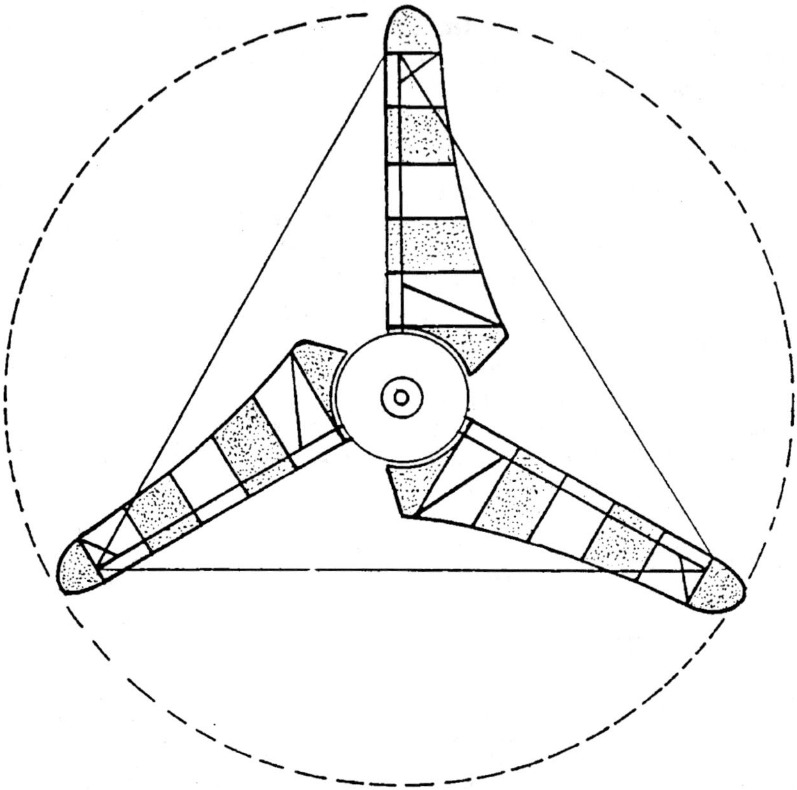

FIGURE 17 Schematic of the Princeton sailwing rotor planform.

The Princeton and Quixote Sailwings

The Princeton sailwing is based roughly on the design of water pumping windmills used on the island of Crete for hundreds of years. It incorporates modern structural materials and aeronautical technology in a machine designed specifically for developing nations.

The basic design (Figures 16 and 17) consists of a 3-blade upwind rotor connected to a single disc-shaped hub at one end of a long horizontal boom. In a technique borrowed from the ancient Cretan windmills, the blades are braced both by a connecting cable and by drag cables connecting each blade tip to a lance protruding forward from the rotor. These cables strengthen the very light rotor. The machine's tail, which resembles that of an aircraft, has trim tabs designed to offset the twisting action of torque produced during operation. An azimuth bearing at the point where the mast (tower) fits the boom permits the tail to orient the rotor freely with the wind.

The sailwing's simple drive train consists of three shafts running through two gearboxes–one at the right angle where the tower meets the boom and one at the right angle at the base of the tower. The mechanical output of the turbine can be used to operate virtually any kind of energy conversion device–water churn, water pump, air compressor, or electrical generator.

Most of the sailwing's structural components are aluminum, and dacron is usually used for the sails. The designer recommends that only the materials specified be used if electrical output is planned.*

"The Quixote" is a modified version of the Princeton sailwing.†

Other Home-Built Systems

Other plans available in the United States include the "Budget Windcharger," a 1.2-kW, horizontal-axis sailwing wind turbine generator and a simple 2–5-kW wind turbine design that comes in 12-ft (3.6-m) and 16-ft (4.8-m) diameters.‡

A good model for low-technology, cottage-industry wind machine production is provided by the many "backyard" manufacturing operations in the United States. One such operation, Dragonfly Wind Electric in Elk, California, is marketing kits and unassembled or assembled versions of a very simple horizontal-axis machine similar to the tilt-axis North Wind 2-kW DOE prototype and the now defunct Parris–Dunn machine. This electrical-output system uses automotive single-pulley alternators or generators with output ranging from 500 to 2,000 W. With a 42-amp Chrysler alternator, the 8-ft (2.4-m) diameter system is designed to produce about 500 W in a wind speed of 17 mph (28 kmph). The major advantage of this kit is its low cost.§

*Plans for the system are available from: Forrestal Campus Library, Princeton University, Princeton, NJ 08540, USA.

†Plans for "The Quixote" are available from: Flanagan's Plans, P.O. Box 891, Cathedral Station, New York, NY 10025, USA. The plans include detailed blueprints, a step-by-step instruction booklet, and a parts list. No welding or complicated metalworking are required.

‡Plans for the "Budget Windcharger" are available from: MI Plans Service, Fawcett Building, Greenwich, CT 06030, USA. Plans for the 2–5-kW design are available from: Environmental Action Reprint Service, 2239 East Colfax, Denver, CO 80206, USA.

§The kit costs $150 without alternator or voltage regulator and with unbalanced blades and unfinished materials. The unassembled version costs $200 (same as the kit, except holes are drilled), and the assembled version, $300 (unassembled blades).

ACCESSORIES AND TOWERS FOR WIND MACHINES

Batteries and inverters are used with many energy systems, not just wind machines, and they are discussed in two later chapters of this supplement, "Energy Storage" and "Solid-State Electric Inverters," respectively.

Instructions for building simple wooden derrick towers were included in the first edition of *Energy for Rural Development*. In addition, guyed or free-standing steel pipes and wooden poles are suitable.

Tower selection becomes much more critical when a high-rpm wind turbine generator is involved. Dynamic interactions among the rotor, yaw system, drive mechanism, and the tower can create harmonically related vibrations that decrease system life and even result in failures. Communications or radio towers that are available commercially worldwide are most commonly used for wind turbine generators. One U.S. manufacturer of such towers, the Rohn Tower Division of Unarco Industries,* is fabricating special towers to accommodate any wind machine. Sectional towers specifically designed for wind machines are also available from Solargy Corporation.†

The octahedral tower, designed by U.S. engineer, designer, and architect Buckminster Fuller, is a relatively low-cost design that offers complete rigidity, an aerodynamically clean profile, and ease of installation. Composed of sections of galvanized steel pipe, the octahedral tower is bolted together from the ground up.‡

WIND POWER APPLICATIONS

Electrical Applications

These applications fall into three categories: interconnected AC, noninterconnected AC or DC, and remote DC.

Interconnected Applications. For this application, total or surplus power from the wind system is fed into utility lines. The wind system must, therefore, produce power that is compatible in frequency and phase with that in those lines. This can be accomplished by either (1) using a synchronous generator with revolutions per minute and resultant frequency regulated by a sophisticated control system, (2) using a nonsynchronous alternator with a line-commutated inverter, or (3) using an induction generator that uses utility power as a reference to produce compatible alternating current.

The first method is generally not appropriate for small wind systems because it requires expensive pitch change and electronic control mechanisms.

*P.O. Box 2000, Peoria, IL 61656, USA.
†17914 E. Warren Avenue, Detroit, MI 48224, USA.
‡The octahedral tower is available from several U.S. distributors, including Natural Power, Incorporated, New Boston, NH 03070, USA.

The second method is commonly used for small wind systems, but it requires the additional cost and complexity of a battery bank, inverter, and the switches and wiring needed to interconnect these components. In addition, losses associated with conversion and reconversion of wind system output reduce considerably the cost-effectiveness of the entire system. The third method appears to be the optimal choice for developing countries as it is the least expensive of the three methods and the least complex. Even though wind turbines with nonsynchronous alternators are generally less expensive, the cost of power conditioning equipment required for interconnection more than offsets this advantage.

Noninterconnected AC or DC Applications. For this application, the wind system is dedicated to one or more loads, and surplus is not fed into power lines. More options are thus open to the user. Many appliances are flexible in terms of the quality of power they will accept, particularly those operated by motors that do not have to have a precise speed. Electric lights have relatively wide frequency operating ranges. DC lights can also be used, and resistance heaters can be operated with virtually any type of AC or DC power.

Examples of noninterconnected AC applications include: electric water pumps for irrigation, residential or industrial applications requiring AC power, heaters and blowers for crop or material drying, and large-scale cooling or ventilation systems. One AC-output wind turbine generator can satisfy both AC and DC power needs at one site. With the installation of a rectifier and a battery bank, energy for DC lights and small resistance heating loads can be stored for use when the wind is not blowing.

DC-output systems are used most efficiently with DC lighting, resistance heating, and appliances with DC motors. An inverter (not line-commutated) can be used to operate AC appliances, but higher cost and output losses are the primary penalties. Most DC-output systems have AC generators, the output of which is rectified. This power production mode is dictated by the high cost of DC generators.

Remote DC Applications. A wind machine is backed up by batteries or other storage systems for this application. Examples of excellent remote applications for DC-output systems include telecommunication relay stations, meteorological stations, rail signaling systems, marine and air navigational aids, and remote pumping stations. Machines used for such applications typically have relatively low output ratings of 20 W to 4 kW. In developing countries seeking to install or improve communications networks, remote DC-output wind systems can be used instead of oil- or gas-powered generators, with the advantages of reduced fuel costs and the possibility of unattended operation over several months to a year or more. Such systems could also be used as a heat or light source in remote communities where labor for maintenance is not available. Aerowatt (France) has designed its systems specifically for remote applications. In the United States, the North Wind Power Company, Inc. has begun to market a 2-kW system developed under contract to DOE.

Mechanical Applications

Such traditional mechanical applications as water pumping for irrigation or domestic water use, pond aeration, and grain grinding are still among the most economical uses of wind power, using any high-torque, low-rpm wind system. The most promising new application, the water churn heating concept, involves heating water by mechanical agitation.

Because combustion fuels can heat water more efficiently, the age-old principle of raising the temperature of water by means of a mechanical brake or churn has not been applied. With the use of wind power, however, mechanical means are the most direct way of heating water to usable temperatures. Furthermore, mechanical heating can be used with virtually any wind machine now available. High-rpm systems are preferable, but high-torque water pumpers can also power water churn systems with the aid of speed-increasing gearboxes.

The range of applications of water churns, powered by windmills or water wheels, covers all processes that require water heating, including producing steam for industrial processes, supplementing home or community heating systems, and producing hot water for use in grain drying and water desalination.

One of the primary advantages of the water churn is its excellent applicability to a wind system. Because both devices involve fluid processes that operate with a cubic power/speed relationship, the water churn's power output will be matched to the wind turbine's output at any wind speed. As a result, the conversion efficiency of such a system is almost 100 percent, with the exception of frictional losses. Heat losses can be kept to a minimum with adequate insulation.

The Water Twister Model 12, suitable for use with small wind machines, has been developed by the All-American Engineering Company (Figure 18). This model has a rated output of 6 hp at 500 rpm and 47 hp at 1,000 rpm. Costs for this system ($3,000 in 1979) are already comparable with those of electrical generators of similar output. The Water Twister is now produced in limited quantities, but the manufacturer estimates that increased production may reduce costs to under $1,000 per unit (1979 U.S. dollars).

Water churn research is now being conducted at Cornell University (Ithaca, New York) within a DOE-funded project entitled "Development of a Wind-Power Water Heating System for Daily Application." A churn simpler in design than the Water Twister is being used in this project.

Considerable work on the water churn concept has also been performed in Denmark. In Skive, Jutland, a water brake powered by a vertical-axis wind turbine was tested in a "low-energy," one-family home in conjunction with passive solar heating and flat-plate collectors. The churn used was a simple

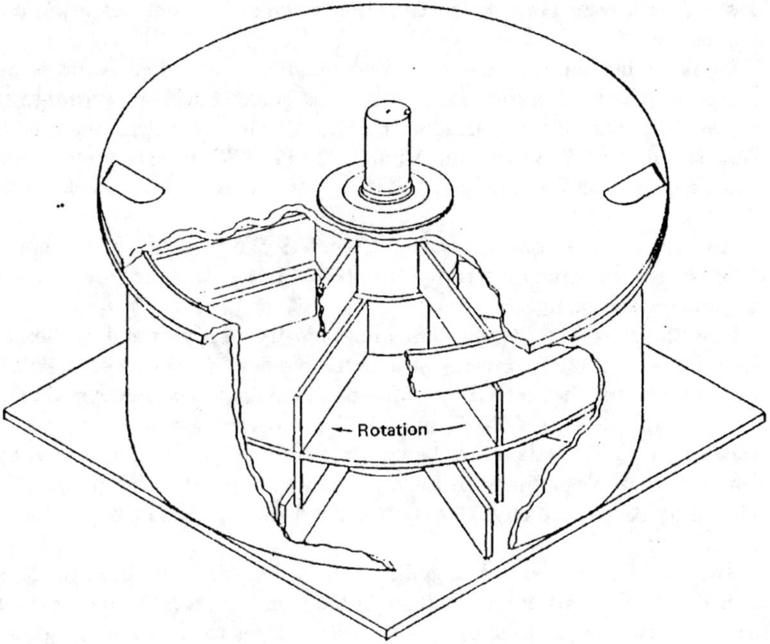

FIGURE 18 Cutaway schematic of the Water Twister manufactured by the All-American Engineering Company (USA).

star-shaped paddle wheel, .45 m (1.5 ft) in diameter. Coupled to the wind turbine through a 1:15 gearbox, the water churn produced 2 kW of power at 225 rpm (see Appendix B for a list of sources of information on Danish research efforts).

CURRENT ACTIVITY IN THE WIND INDUSTRY

United States

A number of new systems have emerged since publication of *Energy for Rural Development* in 1976 (see Appendix A for a complete list of manufacturers). In the 1-2-kW output range are the Enertech 1500, the Whirlwind Model A, the Sencenbaugh Model 1000 series, the Aero Power SL 1500, and the Kedco 1200 series. All but the Sencenbaugh are designed for utility interconnection and have induction generators designed primarily for that purpose.

Somewhat larger machines (2-5 kW) are the Astral/Wilcon AW 10-B, the Product Development Institute Wind Jennie, Kedco 1600 series, Dakota Wind & Sun, Dynergy Corporation 5-m Darrieus, the Pinson Energy Corporation

Cycloturbine Model C2E, and the Independent Energy Systems Skyhawk. Most of these systems are designed for utility connection, but can be adapted to other uses.

Probably the most important development in the United States is the emergence of several larger wind systems that could provide power for small communities and villages, including the Millville 10-kW system, the Dakota Wind & Sun 10-kW system, the Windgen 25 (25 kW) manufactured by the Wind Engineering Corporation, and the Jay Carter Enterprises Mod 25 (25 kW).

Mechanical-output systems available from U.S. firms include the Aermotor, Heller–Aller, and Dempster fan-type water pumpers, and the Bowjon compressed-air water pumper.

New firms are entering the U.S. wind industry each year and established firms are continually improving old systems and developing new ones. Within the next 2 years, the most likely additions to the list of available systems will be several of the 1–2-kW, 8-kW, and 40-kW prototypes developed under DOE contracts, a 20-kW system now being developed by U.S. Wind Power for the California State Department of Energy, a series of Darrieus wind turbines developed by Alcoa, and an 8-kVA system developed by the Jacobs Wind Electric Co., Inc.

The U.S. Federal Wind Energy Program, within DOE, has been the most diverse and active government effort in the world in recent years. Development efforts range from small (1–100-kW) systems to giant multimegawatt systems. Test and evaluation efforts for small systems are conducted under contract by Rockwell International at Rocky Flats (Golden), Colorado. Farm application tests are managed by the U.S. Department of Agriculture's Research Service in Beltsville, Maryland. Field tests of 100-kW, 200-kW, 2-MW, and 2.5-MW systems developed under contract to the U.S. National Aeronautics and Space Administration's Lewis Research Center in Cleveland, Ohio, are either under way or planned at locations throughout the United States. Candidate sites for large systems are selected and evaluated with the assistance of Battelle–Pacific Northwest Laboratories, which is also conducting research on wind characteristics. Analytical research and coordination of the dissemination of information are performed at the Solar Energy Research Institute in Golden, Colorado.

Canada

In 1977, about $500,000 was allotted for wind energy research and development to be dispensed by Canada's National Research Council, the designated lead agency in the field of renewable energy research. In 1979, the total national expenditure in Canada, including that by industry, was approximately $2 million.

Dominion Aluminum Fabricating, Ltd. has been involved in the development and marketing of Darrieus wind turbines for several years. Bristol Aerospace, Ltd. has developed a 9-ft (2.7-m) diameter vertical-axis wind turbine, which has successfully powered a remote unmanned weather station in the Arctic Sea for over a year without maintenance. Developed in cooperation with the University of Manitoba, the Bristol systems are self-starting and are equipped with low-speed multiple alternators.

Winflo Power, Ltd. is now manufacturing horizontal-axis wind turbines that range in output from 300 W to 2.5 kW. The 2.5-kW machine has a cut-in wind speed of 7 mph (11 kmph) and produces its rated output at 22 mph (35 kmph).

Crowfield Associates, Ltd. is planning to produçe a wind power heating system utilizing a 12-kW (23 mph or 37 kmph) vertical-axis wind turbine with two straight blades.

Europe

Over the past 5 years, Europe has experienced a renewed interest in the use of wind energy. Most countries now have government-sponsored wind energy programs.

Denmark's wind energy program, instituted in 1975, is currently emphasizing large wind turbine generators. However, a second stage that includes small WTG development was just getting under way in 1979. The Danes have constructed the Tvind-WTG, a 3-blade machine with a downwind rotor, 54 m (177 ft) in diameter, and a rated output of 2 MW.

The Federal Republic of Germany has initiated the GROWIAN (Grosse Windenergie-Anlage) Program as a main line for research and development. Several projects are under way aimed at adapting small wind energy conversion systems to special types of utilization, including development of a 20-kW wind energy conversion system in cooperation with the Centro Técnico Aerospacial in Brazil.

The Netherlands began a 5-year program in 1976, designed to study the feasibility of wind energy conversion. The Dutch have also initiated a research and development program that has produced a 5-m (16-ft) vertical-axis wind turbine for experimental testing.

Sweden is now testing a 75-kW wind turbine manufactured by Saab-Scania as part of the National Swedish Wind Energy Program managed by the National Board for Energy Source Development. However, the major thrust of this program is aimed at developing large wind energy conversion systems.

Several manufacturers of wind machines are located in Europe (see Appendix A). The French company, Aerowatt, currently manufactures wind machines ranging in output from 30 W to 100 kW. Lubing Maschinenfabrik (FRG) is manufacturing several types of wind pumps, as well as a wind turbine generator with an output of 400 W. Elektro GmbH Winterthur of Switzerland has models available that produce outputs of from 50 W to 10 kW. P.I. Enquiries, Ltd. (Great Britain) is producing the Variable Geometry Vertical-Axis Wind Turbine in 4.5-m (1-kW) and 6-m (4-kW) sizes. In Denmark, manufacturers emphasize heating applications for their wind systems. A 5-blade vertical-axis machine manufactured by Dansk Vindkraft Industri Aps, a 10-kW fan-type system manufactured by S-J Wind-Power ApS, and 22–55-kW systems manufactured by Riisager El-Vindmolle are all major models available. Systems with outputs of 20–200 W designed for remote DC applications are produced by Naestved Maskin Elektro. The Swedish firm Ingenjorsfirma Ture Ekenburg AB manufactures a 3-blade upwind system rated at 10 kW in winds of 43 kmph (27 mph). Finally, Domenico Sperandio & Ager of Italy markets electrical-output systems rated at up to 1 kW. These include the Garbino 250 W (12 volts), the Turbine 500 W (24 volts), and the Monsone 1 kW (48 volts).

USSR

Although accurate information is difficult to obtain, it appears that the USSR is planning to embark soon upon an ambitious wind energy project that will take the form of a 700-mi (1,126-km) network of large wind machines to generate electricity in northwestern Russia. The plan involves erecting 238 wind generators with a rotor diameter of over 50 m (164 ft). Each machine will have a generating capacity of 1 MW. Several smaller 100-kW machines have been installed to produce power locally along the Baikal–Amur railway line. The USSR Enercgomashexport and the USSR Institute for Farm Electrification I-U Veshniakovski have both been listed as producing and developing wind machines, including numerous models from 500 W to 25 kW.

Costa Rica

The Costa Rican Technological Institute (Instituto Tecnológico de Costa Rica) has established an investigation center, which, among other things, is designed to help small villages or communities develop sources of energy (wind power, primarily) by making maximum use of local materials and labor. Uncomplicated construction and maintenance are emphasized.

Australia and New Zealand

In Australia, Davey-Dunlite manufactures wind machines with outputs of 1–5 kW, which are distributed worldwide.

In New Zealand, a Wind Energy Task Force was established in 1974 to coordinate wind engineering, power systems, and meteorological expertise. Lincoln College in Canterbury has built a 6-m (19.6-ft) diameter, 1.5-kW, Cretan-type sail windmill for test and demonstration purposes. A German

"Flip-Flop" (Schlagflugelwind-pumpe) wind machine design is also being studied.

Japan

Japan Wind Power Generator Co., Ltd. is manufacturing wind machines with outputs ranging from 6 W to 1 kW and up. According to the manufacturer, these machines can be used to provide power for remote areas, meteorological stations, and ships.

REFERENCES

Park, J. and D. Schwind (1978) *Wind Power for Farms, Homes and Small Industry*. Nielsen Engineering and Research, Inc. Available from the National Technical Information Service, Order No. RFP-2841/1270/78/4, 5285 Port Royal Road, Springfield, VA 22161, USA.

Wegley, H. L., *et al.* (1978) *A Siting Handbook for Small Wind Energy Conversion Systems*. Battelle–Pacific Northwest Laboratories, PNL-2521. Available from the National Technical Information Service, 5285 Port Royal Road, Springfield, VA 22161, USA.

Energy from Falling Water

Hydroelectricity and isothermal hydraulic air compression, two technologies dependent in different ways upon falling water, were not discussed at length in the original volume of *Energy for Rural Development* as these are well-established technologies for which there have been few recent new developments. However, these technologies are in many ways now being "rediscovered" for use in developing countries and are being widely applied there. (See Appendix C for suggested reading on hydropower and isothermal hydraulic air compression.)

SMALL-SCALE HYDROELECTRIC SYSTEMS

The development of hydraulic turbines has slowed since the 1960s, and current work has been directed more toward the design of hydroelectric equipment packages for cost-efficient systems. Small-scale hydroelectric systems generally fall into two categories: micro-hydro, 0–100 kW, and mini-hydro, 100–1,000 kW. Whether an installation falls into the category of micro-hydro or mini-hydro is determined primarily by the amount of energy available at the site. The physical principles involved are the same for all hydroelectric systems.

As the energy output of a hydro system becomes smaller, every aspect of the system must become simpler and faster to implement. This not only applies to a system's design, civil works, and plant distribution, but also to the organization and techniques used to install and operate the equipment over its life.

With the availability of new materials (i.e., bearings, generators, and control electronics), it has become possible to obtain small-scale hydroelectric systems that have an installed cost ranging from $800 to $1,800 per kilowatt and require less than 5 percent of invested capital per year for operation and maintenance. Such systems can have useful lives of 30–40 years and can be operated automatically from remote locations. These systems can be easily installed in a matter of months or less by unskilled labor. The small-scale approach to hydropower can best be illustrated by the following description of the installation of a typical 5-kW hydroelectric system.

Example: Installation of a 5-kW Hydroelectric System

The water source for this system is a mountain spring located 470 ft (143 m) above the turbine installation in the valley below. The water diversion system (Figures 19 and 20) consists of a primary and secondary sump and screening system that allows the water velocity to decrease and precipitate heavy suspended solids. Then, 130 gal of water per minute (492 1/min), as shown in Figure 21, are drawn up a 6-in. (15-cm) syphon through a screen and carried 2,500 ft (762 m) down the mountain to the power house where a pressure of 173 lb/in.^{2} (12 kg/cm^2) is developed at the nozzle of the 4-in. (10-cm) pitch diameter impulse turbine (Figure 22). The turbine is directly connected to a 5-kW, 3,600-rpm, 120/240-volt, single-phase, 60-Hz brushless generator with permanently lubricated bearings. Voltage and frequency are controlled by the

FIGURE 19 A 3-in. (7.6-cm) PVC pipeline carrying 130 gal of water per minute (492 l/min) over a distance of 2,500 ft (762 m), developing a net head of 400 ft (122 m), and producing 5 kW of electrical energy (see the nomograph in Figure 25). (Photograph courtesy of Independent Power Developers, Inc., Sand Point, Idaho)

FIGURE 20 Close-up of water intake showing a 6-in. (15-cm) syphon that allows the silt to settle out before being drawn into the pipeline. This line provides the water supply for domestic use, livestock, gardens, fire protection, and electrical generation. (Photograph courtesy of Independent Power Developers, Sand Point, Idaho)

electronic governor (Figure 23), which together with the controller allows automatic operation with a wide combination of turbine generator sets or DC to AC conversion.

The electronic governor makes it possible to keep the alternator at essentially full load to prevent the turbine from overspeeding. This device can be designed in such a way that it selects loads to match the alternator output according to a predetermined set of priorities. Thus if the water supply increases or decreases, the governor keeps the speed essentially constant, adjusts the alternator output for maximum possible power output, and selects the priority loads to which that power is delivered.

The two nomographs shown in Figures 24 and 25 show the relationship between the various parameters. The lines drawn on the nomographs are for the example just presented. These nomographs can be used to obtain preliminary design numbers for a wide variety of hydroelectric installations.

The importance of keeping the load factor high is illustrated in the simple nomograph devised by R. G. Ramakumar of Oklahoma State University and shown in Figure 26. The lines emanating from the lower left represent constant load factors. Load factor is simply the average power output of the generating equipment, divided by its rated power output. The more electrical power used, up to the rated capacity of the system, the closer the system load factor comes to 1.0 and the lower the cost per kilowatt-hour. Micro-hydroelectric systems normally have a load factor range of between 0.6 and 1.0.

As an example, if an installation costs $1,100 per installed kilowatt and has a load factor of 0.7 (that is, the unit delivers average power equivalent to 70 percent of its full load capacity), a vertical (solid) line is drawn from $1,100 on the horizontal axis to the 0.7 load factor line. A horizontal (solid) line is

FIGURE 21 One hundred and thirty gallons of water per minute (492 l/min). See the nomograph in Figure 24. (Photograph courtesy of Independent Power Developers, Sand Point, Idaho)

then drawn over to the right-hand vertical axis to get a cost per kilowatt-hour of 21 mills or \$.021. If the load factor were 0.3 (dashed lines), then the cost would be \$.046 and so on. In each case, this is that part of the cost attributable to capital investment and operation and maintenance. Since there is no fuel cost in micro-hydroelectric systems, this becomes essentially the total cost. The nomograph in Figure 26 assumed a 7.5 percent interest rate, an operation and maintenance (O&M) cost of 3 percent of capital costs, and a 35-year amortization. Similar nomographs have been worked out for a wide

FIGURE 22 Close-up of the impulse turbine and nozzle on the 5-kW turbine generator set, which weighs 130 lb (59 kg), including the speed control. (Photograph courtesy of Independent Power Developers, Inc., Sand Point, Idaho).

FIGURE 23 Electronics for speed governor and voltage control. A malfunctioning card is simply replaced by a spare and then mailed to the nearest repair station. (Photograph courtesy of Independent Power Developers, Inc., Sand Point, Idaho)

variety of interest, operation and maintenance, and amortization figures.* Placing real economic value upon the electricity produced, of course, depends entirely on having a beneficial and valuable use as much time as possible for as close to 100 percent of the energy available as possible.

In any economic analysis of hydroelectric systems, it is important to look at the proportion of the budget spent on civil works, generating equipment, transmission, and operation and maintenance. A study of the economics of micro-hydro systems in both developed and developing countries has shown invested capital to be allocated as follows: civil works, 24–52 percent; generating equipment, 30–40 percent; transmission, 10–30 percent; and operation and maintenance, 3–8 percent. These percentages were derived from installed cost data in Nepal, Bolivia, Peru, Panama, Guatemala, and the United States, and the sample included 63 installations with powers ranging from 3 kW to 100 kW. Typically, the lower the head (elevation) the higher the installed cost per kilowatt. Of course, all of these percentages can vary widely depending on individual circumstances.

*These can be obtained from the Engineering Energy Laboratory, Oklahoma State University, Stillwater, OK 74078, USA.

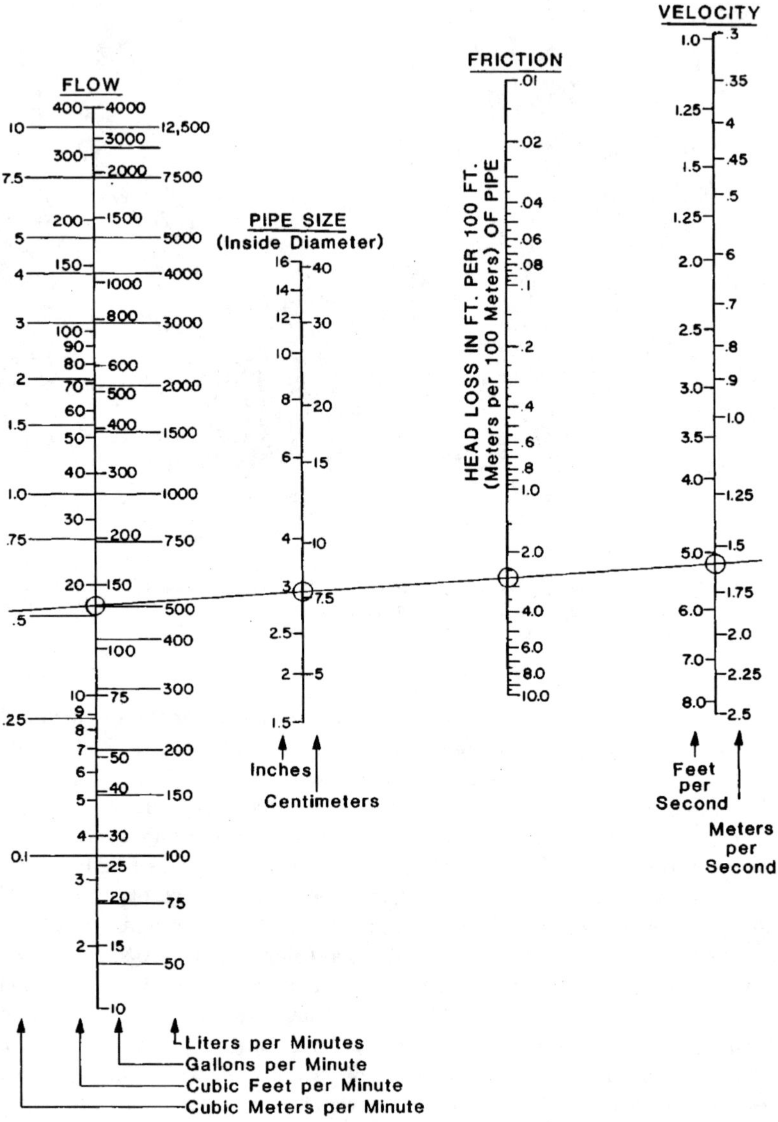

FIGURE 24 Hydroelectric system sizing nomograph relating flow, pipe size, frictional losses, and velocity.

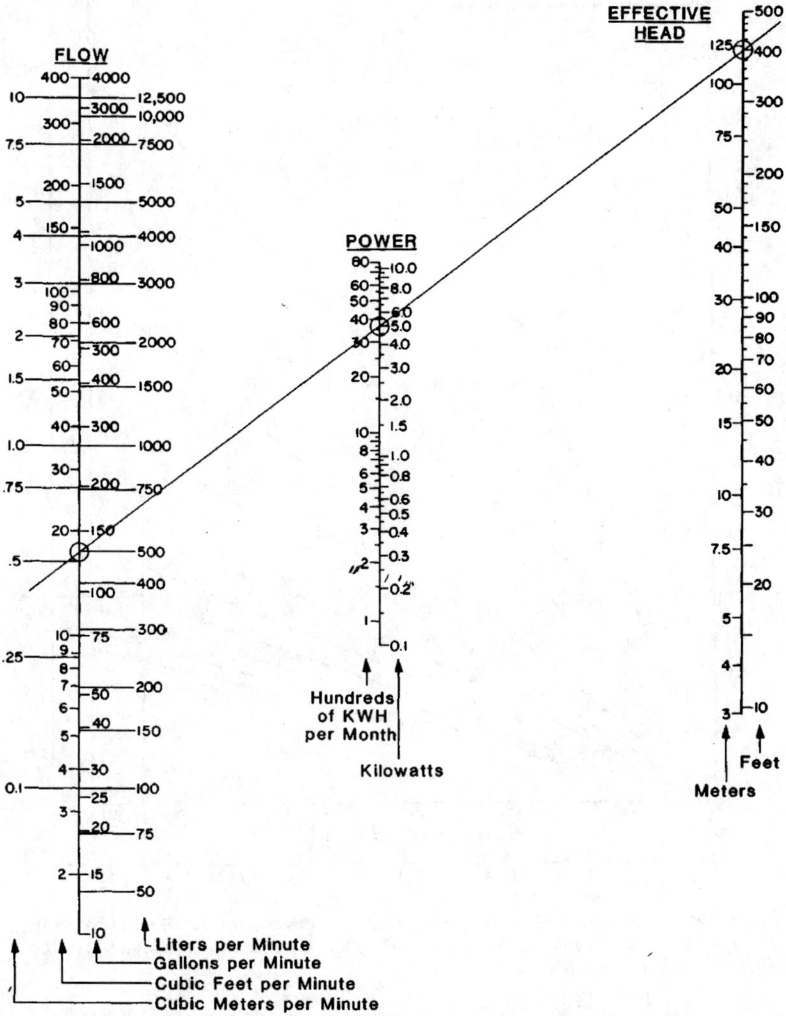

FIGURE 25 Hydroelectric system sizing nomograph relating flow, power available, and effective head. For larger hydroelectric power requirements, one need only consider the fact that for any given head, power availability is directly proportional to water flow.

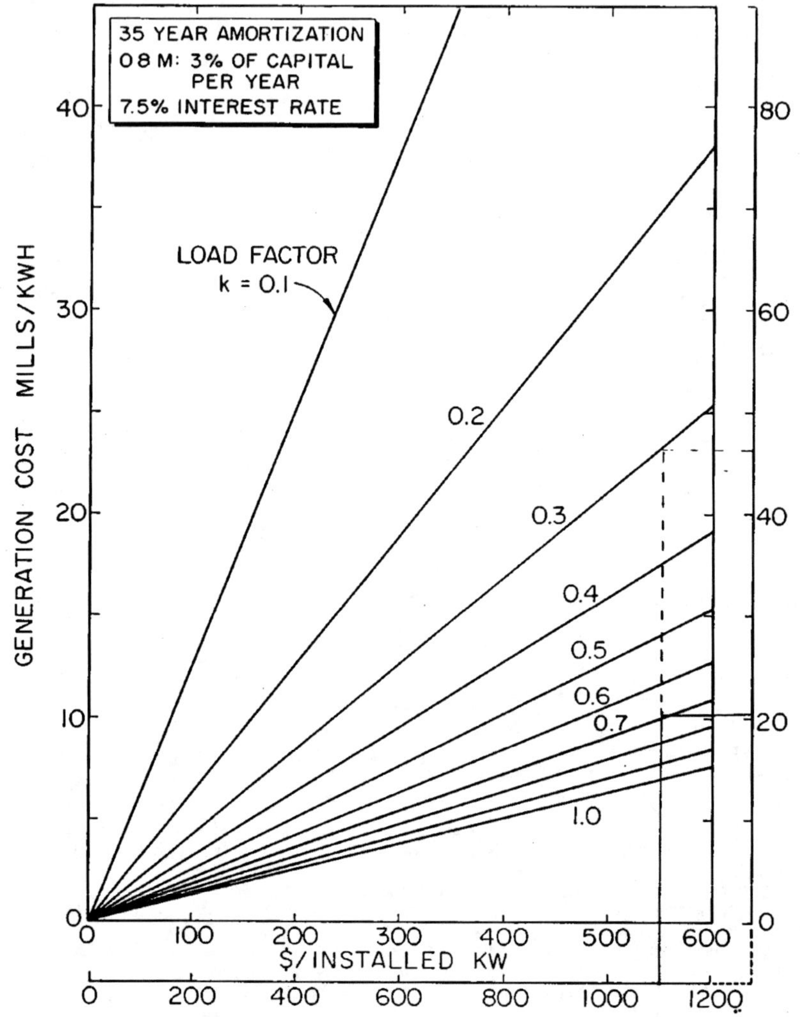

FIGURE 26 Nomograph for determining cost per kilowatt-hour attributed to capital, operation, and maintenance costs (devised by R. G. Ramakumar, Oklahoma State University, Stillwater, Oklahoma).

Sources of Information on Small-Scale Hydroelectric Systems

A number of organizations located both in the United States and in other countries publish information on this technology, and they could be of assistance in contacting suppliers of hydropower equipment (VITA 1980). In the United States, Volunteers in Technical Assistance (VITA) and the National Center for Appropriate Technology (NCAT)* have published material that enables one to determine the power potential of a water resource and measure the head and flow rate, and gives instruction for the actual installation of micro-hydro systems (VITA 1967, n.d.; NCAT 1979). The Swiss Center for Appropriate Technology (SKAT) will answer inquiries from developing countries on micro-hydro and other types of water power systems (SKAT n.d.). In Great Britain, the Intermediate Technology Development Group (ITDG)† distributes publications on small-scale water power development and responds to inquiries on micro-hydro and other aspects of hydropower. Groupe de Recherche et d'Echanges Technologiques (GRET)‡ in France has a number of documents on micro-hydro and other forms of water power. In Colombia, the Center for Integrated Rural Development of Las Gaviotas (Universidad de los Andes, Bogotá) does research, development, and testing and training in water resources and other areas, and distributes hydropower technical reports in Spanish. In Papua New Guinea, the Appropriate Technology Development Unit (ATDU)§ of the University of Technology conducts research on and does pilot production of micro-hydro systems. (See Appendix A for a list of manufacturers of small-scale hydroelectric systems and Appendix C for a list of suggested reading on this subject.)

HYDRAULIC AIR COMPRESSION

Prior to the widespread use of electricity and the internal combustion engine, compressed air was considered a source of primary power, and falling water was considered a prime source of power for air compression and storage. This technique was used extensively until the beginning of the 20th century when

*Volunteers in Technical Assistance, 3706 Rhode Island Avenue, Mt. Rainier, MD 20822, USA; National Center for Appropriate Technology, P.O. Box 3838, Butte, MN 59701, USA.

†Intermediate Technology Development Group, 9 King Street, London WC2E NH, England.

‡Groupe de Recherche et d'Echanges Technologiques, 34 Rue Dumont d'Urville, 75116 Paris, France.

§Appropriate Technology Unit, University of Technology, P.O. Box 793, Lae, Papua New Guinea.

less capital-intensive and more easily acquired sources of energy became available. Since that time, the technique has become a forgotten art, and most scientists, engineers, and technicians educated in the last half of this century are simply not familiar with the fact that air can be isothermally compressed using falling water. (See Appendix C for a listing of recent literature on this subject.)

The Process

This relatively simple process is illustrated in Figure 27 (in the interests of simplification, some details have been omitted). When water is allowed to fall through a long, enclosed column (such as a vertical pipe or even a mine shaft in rock), it will capture a significant amount of air at the top of the column. As the completely enclosed column of water falls, the bubbles are carried to the bottom of the column and compressed as they go down. The heat of the compression that results is immediately absorbed from each bubble by the surrounding water. Since the specific heat per unit volume of water is many times that of the compressed air, the temperature of the water does not rise significantly and the effect is equivalent to isothermal compression of the air.

At the bottom of the column, the water passes slowly through an enclosed chamber in which the bubbles have time to rise. Since the air in the bubbles is at the head pressure of the water, the enclosed chamber is thus filled with air at a pressure equal to the water head pressure.

For safety reasons, the bottom level of the compressed air must never be allowed to reach the bottom water inlet or outlet level in order to prevent

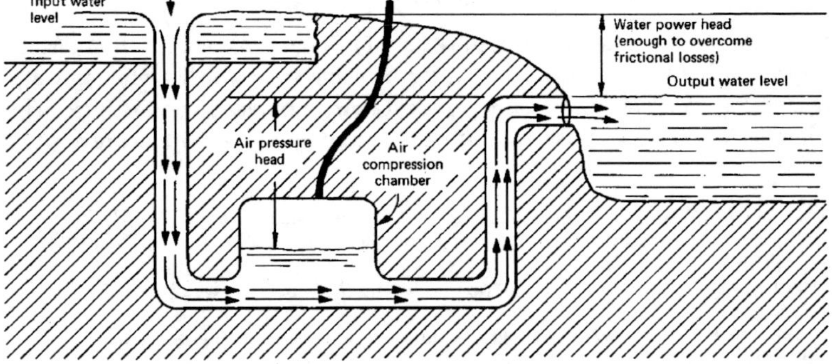

FIGURE 27 Simple illustration of a hydraulic air compressor.

backflow of air up the intake. A simple blow-off air line must be installed just above the storage tank inlet and outlet.

Such compressors have several desirable operating characteristics: (1) air is compressed isothermally, which means high thermodynamic efficiency; (2) compressors have no moving parts, and thus maintenance is minimal and service life maximal (at least 50 years); (3) within limits, the pressure attainable is a design feature determined solely by the placement of the air-water separator below the tailrace elevation; and (4) as should be expected, the higher the flow head (the distance between the elevated source of water and the tailrace), the more air can be induced and compressed, and hencc the more power available.

Analysis of a large number of old compressors reveals a very simple relationship between the flow head available, the quantity of water flowing, and the quantity of air induced and compressed. If the flow head and quantity of water flowing are known, one can determine the quantity of air that can be compressed to a predetermined pressure and, in turn, the amount of work that can be performed.*

Applications of Hydraulic Air Compression

Compressed air can be used as a means of energy storage; to run a wide variety of pneumatic tools, machinery, and control equipment; or to turn a turbine wheel and provide rotary mechanical power, which, in turn, can be used to generate electric power. When compressed air is expanded through a turbine

*The simple functional relationship between the head, quantity of flowing water, and amount of air compressed is as follows in both English and metric units:

(English)

$$\frac{Q_A}{Q_W} = \frac{h}{100}$$

and

$$W_A = \frac{h}{1,307} Q_W$$

(Metric)

$$\frac{Q_A}{Q_W} = \frac{h}{30.48}$$

and

$$W_A = \frac{h}{24.82} Q_W$$

where

h = head in ft (m)
Q_A = air induced in ft^3/sec (m^3/sec)
Q_W = water rate in ft^3/sec (m^3/sec)
W_A = lb of air/sec at 59°F (kg of air/sec at 15°C)

Once the actual amount of compressed air is determined for a system, the theoretical energy available from reversible adiabatic (isentropic) expansion can be calculated in and is actually doing work, the temperature of the exit air drops significantly, thus providing refrigeration as a side benefit (or perhaps in many cases the main benefit). If compressed air is heated prior to passing through the expander turbine, the energy output of the turbine can be increased significantly, thus converting low-grade (heat) energy to high-grade (mechanical) energy. If a significant amount of heat is added, the hydraulic air compressor could be used, for example, to provide input air for a large gas-fired turbine electric generator. Under proper design conditions, the fuel consumption of such a gas turbine generator could be reduced to perhaps half of that required (for any given amount of electricity produced) if the gas turbine had to compress its own air. The net amount of electricity produced could be higher than that produced by a hydroelectric system using the same falling water plus the gas turbine operating alone.

Whether low-grade heat is added or not, the exit air can be relatively cool (or even cold) and the refrigeration benefit need not be lost. Since its operating temperature would be fairly low, the turbine can be made of a variety of common metals. Such turbines can be manufactured at relatively low cost, and could be fabricated in some developing countries.

In many countries in which hilly or semi-mountainous land is farmed, irrigation can be a problem. In combination with an air lift pump, the hydraulic air compressor can be used to lift water to any reasonable height.

Finally, hydraulic air compressors are useful in mining operations (their original purpose), including mine ventilation and the operation of mining equipment. While the original application was in hardrock mining, the system

terms of the fundamental relationships between temperature, pressure, and volume as follows:

$$W = \frac{mR}{0.4}(T_1 - T_2) = \frac{P_1V_1 - P_2V_2}{0.4}$$

$$\frac{T_2}{T_1} = \left(\frac{V_1}{V_2}\right)^{0.4} = \left(\frac{P_2}{P_1}\right)^{0.2857}$$

where subscripts indicate initial and final values for temperature, pressure, and volume, and

m = mass of air captured and compressed in lb (kg)
R = gas constant, 1545 ft-lb/°R (8.314 J/°K)
W = work done in ft-lb (J)
P = pressure in lb/ft^2 (Newtons/m^2)
T = temperature in °R (°K)
V = volume of air in ft^3 (m^3)

Any figures for work or energy available calculated from the above equations must, of course, be multiplied by the efficiency of whatever turbine or other expander is used.

could be used to pressurize water for placer mining or to elevate water for gravity washing in small applications.

All of the above applications require a rudimentary system with a minimum of moving parts, minimal maintenance, and a long service life.

REFERENCES

NCAT (1979) *Micro-Hydro: Reviewing an Old Concept*. National Center for Appropriate Technology, Butte, Montana, USA.

SKAT (n.d.) *Harnessing Water Power on a Small Scale*. Swiss Center for Appropriate Technology, Institute for Latin-American Research and for Development Cooperation, St. Gall University, Switzerland.

VITA (1967) *Low Cost Development of Small Water Power Sites*. Volunteers in Technical Assistance, Mt. Rainier, Maryland, USA.

VITA (1980) Micro-Hydro Power Projects. Energy Fact Sheet, Number 4. Volunteers in Technical Assistance, Mt. Rainier, Maryland, USA.

VITA (ñ.d.) One Kilowatt River Generator. Technical Bulletin No. 1, Volunteers in Technical Assistance, Mt. Rainier, Maryland, USA.

Energy from Biomass

INTRODUCTION

Energy can be derived from a wide range of biomass products, but with varying implications for agriculture. Adding the pressures of energy demand to intensive crop production can aggravate such problems as depletion of soil nutrients and of organic matter, which is critical to the maintenance of soil tilth, water infiltration and holding capacity, and a range of other physical and chemical properties.

In contrast, biomass energy recovery systems that include crop material, utilize animal wastes, provide animal feed, or return nitrogen, phosphorus, and potassium to the land can greatly increase the efficiency of many small farms in developing countries by encouraging integration of the components that result in higher production efficiency.

The intensive agricultural systems so common to developing nations are not totally lacking in energy potential, despite the problems just cited. For example, the lowland rice paddy is a biologically stable crop cultivation system (Geertz 1963): its anaerobic conditions create a chemical balance that makes low-concentration soil nutrients more readily available; considerable nitrogen is fixed in the paddy water by free-living organisms (IRRI 1977); and little soil erosion occurs. Rice straw can be converted to energy in some situations in which other uses do not take priority. Sugar cane is another relatively stable biological system, and it can be cultivated solely for energy use. There are, however, far greater overall benefits in intensive agricultural systems when energy recovery is a by-product.

Perennial cropping systems–shrubs and trees–have high untapped potential for the use of new technologies that convert sunlight to usable energy via biomass. Such systems would be concerned primarily with the production of cellulose rather than the protein–starch–carbohydrate production of intensive agriculture. Conversion methods include combustion, charcoal production with a heat by-product, the various pyrolysis processes, or enzymatic hydrolysis to sugars and fermentation to alcohol.

Conversion technologies for producing fuels from biomass can be divided into two categories: biological and thermochemical. Biological conversion processes, as applied to developing countries, were not only treated in the original volume of *Energy for Rural Development*, but also in *Methane Generation from Human, Animal and Agricultural Wastes* (National Research Council 1977) and a forthcoming publication of the National Academy of Sciences, *The Potential for Alcohol Fuels in Developing Countries* (working title). These processes are covered in the following sections on "Anaerobic Digestion," "Energy Farms Utilizing Aquatic Plants and Wastewater," and "Ethanol (Alcohol) Production." Thermochemical conversion processes are discussed in a section that focuses principally on direct combustion and pyrolysis, which are near-term technologies of importance to developing countries. (See Appendix C for an extensive list of suggested reading on energy from biomass.)

ANAEROBIC DIGESTION

It is only logical that developing countries, which are largely rural, have taken a great interest in biogas, the gas produced as a result of the anaerobic digestion (biomethanation) of biomass. Agricultural and animal wastes, the major raw materials for biogas production, abound in the rural areas, and the rural populace, who are especially vulnerable to the price and supply fluctuations of conventional fuels, can benefit directly from family- and community-size biogas systems.

Microbiological conversion of organic material into methane gas through anaerobic digestion has been studied extensively, and in recent years has been of particular interest in countries facing shortages of coal and petroleum. The basic principles involved in this process are thoroughly discussed by Singh (1971); in the National Academy of Sciences (NAS) publication, *Microbial Processes: Promising Technologies for Developing Countries* (National Research Council 1979); and in a forthcoming NAS publication, *Food, Fuel, and Fertilizer from Organic Wastes*. Diagrams for simple digesters, particularly for animal wastes, can be found in these books. Because this process is discussed in such detail elsewhere, only a brief summary is presented here. (See Appendix A for a list of manufacturers and designers of methane digesters.)

The Process

Biomethanation is the anaerobic digestion (without free oxygen) of organic material. This process produces biogas, which consists of carbon dioxide and methane with lesser amounts of nitrogen and traces of hydrogen sulfide and hydrogen. The percentage of methane in the total volume of biogas produced during this process is highly dependent on the control of certain parameters—substrate, carbon-to-nitrogen ratio, temperature, the pH, organic loading rate, and solids concentration—all of which influence the delicate microbial population balance between the acid-forming bacteria and the methanogenic bacteria. In brief, these parameters should be adjusted as follows in order to generate the greatest volume of methane in the shortest period of time.

Substrate. The initial substrate or organic material must have a high carbon content, mainly in the form of cellulose, hemicellulose, sugars, lipids, and protein. The substrate should also be low in lignin which is nonbiodegradable and hinders the microbial breakdown of the cellulose.

Carbon-to-Nitrogen Ratio. The optimal C/N mass ratio is 30:1. A higher ratio indicates a substrate that is deficient in nitrogen, which would result in the biodigestion ceasing when the nitrogen has been depleted and before all the available carbonaceous material has been converted to methane and carbon dioxide. A lower C/N ratio usually indicates that there will be loss of nitrogen during the digestion process.

Temperature. Sudden temperature fluctuations will slow down or stop the process. The optimum temperature range is generally from 33° to 38°C (91° to 100°F), or the so-called mesophilic range. Biodigestion will proceed at a slower rate below 33°C, essentially stopping at 10°C (50°F), and will proceed in the temperature range of from 54° to 60°C (129° to 140°F) with the so-called thermophilic bacteria. However, this group of bacteria is extremely sensitive to temperature fluctuations. Further, a greater amount of the energy is utilized to maintain the biodigester at this high temperature unless solar energy is used for temperature control.

pH Factor. The optimal pH range is from 7.0 to 7.2, but the pH may vary between 6.6 and 7.6. If the system is upset, the pH usually drops due to excessive acid production, a condition that may be difficult to correct. Careful addition of lime to raise the pH to near 7.0 may restore the system to normal.

Agitation. Some daily mixing is desirable to keep heavier material distributed throughout the system. The amount of agitation required (if any) is highly dependent on the type of feedstock used.

Solids Concentration. Terrestrial plant material and animal wastes generally require dilution with water to reduce the solids concentrations to 7-9 percent. A minimum amount of water should be added to aquatic vegetation since the initial solids content is generally below 7 percent of the fresh weight (see "Energy Farms Utilizing Aquatic Plants and Wastewater").

Organic Loading Rate. Continuous-feed digesters can be easily upset by adding fresh substrate too rapidly. The optimal loading rate varies according to such parameters as digester design, substrate, and temperature. Thus it is difficult to give any general advice as to how to adjust this variable for optimal production of gas.

Recent Technological Developments

In the past 5 years, significant advances have been made in methane production by anaerobic digestion. A paper by Klass (1980) presented to the 1980 Bio-Energy World Congress and Exposition describes recent developments in this area, many of which apply for the moment only to large-scale, municipal-type digesters found in urban areas.

Feedstock Evaluation. Considerable work has been aimed at better understanding the anaerobic biodegradability of organic components and predicting the performance of a particular feedstock under anaerobic digestion conditions without conducting long-term batch digestion tests. Recent work includes studies of the effect of molecular structure on fermentation (Chou, *et al*. 1977), the use of kinetic constants to predict methane production (Chen and Hashimoto 1979), and rapid methods of predicting biodegradability based on substrate composition (Chandler, *et al*. 1979).

Methane yields in an anaerobic digestion process can be increased by pretreating the feed (e.g., acid hydrolysis of cellulose, alkaline treatment to break up lignocellulose complexes) so that the complex organic structures, especially the polymers, are broken down to lower molecular weight species which are then more susceptible to microbial degradation. Gossett, *et al*. (1975) and McCarty, *et al*. (1979) have made detailed studies of the effects of acid and alkaline treatments on methane production.

Digester and System Designs. Stirred-tank reactors, usually in the complete-mix mode, are still the most commonly used digestion systems, but other digester designs are being developed—for example, plug-flow, two-phase, and fixed-film.

Plug-flow digesters, proposed for low-cost, small-scale systems, are reportedly quite suitable for such feeds as dairy cattle manure (Jewell, *et al*., 1979a). When compared to the complete-mix digestion of dairy cattle manure under mesophilic conditions in farm-scale digesters, the plug-flow unit provides higher methane yields and more efficient destruction of volatile solids. The plug-flow tank costs about $10,000 compared to $30,000 for the complete-mix digester. These systems have been in operation on Michigan and Wisconsin farms for several years (Abeles, *et al.* 1978).

In a two-phase digestion system (for insoluble substrates), the acid- and methane-formation phases are physically separated by control of the detention time and the loading rate of the volatile solids (Ghosh, *et al.* 1975; Ghosh and Klass 1978; Norman and Frostell 1977). Two-phase digestion permits optimal operating conditions for each microbial population, which means shorter detention times and more efficient conversion to methane.

Digesters that contain solid supports for anaerobic organisms seem promising for both efficient waste stabilization and methane production (Pitt and Genung 1980; Jewell 1980; Jewell, *et al.* 1979b). Digesters in the experimental stages include a 5,000-gal/day (18,926 l/day), fixed-film, upflow bioreactor (ANFLOW) containing a packed bed of ceramic Rasching rings; settled sewage sludge and upflow bioreactors containing small particles that behave like fluid beds; and anaerobic, attached-film, expanded-bed bioreactors (AAFEB), which reportedly are effective for treatment of primary settled sewage. A 50,000-gal/day reactor is now being developed to demonstrate ANFLOW. It is anticipated that methane will be recovered at yields of about 5 ft^3 (.14 m^3) for each pound of BOD (biochemical oxygen demand) destroyed in the process.

One difficulty with anaerobic digestion systems is that the organics are not completely gasified in the process; residuals are always left in the digester effluent. An innovative approach to this problem combines anaerobic digestion and thermal gasification (Chynoweth, *et al.* 1980; Anonymous 1979). The Institute of Gas Technology (USA) has developed the BIOTHERMGAS process in which anaerobic digestion, dewatering, steam-oxygen gasification, and catalytic methanation of the hydrogen-rich gas from the thermal gasifier are combined in that order to produce methane. Thus refractory organics from the digester are completely gasified, and waste heat as well as ammonia and inorganic nutrients can be recycled from the thermal gasifier to the digesters.

Anaerobic digestion and thermal processing can be combined in yet other ways. Four alternatives for conversion of refuse-derived fuel (RDF) to electric power via gas have been analyzed, namely (1) mesophilic digestion with the residue used as landfill, (2) mesophilic digestion with thermal processing of the residue, (3) thermophilic digestion and thermochemical processing of the residue, and (4) mesophilic digestion of RDF's food–waste fraction and thermochemical processing of the fiber fraction. The latter route was found to be the most economically attractive. The cost of producing fuel gas by these processes ranges from $2.46 to $14 per million Btus at a capacity of 2,000 t/day. If efficiency can be optimized and credits for disposal of municipal solid waste of $7.50–$10/t obtained, the cost of using gas produced by these methods could approach present costs for fuel oil (Anonymous 1979).

Because steel tends to corrode over a period of time when used in anaerobic digesters, one recent design has evolved that obviates the need for steel. The bag digester (Figure 28) is a sausage-shaped bag made of 0.55-mm (.02-in.) thick Hypalon, which is laminated with neoprene and reinforced nylon (Skrinde 1978). It is provided with a polyvinyl chloride (PVC) inlet and outlet. Advantages of bag digesters, which originated in Taiwan, are reported to be low-cost mass production capabilities and easy transportability. They are available in both circular and rectangular configurations and in sizes from 5 to 100 m^3 (177 to 3,531 ft^3).

A host of variations in biogas plant configurations has been developed (Figures 29–36), incorporating design features particular to the type of feedstock used, construction materials available, and ultimate uses of the by-products that are produced.

Production Parameters. Loading rate, detention time, and temperature probably have the most influence on the methane generation process. Ideally,

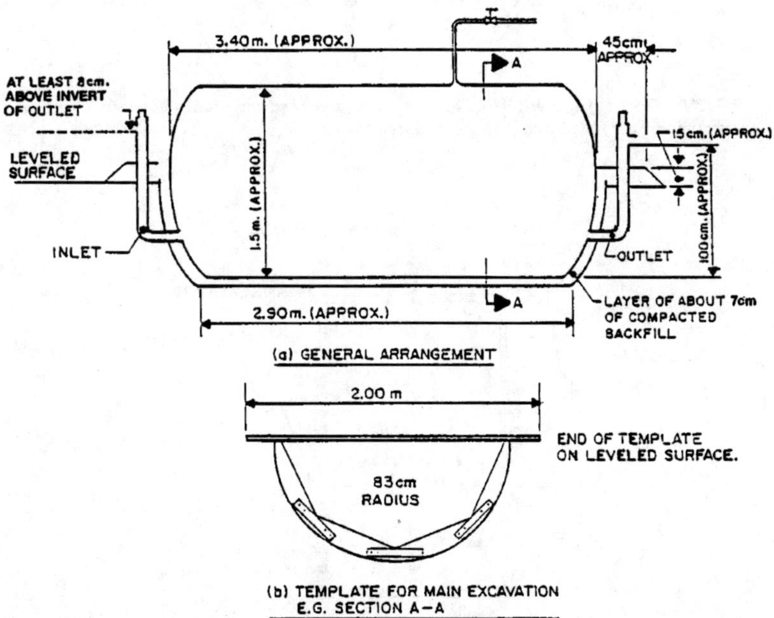

FIGURE 28 A 5-m^3 bag digester. (Source: Skrinde 1978)

digesters should be loaded as much as possible for the shortest period of time consistent with high methane yield and destruction of volatile solids. Many projects have been working toward these goals (see Bryant, *et al.* 1976; Hashimoto, *et al.* 1978; Schellenbach 1980; Wujcik and Jewell 1979).

Additives. Considerable work has been done the last few years aimed at using additives–coal and flyash, carbon, lactobacillus cultures, enzymes, and growth factors such as amino acids–to improve digestion. Radiation, although not strictly an additive, has also been evaluated (see Spencer 1978;

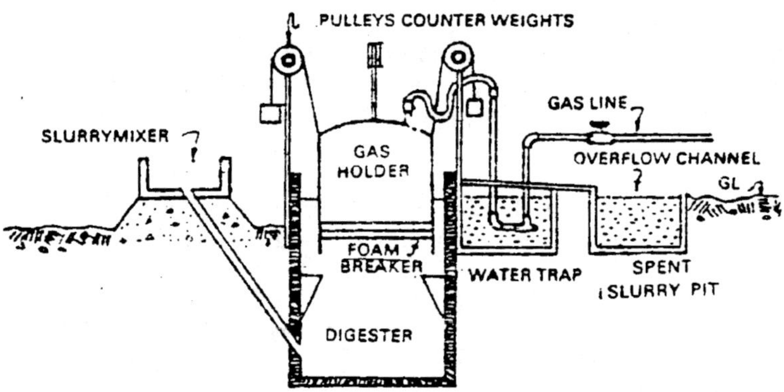

FIGURE 29 Gobar gas plant (Indian Agricultural Research Institute) with overflow outlet, designed by Patel, India. (Source: Maramba 1979)

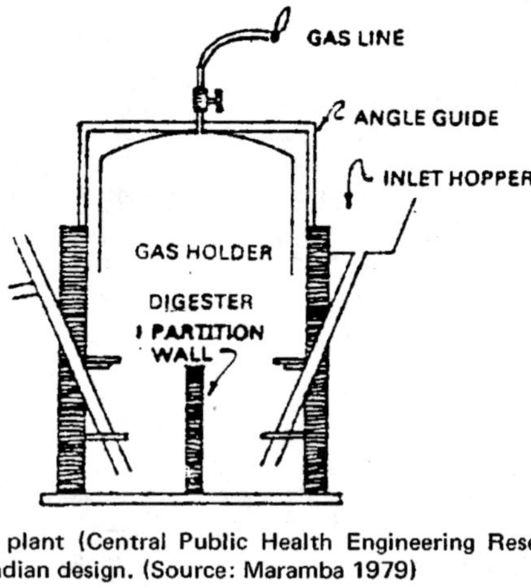

FIGURE 30 Gobar gas plant (Central Public Health Engineering Research Institute, India) with pipe outlet; Indian design. (Source: Maramba 1979)

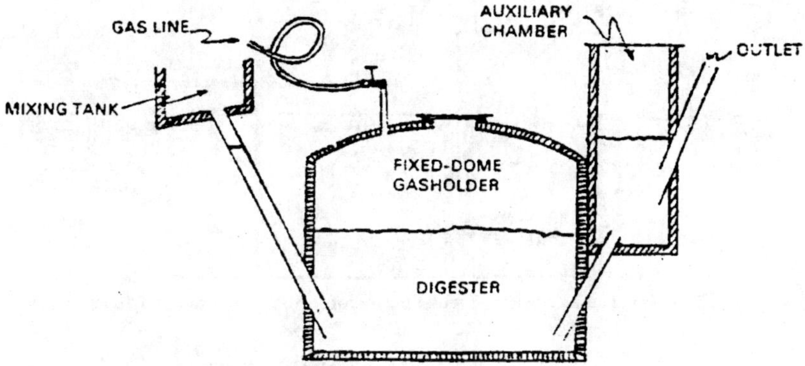

FIGURE 31 Fixed-dome, continuous-feed biogas plant, Maya Farms, the Philippines; Chinese design. (Source: Maramba 1979)

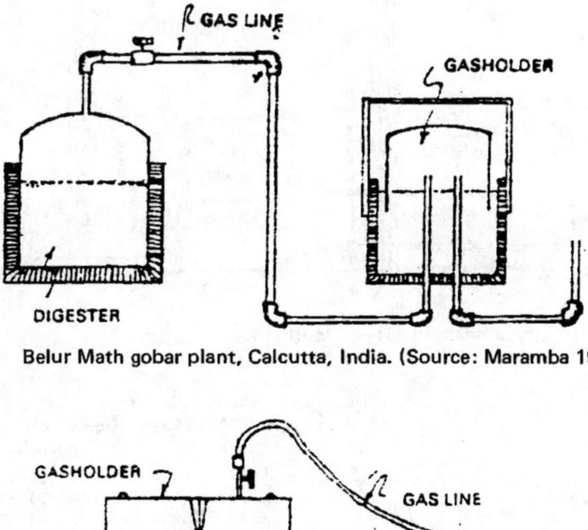

FIGURE 32 Belur Math gobar plant, Calcutta, India. (Source: Maramba 1979)

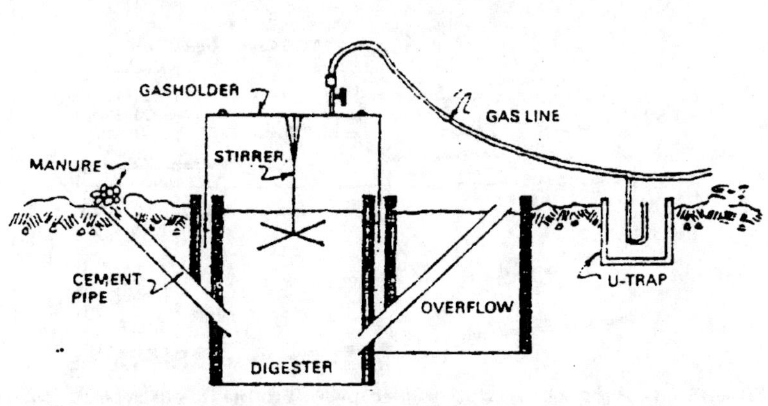

FIGURE 33 Taiwanese single-stage biogas plant. (Source: Maramba 1979)

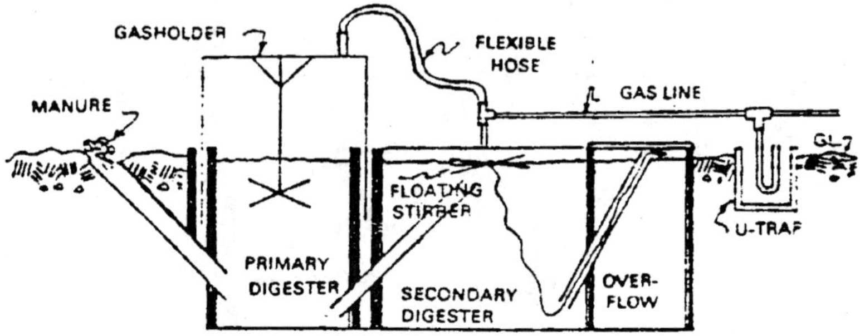

FIGURE 34 Taiwanese two-stage methane generator. (Source: Maramba 1979)

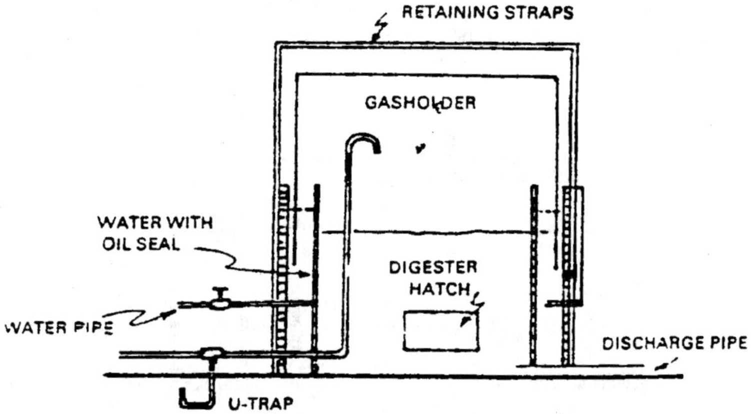

FIGURE 35 Kenyan batch digester. (Source: Maramba 1979)

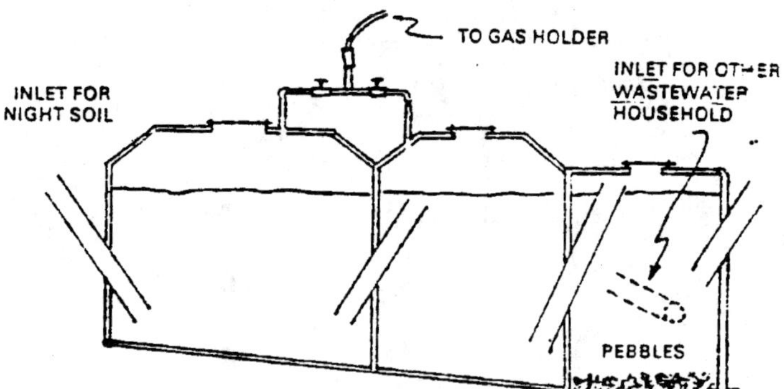

FIGURE 36 Night soil digester, designed by D. Bautista, the Philippines. (Source: Maramba 1979)

McConville and Maier 1978; Ghosh and Klass 1980; Bullock, *et al.* 1980; Novak-and Ramesh 1976; Gottschall, *et al.* 1978). Refinement of the additives as well as treatments that can improve digestion performance and careful economic analysis should facilitate selecting practical methods for commercial use.

Applications in Developing Countries

Using known technology to produce biogas to satisfy a significant portion of human needs has been approached differently in various countries and by various organizations (see Skrinde 1978).

In 1974, the Economic and Social Commission for Asia and the Pacific (ESCAP), a United Nations organization, adopted the Colombo Declaration which stated that energy was one of its most urgent priorities for action. Subsequently, ESCAP undertook a program designed to accelerate the development of the biogas industry throughout Asia.

India is carrying out an extensive program of biogas research and plant construction. During the past 30 years, such institutions as the Indian Agricultural Research Institute, New Delhi; the Khadi and Village Industries Commission (KVIC), Bombay; the Gobar Gas Research Station, Ajitmal; and the Planning Research and Action Institute, Lucknow, have accumulated useful information on the anaerobic fermentation of cow manure (known as "gobar" in Hindi). The KVIC began playing a key role in the extension of biogas plants throughout India in the 1960s, and by 1978 the number had reached 50,000, with an ultimate target of 100,000 units (see Sathianathan 1975; Subramanian 1977).

Standard biogas plant designs utilizing concrete tanks, concrete inlet and outlet basins, and steel covers serving as floating gas holders have been developed and published by KVIC (1974). These digesters have no pumps, motors, mixing devices, or other moving parts, and digestion takes place at ambient temperature. Detention time by volume is generally about 30 days. The digesters contain a baffle in the center to avoid mixing new material with digested sludge. These are primarily single-family units.

A number of large-scale biogas plants are also operating in India. The Bharatya Agro Industries Foundation (BAIF) has a 12,000-ft^3/day (340-m^3/-day) plant operating at the village of Uruli Kanchan (near Poona), using the dung from a cattle-breeding operation (approximately 600 cows). The gas produced is used for cooking, lighting, generating electricity, and pumping water.

Under the direction of the Gobar Gas Research Station in Ajitmal, a community biogas plant has been installed and is operating successfully in the village of Fatah–Singh–Kapurva (Uttar Pradesh). This plant, operating on dung supplied by the villages, provides cooking gas, village lighting (electric), water pumping, and grain milling.

Several other community biogas plants are being installed throughout India, including one designed by A. K. N. Reddy and the ASTRA (Application of Science and Technology to Rural Areas) group at the Indian Institute of Science, Bangalore. This plant will supply cooking gas, only during cooking hours, to all 60 houses in a village. The gas will also be used to pump drinking water for the village, to provide three hours of electric lighting in each house, and to mill rice husk and lime for cement. The proceeds of the sale of the cement will pay the wages of the five people needed to operate the system. The residue (fertilizer/soil conditioner) will be distributed to farmers on a pro rata basis, determined by the amount of dung contributed.

In view of the economic advantages inherent in large-scale community biogas systems (National Research Council 1977), the progress of these systems in India will be of great interest.

In the Republic of Korea, considerable research and development work has been carried out under the direction of the Organization of Rural Development (ORD), and 25,000 digesters have been installed. The severe Korean winter and the fact that Korea does not have a cattle-based economy have influenced the biogas program. From 1969 to 1976, ORD conducted a major campaign to encourage the construction of biogas plants that digest both animal and human wastes. The Institute of Agricultural Engineering and Utilization and the Rural Guidance Bureau are in charge of development and application, respectively.

The Korean government was subsidizing 33–50 percent of the cost of new biogas plants, but this subsidy has been discontinued, and the number of new units being constructed has declined. Only 4,000 new units were constructed in 1975, and in 1976, ORD discontinued the installation of demonstration family digesters. Shortage of animal wastes, difficulty in collecting wastes, and the rapid urbanization of rural areas have resulted in discontinuing the use of a number of small digesters.

In Pakistan, nearly 100 biogas plants have been built, and there is growing interest in the biogas industry. The government has been supplying free gasholders, provided that farmers build their own digesters. Extension of the program has been hampered by low temperatures from November to March and the high cost of imported steel for fabrication of gasholders. The Appropriate Technology Development Organization of Pakistan has recently been engaged in the design and construction of 10 m^3-capacity (353 ft^3), fixed-gasholder digesters, based on Chinese technology. One such unit was constructed near Karachi in 1976 at a cost of $590.

In the Philippines there is a growing interest in biogas systems, arising essentially from their use for pollution control and their public health benefits. Some biogas units have been installed through a program of the National Institute of Science and Technology (Maramba 1979). At Maya Farms, where an integrated agroindustrial complex of 7,500 hogs* has been established, 48 digesters of 22 m^3 (777 ft^3) each are operating on a batch cycle of 30 days. Stirring is carried out for two minutes each day, and digested sludge from finished batches is used as a starter. The gas produced is used within the farm-processing complex as a source of fuel for canning, rendering, cooking, generating electricity, deep-well pumping, and other purposes. Digesters using hog wastes are also located at the Daughters of Charity Mission in Rizal; the University of the Philippines, Los Baños; and the University of the Philippines, Quezon City.†

Biogas utilization in Thailand started on a small scale 5 years ago, and now there are 225 units in the country. Biogas plants have been promoted by the Agricultural Economics Department, the Sanitation Division of the Health Department, the Applied Scientific Research Corporation, and the Animal Husbandry Department of Kasetsart University, with subsidies from the government for the gasholders. Plans call for building 50 digesters every year, but with no additional subsidy from the government. The present emphasis of biogas plants in Thailand is on waste treatment rather than on energy production.

Approximately 12 biogas plants are operating throughout Indonesia, with numerous experimental units being evaluated to reduce construction costs.

Seven million plants have reportedly been built in the People's Republic of China, and emphasis has been placed on a design that destroys bacteria and parasitic eggs in human and animal wastes so that the digested slurry can be safely applied to crops (McGarry and Stainforth 1978).

Integrated Farming System (IFS)

In an integrated farming system (Figure 37), the anaerobic digestion of biomass provides pollution control, biogas production, nutrient and fertilizer effluents, and feed supplements for rural areas. Optimization of the IFS to provide fuel, food, and fertilizer for rural areas in developing countries is thought to be accomplished most easily as follows. Raw wastes are fed on a continuous basis (twice a day) to a digester with a specified volume of water

*The operation is now undergoing expansion to handle ultimately 18,000 animals.

†One of the more interesting aspects of the Maya Farms approach is the integrated recycling of the digested sludge and the liquid effluent from the digesters. The sludge is dried in open "ponds" and the dried material is mixed with pig feed at a rate of 10 percent. Aside from the saving in feed this represents, farm managers have found that vitamin B_{12} and other unidentified growth factors in the sludge accelerate growth rates by 15 days. The total saving in feed repaid the cost of the biogas installation in 2 years.

Liquid effluent from the digesters, together with pig sty washings, are sent to a pond where they supply nutrients for the growth of algae used as food by a colony of ducks. The droppings of the ducks in turn provide food for tilapia that are grown in the pond. The fish are sold as food to farm workers and others in the area, and the ducks are slaughtered and canned, providing the operation with additional income.

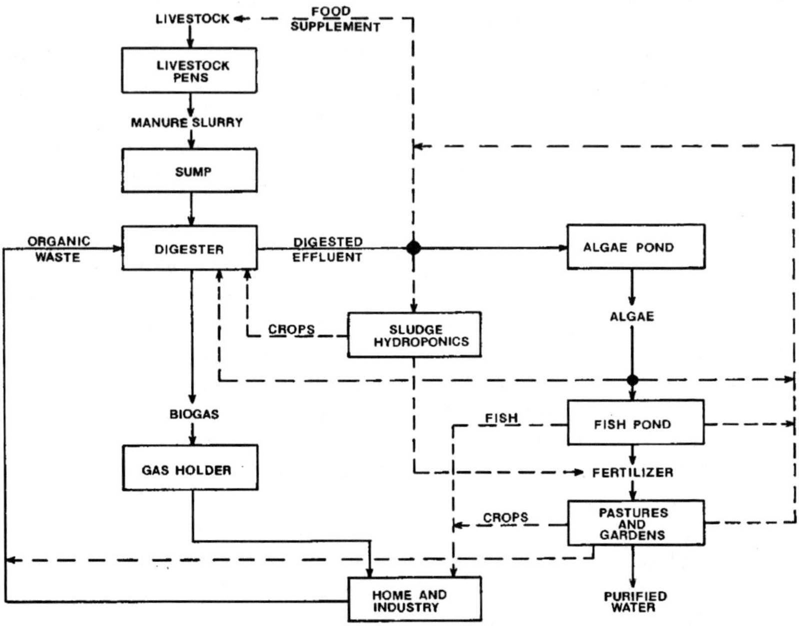

FIGURE 37 Integrated farming system.

for proper dilution. If the digester is covered with a greenhouse structure or other material, digester temperatures can be maintained at a particular level to assure maximum gas production for the type of waste used and surrounding environmental conditions.

Effluent (or sludge) from the digester is then sent to algae basins which are only a few inches deep in order to promote maximum algal growth and water stabilization. Some of the algae may be harvested to serve as a protein supplement in animal feed, or may be returned to the digester as raw feedstock. Effluent from the algae basin or pond then flows into deep fish ponds where fish and ducks can feed on plankton in the pond.* Mineralized nutrients in the pond waters provide a favorable aquatic habitat for the growth and eventual proliferation of a natural, biological aquatic ecosystem. The effluent from the fish ponds is then used in vegetable gardens and on crops as irrigation water containing sufficient nutrients to promote maximum growth of food, animal feed, and fiber. If the system is designed properly, it will

*Certain types of fish will also feed on the algae.

work by gravity flow, so that loading the digester with the raw waste slurry is the only required task. If animal wastes are to be used, this task could be conveniently accomplished by installing slatted floors in animal pens or barns to permit the wastes to be dropped directly into a digestion mixing chamber.

An integrated approach to waste recycling such as the one mentioned above can be economically viable, particularly if the main objective is self-reliance at the village level. Case studies of the application of such technology in rural areas of India, Korea, and the Philippines (as described in the previous section) have produced favorable economic evaluations of the process, based solely upon direct economic benefits of the system's end-products. In India and Korea, the anaerobic digestion process has produced methane with a higher economic value (resulting from the conservation of other domestic energy sources such as rice straw, wood, coal, or oil) than the cost of its production, while at the same time, a safer, more efficient fertilizer is produced than otherwise would be available. In the Philippines, the integrated farming system using sludge recovery and processing to provide pollution control, production of animal feed supplements from sludge solids, and liquid effluent to grow algae, fish, and aquatic vegetation for food and energy, has produced direct benefits well in excess of production costs (Maramba 1979).

ENERGY FARMS UTILIZING AQUATIC PLANTS AND WASTEWATER

In the last few years, the concept of using vascular aquatic plants for biological waste treatment and nutrient reclamation systems has become very popular in the United States and other industrialized countries. Highly productive vascular plants (such as the water hyacinth, *Eichhornia crassipes*), instead of algae, growing in wastewater ponds, are used to absorb the nutrients (and pollutants) and produce biomass (organic material) which can be harvested, processed, and returned to the environment in the form of fertilizer, feed, or energy. In a sense, this form of waste treatment can be considered an "energy farm" when plants are maintained to achieve maximum productivity, harvested, and then anaerobically digested to produce biogas.

The recognized techniques of producing biogas from the harvested plant material have not changed much in the last few years. However, a single large-scale system that encompasses both waste treatment/nutrient reclamation using vascular aquatic plants and energy production from the harvested plant material has yet to be devised. Vascular aquatic waste treatment systems are located at Coral Ridge Development in Coral Springs, Florida; the National Space Technology Laboratories (NSTL) of the U.S. National Aeronautics and Space Administration (NASA) near Bay St. Louis, Mississippi; Rio Hondo, Texas; and Disney World, Florida. A demonstration project at San Diego, California, is planned, and the waste treatment part of this system was due to be installed by late 1980. Energy production as an integral part of several of these systems is being considered, but a definite time schedule for this aspect of these projects has not yet been adopted.

According to a publication of the National Academy of Sciences, *Making Aquatic Weeds Useful: Some Perspectives for Developing Countries* (National Research Council 1976), "The recovery of fuel from aquatic weeds, even if on a small scale, has interesting implications, especially for rural areas in developing countries. As many developing nations have an apparently inexhaustible supply of aquatic weeds within their borders, this potential energy source deserves further research and testing."

Due to widespread interest in the United States, this new waste treatment method should be well developed and perfected in the next 5 years. Designing and testing an optimum harvesting system is still the biggest single problem; however, this may not be a problem in developing countries where small, individual systems may be more practical and hand harvesting is a feasible alternative.

Vascular Aquatic Plants

A variety of aquatic plants has been studied for use in aquaculture waste treatment systems, and thus, potentially, energy farms. The water hyacinth has received most of the research attention in tropical and subtropical climate zones (Figure 38), and a large research program in this field has been sponsored by NASA (Wolverton and McDonald 1976, 1979a, 1979b; McDonald

FIGURE 38 Water hyacinths (*Eichhornia crassipes*) growing on a domestic sewage lagoon.

and Wolverton 1980). Other research projects have been conducted by Dinges (1978) and Cornwell, *et al.* (1977). Emergent aquatic plants such as the bulrush (*Schoenoplectus lacustris*) and submerged plants such as elodea (*Elodea canadensis*) have been studied by Seidel (1976) and McNabb (1976), respectively, for wastewater reclamation in northern climates. Since most developing countries lie within tropical and subtropical climate zones, further discussions will be confined to aquatic species that can be used in these areas.

Aquaculture waste treatment systems used in conjunction with energy farms should be designed to maximize biomass production; consequently, nutrient removal would also be maximized. To date, several designs for a total system have been suggested, but each has its advantages and disadvantages. Total daily wastewater volume, nutrient and organic concentration in the influent waters, climate, availability of land, mode of harvesting, and degree of purification and nutrient removal are some of the more important variables that must be carefully considered in choosing a suitable design.

Wolverton (1979) recently published design characteristics for small wastewater treatment systems applicable to U.S. needs. A more comprehensive engineering handbook compiled under contract with NASA is now available (Wolverton and McDonald 1980). Other pertinent references on aquaculture and wastewater treatment are listed in Appendix C.

Following is a brief description of some of the system management problems that have to be considered for this method of treatment regardless of whether it is carried on in a developed or developing country.

Harvesting. An aquaculture system designed for maximum biomass production will generally need harvesting every 3-4 weeks to maintain optimal growth conditions. The need for mechanical harvesting depends on the size of the system and the availability of inexpensive labor for hand harvesting as a substitute. Mechanical harvesting would not be economically practical for small aquaculture systems servicing only a few families.

A variety of harvesters has been built to harvest aquatic plants, and most were designed to remove water hyacinths (see National Research Council 1976; Wolverton and McDonald 1979c). Two of these harvesters are shown in Figures 39 and 40. Water hyacinth harvesters can generally remove anywhere from 9 to 90 tons of plants per hour depending on plant density, lagoon configuration, and operator skill in using the harvester.

Prevention of Spreading of Aquatic Weeds to Unwanted Areas. If water hyacinths or any other aquatic plants used for waste treatment are already established in the waters receiving the discharge, no special care is needed to prevent the plants from spreading. However, if this is not the case, screen barriers at the discharge point should be constructed to prevent uncontrolled growth in areas where plants are not wanted. The levees or retaining banks of

FIGURE 39 Water hyacinth harvester consisting of a rotary pick-up head, chopper, and conveyor belt.

lagoons should be sufficiently high to prevent any flood waters from sweeping plants into uncontaminated areas.

Odor Control. Aquatic plant wastewater treatment systems receiving light to medium daily organic loads (generally expressed in units of kg BOD_5/ha/day) present no odor problems. The BOD_5, or 5-day biological oxygen demand, measures the amount of oxygen required by aerobic bacteria to stabilize the organic matter in 5 days. If sufficient oxygen is not supplied to the waters to keep the system aerobic, or at least the upper portion aerobic, then anaerobic bacteria will also be present. These bacteria cause gaseous chemicals such as hydrogen sulfide to be liberated, which can create the foul odors normally associated with overloaded wastewater treatment systems.

A system receiving approximately 50 kg BOD_5/ha/day (44.5 lb BOD_5/a/day) or less will fall into the light- to medium-loaded range. In general, if this system is covered with aquatic weeds and harvested when needed, it will not emit odors. A system receiving a heavier load and covered with aquatic plants will emit odors at night when the plants are photosynthetically inactive. This problem can be corrected by either adding supplemental mechanical aeration at night or keeping the plants thinned out enough to allow some algal growth and natural aeration from wind action. A recent study of a duckweed-covered system (*Spirodela, Lemna*, and *Wolffia*) proved that this type of system receiving a heavy organic load could be odor-free (Wolverton 1979). The duckweed was so thick that it acted as a natural sealant and prevented the hydrogen sulfide from being emitted from the lagoon and causing foul odors.

FIGURE 40 Conveyor harvester receiving plants from a pusher boat in foreground.

Mosquito and Insect Control. Aquaculture systems that are not kept totally anaerobic may be stocked with mosquito fish (*Gambusia affinis*). These small fish feed on the mosquito larvae and serve as effective, natural "larvicides."

Water hyacinths and other aquatic weeds may occasionally be damaged by insects, especially the spider mite (*Bryobia praetiosa*). If damage is severe, a light application of an insecticide such as Malathion may be required.

Nutritive Value of Sewage-Grown Aquatic Plants

The nutritive value of aquatic plants generally reflects the fertility of the waters in which they are grown. This phenomenon can be easily seen from the difference in the two plants shown in Figure 41. In one study by Wolverton and McDonald (1978) in which water hyacinths were grown in four wastewater treatment systems receiving nutrient loads from very light to very heavy, the crude protein in the whole plants varied from 9.7 percent to 23.4 percent of dry weight, and phosphorus content ranged from 0.31 percent to 0.89 percent of dry weight. Tables 3 and 4 depict the moisture content of vascular plants—water hyacinth (*Eichhornia crassipes*), duckweed (*Spirodela, Lemna*, and *Wolffia*), and water pennywort (*Hydrocotyle ranunculoides*)—

FIGURE 41 Visual comparison of plant differences when grown on a nutrient-rich medium such as a domestic sewage lagoon and on an unpolluted lake. The water hyacinths were approximately the same age and collected on the same day.

TABLE 3 Proximate composition (dry weight basis).

Plant	Crude protein (%)	Fat (%)	Crude fiber (%)	Ash (%)	Total carbohydrates (%)
Water hyacinth	23.4	2.20	17.1	20.4	36.9
Water pennywort	22.4	2.19	11.8	17.4	58.0
Duckweed	37.1	3.40	15.6	12.5	47.0

TABLE 4 Nutrient and mineral composition (dry weight basis).

Plant	Nitrogen (%)	Phosphorus (%)	Potassium (%)	Sodium (%)	Calcium (%)	Copper (ppm)	Zinc (ppm)	Iron (ppm)	Sulfur (%)
Water hyacinth	3.74	0.85	3.45	1.63	0.56	42.0	48.0	2260	0.40
Water pennywort	3.58	0.42	3.33	1.34	0.77	11.5	59.4	7490	0.31
Duckweed	5.94	1.01	2.13	0.74	0.88	14.1	1885.0	1450	0.85

that have been grown in highly enriched, domestic sewage lagoons. These plants, like many other aquatic species, would be excellent sources of protein and minerals. The sludge remaining after biodigestion would contain all of the minerals and most of the nitrogen and sulfur present originally in the plants. This sludge, which is completely stabilized, could be air-dried and used as a natural fertilizer for crops or even for mineral recycling in a "perpetual" biomass system for energy production.

Producing Biogas from Aquatic Plants

The simple biodigester designs currently available are particularly suited for digesting animal manure. Extensive use of this method on an individual home basis is found in India. However, even the simplest of biodigesters requires some daily attention and knowledge of biodigestion in order to generate a sufficient volume of methane safely and reliably to make the effort worthwhile. Simple batch digesters require the longest digestion times. (See the preceding section on anaerobic digestion for a more thorough discussion of energy from biogas and the NAS publication, *Methane Generation from Human, Animal and Agricultural Wastes* [National Research Council 1977].)

If fresh plant material is used, the plants should be chopped, crushed, or blended in some manner prior to loading in a digester in order to shorten the digestion time by increasing the surface exposure for bacterial attack. Choppers that can be adapted to this use are commercially available for grinding a variety of agricultural products (see National Research Council 1976). However, the initial investment, operation, and maintenance costs of these machines could only be justified for large installations.

A new approach to anaerobic digestion is currently under development by NASA's National Space Technology Laboratories, incorporating an "anaerobic filter" into the digestion system (Wolverton and McDonald 1979b). As depicted in Figure 42, the first chamber holds all the pulp and plant juice. The anaerobic filter in this case consists of small rocks (pea gravel) which provide a large surface area for bacterial growth. The plant juice is periodically circulated through the anaerobic filter and exposed to a mixed culture of the acid-forming and methanogenic bacteria. Preliminary results with this technique indicate that complete digestion of fresh plant material such as water hyacinth, duckweed, and kudzu (*Pueraria lobata*, a prolific terrestrial plant) can be accomplished in an average of 23 days with no loss in total methane production over the old batch or continuous-feed methods.

ETHANOL (ALCOHOL) PRODUCTION

In the years since publication of *Energy for Rural Development*, a growing body of information and practical experience has been gathered on ethanol production in the United States and elsewhere. Currently, a small but rapidly growing number of plants are producing ethanol on a daily basis, making available, probably for the first time, hard data on the costs of and energy balances for producing 150-200-proof alcohol (75-100 percent alcohol, 25-0 percent water by volume). This experience also provides information on the training necessary to operate such plants successfully.

Because of the greater availability of information (rather than the discovery of any new knowledge), many concerns that have previously been the subject of much disagreement can now be scrutinized in the light of experience.

One concern about ethanol production is whether the amount of energy obtained from alcohol sufficiently exceeds that used in the planting, growing, harvesting, transportation, crushing, heating, and distillation needed to produce it. Unfortunately, the response to this concern is not at all clear-cut and depends a great deal on the individual situation.

Many things can be done, however, to improve this energy balance. By locating a distillery near its feedstock source, for example, transportation energy can be saved. Heat for cooking (only required for starches used as feedstock, not sugars) can be provided from wood, wheat straw, or other waste cellulose and lignin materials. Solar energy can also be used for cooking and for distillation, and a simple heat recovery system can be employed after fermentation (to preheat the liquid or so-called "beer"), using heat from the condensed stillage. Besides improving the energy balance, such techniques mean that low-grade (and perhaps otherwise unusable) energy is used to produce high-grade energy.

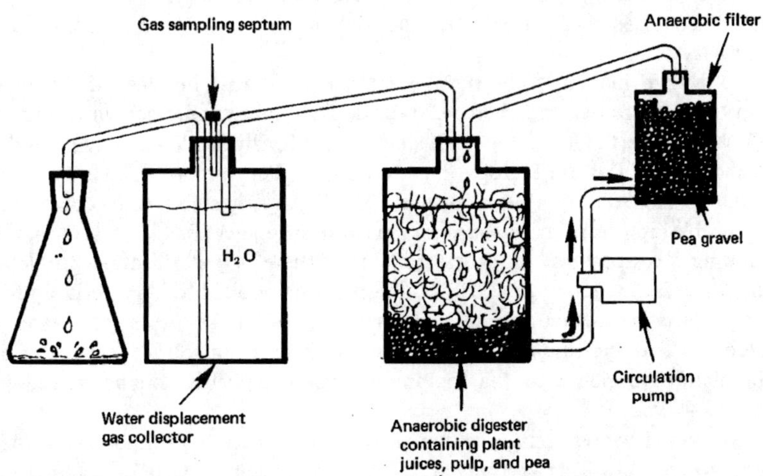

FIGURE 42 Two-stage anaerobic digester system for producing methane.

Another concern about alcohol production is that fermentation of starch, sucrose, or glucose consumes feedstock materials that could serve as animal fodder or human food. In a world facing chronic food shortages, some individuals question the wisdom of adding still another consumer to the food chain. Thus each developing (and developed) country will have to study its basic needs and resources carefully, and then choose the energy paths that best serve those needs. Alcohol production may fit well in some countries and not at all in others.

Operation and Production Rates of a Typical Fuel Alcohol Production Plant

Although alcohol can be produced by individuals with the crudest of equipment, alcohol production can probably be carried out most efficiently in developing countries at the local village or farm cooperative level. The number and size of alcohol production plants would then be balanced against the problems of transporting the feedstock.

This section will describe briefly the operation of an alcohol plant (batch process) that makes about 1,250 gal (4,731 l) of 190-proof alcohol per day (see Figure 43). This is not a general description of all plants, as plants vary according to available feedstocks, process energy sources, enzyme sources, etc. Such a plant will require about 15 t (13.6 metric tons) of dry grain (such as corn, wheat, barley, etc.), although anything with a considerable amount of sugar or starch is a possibility. Perhaps up to half of the grain can be salvaged for other uses in some cases. If the input feedstock is a root vegetable or a sugar source, the amount of material needed is determined by the starch or sugar, cellulose, and water content. Before an alcohol plant is constructed, indigenous feedstocks should be analyzed in order to forecast alcohol productivity.

As shown in Figure 43, the input stock must first be cleaned by some appropriate process and then pulverized. The grain grinder shown in Figure 44 can handle the 15 t of grain a day needed for the 1,250-gal/day plant described here. For root crops it is desirable, where practical, to remove as much of the cellulose fiber as possible, leaving mostly starch to be converted.

If the input is primarily sugar (as in sugar cane juice, fruit juices, etc.), the cooking step may not be necessary, unless there is a sterilization problem. However, input that is primarily starch must be cooked to help break up the long molecules so they can be more easily fermented. In such a case, water is added so that the resulting mixture is about 20 percent solids by weight. For the plant described here, the cooking vat would probably hold about 12,000 gal (45,423 l).

The starch/water mixture is then held at 212°F (100°C) for about one-half hour to help break down the starch and to kill bacteria. Often a commercial

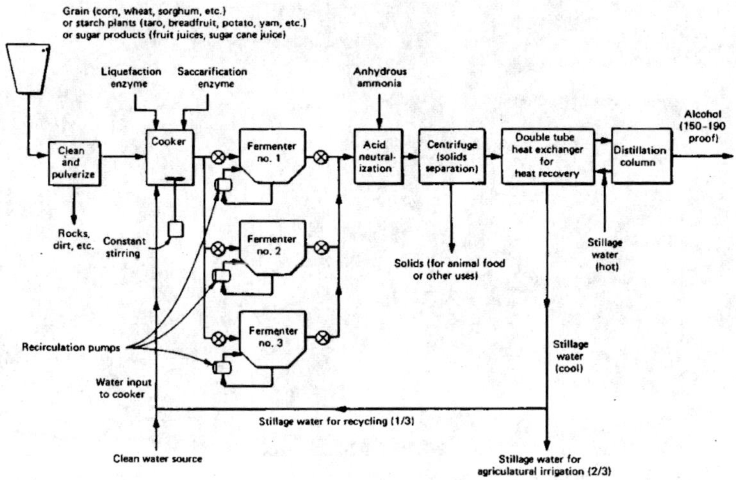

FIGURE 43 Flow diagram of a batch alcohol plant.

liquefaction enzyme is added at this stage to help break down the long starch chains. The cooking process is more effective if the mixture is stirred continuously.

The heat for cooking (and later for distillation) is often provided by a steam boiler, which can be fired by a variety of waste agricultural materials as well as by oil or natural gas. Figure 45 shows a steam boiler that uses 7 rolls of wheat straw (about 7,720 lb or 3,500 kg) per day to produce 1,250 gal of alcohol per day, all cooking and distilling energy included.

After cooking for one-half hour, the mixture is cooled to 170°F (77°C) and more liquefaction enzyme is added. The mixture is then cooled to 120°F (49°C) and a saccharification enzyme is added to convert the starch to sugar. The batch is then cooled further to 90°–95°F (32°–35°C), when it is pumped into one of several fermentation tanks. A proper yeast is then added.

It is essential that the fermentation tank be extremely clean before the batch to be fermented is pumped in, in order to destroy competing microorganisms. Scrubbing a tank down with soda ash between batches and keeping the tank completely full during fermentation will generally restrict the production of vinegar. For safety's sake, any person scrubbing down a tank must wear a sealed air mask with a fresh air supply. Because it is heavier than air, the carbon dioxide produced by the fermentation tends to stay in the tank, which could lead to asphyxiation.

Fermentation takes two or three days, depending on temperature. A small, continuously operating pump could serve both to slowly remove the heat

FIGURE 44 Grain grinder for 1,250-gal/day plant.

generated during fermentation and to stir the mixture. About three tanks are needed if the distillery is to run continuously.

After fermentation stops (determined by the fact that no more carbon dioxide is being produced), the so-called "beer" is ready for distillation. The input mixture ratio (20 percent starch by weight in this case) should have been adjusted so that all of the sugar was used up before the alcohol concentration became high enough to kill the yeast. Since the beer is slightly acidic at this point, it is desirable to add a small amount of anhydrous ammonia or some other appropriate base. This allows all subsequent plumbing to be made of relatively inexpensive carbon steel rather than expensive stainless steel. (Because the only high temperatures are found in the cooker, inexpensive plastic pipe can be used throughout the system until the heat recovery system just before the distillation column is reached.)

The solids must be removed before distillation, which is most easily done with a centrifuge. These solids (from grain feedstock) have significant nutritional value as an animal feed supplement and thus will not be wasted. Generally, any economic analysis of alcohol production costs should include a credit for these by-products of fermentation.

The liquid is next passed through a double steel wall heat exchanger, heated by the just-condensed water returning from the distillation column. This heat recovery step significantly reduces the input fuel required for distillation.

The kind of distillation column (and distillation energy) required is determined in part by the concentration of alcohol needed. The management of the distillation column temperature is critical. At the bottom, the input temperature is 212°F (100°C) and at the top 173°F (78.5°C). If only fuel is desired, and not drinking alcohol, one distillation will suffice (even if alcohol as pure as 190 proof is required), although generally two separate columns will be used. Figure 46 shows a distillation column for the kind of plant just described, which is really two columns in tandem. Although a variety of distillation columns can be used, a large condensing surface is required. Alcohol of 100 proof, or possibly higher, has been obtained with a simple distillation column consisting merely of enclosed tubes packed with rocks.

If the alcohol produced is used as fuel directly (without mixing it with gasoline), 150 proof is adequate, and considerably less distillation energy is required per gallon (or equivalent Btu) output.

As shown in Figure 43, some of the plant's stillage is used as make-up water and the remainder as irrigation water. This takes advantage of the nutrients as well as the water. Research is being done in several countries (e.g., the United States and the Philippines) to determine the possibility of using stillage as a human food supplement.

Many variations are possible in the above process. Alcohol production has become very active in the last 2 or 3 years, and will undoubtedly remain so for the intermediate future.

Adapting Alcohol Processes to Developing Countries

Much can be done to simplify the process just described in order to adapt it (or similar processes) to situations in developing countries. Although the special materials required for this process (e.g., liquefaction and saccharification

FIGURE 45 Steam boiler designed to operate from rolls of wheat straw (in background).

FIGURE 46 Distillation column (tall column in background) for 1,250-gal/day plant.

enzymes, anhydrous ammonia, yeast, caustic soda) may have to be imported initially by some countries, local capabilities for producing these materials could eventually be developed where they do not yet exist.

Concerning the necessary equipment, a sophisticated steam heating system for the cooker and distillation column may not be required if indigenous fuels or perhaps even solar energy can be utilized. The cooking vessel itself could be ceramic. Effective distillation columns could be made from rocks, although a second column with steel condensation plates may be required to achieve the desired 150-proof minimum. When cheap indigenous fuels are available, there is little need initially for a double tube heat exchanger for heat recovery. It would be one of the more expensive and difficult items to obtain or fabricate locally in some countries. Fermentation tanks can be made of hardwood as they have been for centuries. The type of wood required would be less critical than that needed for producing drinking alcohol. Finally, a centrifuge is only one means of removing solids. Settling tanks can also often be used for nonsuspended solids, with simple filter presses used to improve liquid recovery. In some cases, the fermented mixture can be simply heated directly (with the alcohol and some water being driven off) and the high-nutrient residue can then be cleaned out later.

In short, alcohol production plants of reasonable size can be built and operated as a village or farmers' cooperative project. Although such plants may not be as efficient in their use of labor, these considerations can often be secondary to the problems of providing employment and producing fuel locally.

Use of Ethanol as a Substitute Fuel for Internal Combustion Engines

Ethanol can be used as a fuel for internal combustion engines, and direct or gas-fired turbines can also use alcohol as a fuel with few problems.

The use of fuels consisting partially or totally of alcohol to run the spark ignition-type engine has received much attention in recent years, particularly in Brazil and the United States. The practice of mixing ethanol (of at least 190 proof) with gasoline has now become widespread in both countries. Mixtures of up to 20 percent alcohol and gasoline have been used. In the United States, the current standard mixture (so-called gasohol) is 10 percent ethanol, and experience indicates that most automobiles can run on this mixture, without carburetor adjustment, provided the alcohol remains well above 190 proof. At lower proofs, water separation can occur in certain cold weather situations, which sometimes results in difficult starting and rough-running engines.

Careful dynamometer tests run by several laboratories, including John Deere and Company (USA), indicate that with careful tuning, spark ignition engines running on gasohol can be operated with *slightly higher* fuel efficiencies than when running on gasoline alone. Thus, from a technical point of view, a 10 percent mixture of alcohol and gasoline is quite feasible. There has not yet been enough long-term experience to indicate whether problems with corrosion or undue engine wear will occur, but no such problems appear to be developing with any regularity. A few automobiles have required minor carburetor adjustments to achieve satisfactory performance.

When the spark ignition engine is fueled by ethanol alone, alcohol as low as 150 proof can be used and still allow satisfactory engine operation. However, the carburetion must be adjusted, including changing the jets, and certain types of rubber gasket material that may be damaged should be replaced. A technique of simply vaporizing the alcohol with heat from the exhaust manifold and feeding the vapors into the intake manifold has met with considerable success in several places. Thus the conventional carburetor is eliminated altogether.

An engine operating on ethanol alone can be difficult to start, particularly in cold weather, and this problem worsens as the proof of the fuel decreases.

It is often overcome by starting the engine (in cold weather) on gasoline, warming it up, and then switching to the alcohol fuel. (Some operators prefer to switch back to gasoline before shutting it off.) Another possibility is ether injection upon starting, although this must be done with considerable care or the engine can be severely damaged.

Significant attention has also been given to operating diesel engines partly on ethanol. Ordinary diesel fuel has significant lubricating as well as combustion properties, and diesel engines depend upon the former to obtain injection pump lubrication. Therefore, it is not practical to use plain alcohol, which has no lubricating capability. However, alcohol can be mixed with the diesel fuel before it passes through the fuel injection pump. Tests at the John Deere Laboratories have indicated that a diesel engine will run quite satisfactorily with mixtures of up to 30 percent alcohol and 70 percent diesel fuel without obvious adverse effects on the engine. Dynamometer tests indicated that engine efficiency remained essentially unchanged from its efficiency on diesel fuel alone.

The practice of directly mixing fuels for diesels is, however, potentially more damaging than would be the case in spark ignition engines. The alcohol proof must be kept extremely high (perhaps 195 or better) because of the injection pump problem. If there is any water separation, as could occur easily in cool or cold weather, the lubricating capability of the fuel mixture can break down, thus destroying the costly injection pump. However, if diesel fuel is injected by the fuel injection pump in the normal way, and an alcohol spray is injected into the air intake, all lubrication requirements are met, the fuel injection pump is protected, less diesel fuel is required, and the alcohol can be as low as 100 proof and still achieve adequate operation. The higher alcohol proofs (around 150 or more) probably give more efficient operation. With 150 proof or better, engines have been run on mixtures as high as 85 percent alcohol and 15 percent diesel fuel. It is desirable to start the engine on pure diesel fuel and warm it up before beginning alcohol injection, however.

For turbocharged diesel engines, work has been done on alcohol/water injection both before and after the turbocharger. In the former case, the input air draft is adequate, but fuel metering is difficult. If there are droplets of any size in the spray, the turbocharger can be severely damaged or destroyed. For injection after the turbocharger directly into the manifold, a separate injector pump will be required. However, it need develop pressures of only two or three atmospheres for a relatively small volume flow (of the alcohol/water spray). This system inherently provides an easy way of metering fuel, and since it cools the air charge before it enters the combustion chambers, the engine may actually be under somewhat less heat stress in its overall operation.

Work will continue on ways to operate engines on alcohol, but most of the problems have been identified. There do not appear to be any technical reasons why this technology could not be applied widely in the near future; however, such widespread application will be determined more by economic and political factors than technical ones.

This subject will be covered more thoroughly in a forthcoming National Academy of Sciences publication, *The Potential for Alcohol Fuels in Developing Countries* (working title).

Chemical or Enzymatic Hydrolysis of Cellulose

Hydrolysis is a process in which cellulose molecules can be reduced to constituent sugars which can then be fermented. Enzymatic and chemical (acid) hydrolysis of cellulose are two lines of current research and development that could have a significant impact on the future of ethanol as a fuel. If hydrolysis by either method becomes practical, then essentially all plant biomass becomes a potential feedstock for alcohol production, and the threat of alcohol production competing with the food chain becomes far less serious.

Cellulosic waste materials are broadly composed of cellulose, which can be hydrolyzed into glucose; hemicellulose, which can be hydrolyzed into its constituent five carbon sugars; and lignin, which bonds the cellulosic materials together. To decompose cellulosic wastes, it is necessary to remove the lignin and break the crystalline structure of the cellulose so that it can be hydrolyzed either chemically or biologically.

Biological decomposition of cellulose has been the subject of research for a number of years by the U.S. Army Natick Research and Development Command (Spano 1976). The Natick laboratory has isolated a fungus, *Trichoderma reesei*, certain strains of which are highly efficient in the saccharification of cellulose into glucose. The problem with this process, however, is efficiently removing the lignin and breaking the crystalline structure of the cellulose so that the *Trichoderma reesei* can perform hydrolysis.

A University of Arkansas research program uses a mutant strain of *Trichoderma reesei* that is grown continuously to produce a complete cellulolytic system. This process features a simultaneous saccharification and fermentation process that results in approximately 60 percent conversion of the cellulosic waste within a 48-hour period, more than twice the conversion obtained by saccharification alone. Reportedly, this process is now commercially feasible and available for licensing.

Chemical hydrolysis of cellulosic waste materials, primarily wood, has been the subject of research and commercial application for more than 100 years. Until recently, very little commercial interest was generated by this technology because of its high cost of producing sugar as compared with agricultural techniques. However, with the advent of a heavy demand for fuel ethanol, this research has been accelerated.

A process developed by Purdue University (Ladisch, *et al.* 1978) involves a hemicellulose extraction of the five carbon sugars, a solvent pretreatment of the cellulose residue, hydrolysis of the treated cellulose into glucose, and subsequent fermentation of the glucose into fuel ethanol. Lignin is the most significant by-product of this process. Reportedly, a commercial plant is being designed for demonstration of this process.

J. W. Jelks, an Oklahoma inventor, has been issued U.S. patents on processes that treat plant organic matter to improve its digestibility by ruminants. Seven such plants have been built and are operating commercially, demonstrating a high degree of conversion of cellulose and hemicellulose into digestible sugars. Jelks is presently adapting his processes to utilization of the sugars for fermentation to fuel ethanol.

THERMOCHEMICAL CONVERSION OF BIOMASS

Selected types of thermochemical conversion technologies appear to be applicable generally to rural development in developing countries (Table 5). However, here the primary emphasis is on direct combustion (burning or incineration) and pyrolysis which are near-term technologies of importance to these countries. The prospects for the introduction of reliable, small-scale gasifiers in developing countries may be realized in the intermediate term (5–10 years).

Direct Combustion Processes*

Direct combustion, undoubtedly the most widespread process for the thermochemical conversion of biomass, is a direct process for production of bioenergy, but in its most commonly used form it is also the least efficient. An estimated nine-tenths of the people in poor countries heat and cook with firewood which has resulted in serious environmental degradation and increasing economic burdens caused by a growing scarcity of firewood (Eckholm 1975). The firewood energy crisis is the most acute today in the countries of the Indian subcontinent and in the semiarid stretches of Central Africa bordering

TABLE 5 Thermochemical processes for bioenergy applicable to rural development.

Process	Principal bioenergy/biofuel products
Direct combustion	Heat, hot water, steam, electric power
Pyrolysis	Combustible char, pyroligneous oil, and low-Btu gas
Gasification	Low- and medium-Btu gas

*Direct combustion processes are also dealt with in a NAS publication, *Firewood Crops: Shrub and Tree Species for Energy Production* (1980a).

the Sahara Desert. Other regions, such as Latin America, also suffer from the same problem. Wood and charcoal scarcity afflicts most of the Andean region, Central America, and the Caribbean.

Most rural areas of developing countries depend upon notoriously inefficient open fires for cooking and heating. For cooking, perhaps 2-5 percent of the energy produced by burning or direct combustion is used effectively while the rest is wasted. Devices or hardware that harness this energy efficiently for both cooking and heating are needed (see VITA/Intermediate Technology Development Group 1980).

Domestic Heating and Cooking (Family Level). An efficient, simple, wood-burning cookstove should not only confine the fire to the cooking area but conserve the heat produced by minimizing heat loss to the surrounding area, such as the walls of the fire container itself.

Walton and co-workers (1978) investigated simple, wood-burning stoves for use at the individual family level in sub-Sahara Africa, and found that much of the equipment that has evolved outside Africa for wood burning is appropriate, with modifications, for application in developing countries. Figure 47 shows a design that incorporates many of the features of the very efficient, ancient Japanese earthenware cooker, the Kamado, but that could be built with inexpensive raw materials. The cookstove consists of five essential components: (1) a 5-gal paint can and lid, with vent holes cut in the lid

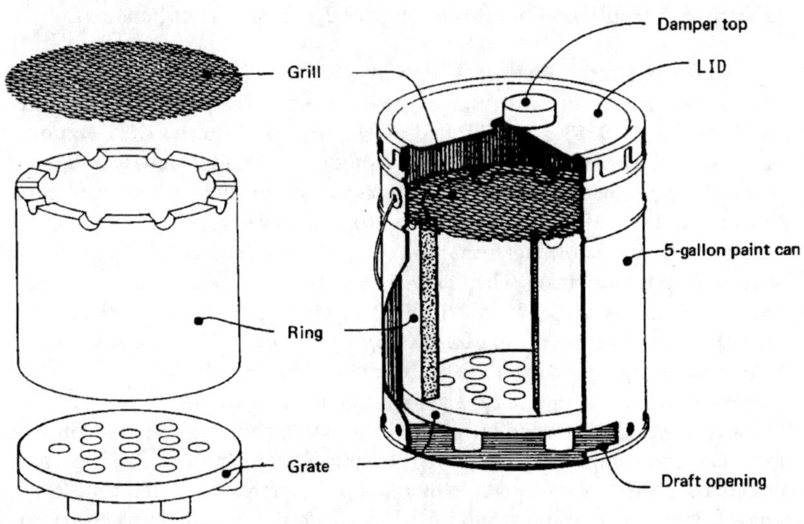

FIGURE 47 Schematic of a wood-burning cook stove constructed from a 5-gal paint can.

and at the base of the can to control draft; (2) two ceramic blocks or stones to support the grate; (3) a ceramic grate to support the fire with holes or air passages to admit air for combustion; (4) a ceramic liner (or shell) to insulate the fire from the can; and (5) a metal grill or other structure to support the food being cooked.

The ceramic components used in this cookstove are easily fabricated, and, as they can be of relatively low strength, they need not be fired in a high-temperature kiln. The 5-gal paint can could be replaced by a simple sheet metal shell which could be rolled and used as an open-ended support structure.

McGill University's Brace Research Institute (1977) has also developed a simple and cheap ($1.50) wood-burning cookstove for family use which is based on a simple, mass-produced galvanized bucket.

Closed wood stoves have been improved in Western countries to a point where efficiencies range from 25 to 60 percent. Stoves can be better controlled, and there is less leakage of air into a stove, more control of air input, and better use of the flue gas heat. By use of air heat extractors, additional heat can be extracted from flue gases above that normally gained through the stove and stovepipe. Increased energy efficiencies of from 3 to 20 percent are reported, which correspond to a reduction of 6–40 percent in the amount of fuel consumed for the same useful heat output (Shelton and Shapiro 1978).

Wood-burning furnaces are gaining increasing attention for use in larger buildings and are equivalent in efficiency to wood stoves. Their application in developing countries at the family level is questionable, although they might be adapted for multi-unit housing or industrial or government buildings.

Centralized Energy Systems. Electricity, of course, can be produced by direct combustion of biomass residues, as can steam. The Forest Products Research Society (1979) has published a series of case histories of operational biomass combustion systems for energy production, and Karchesy and Koch (1979) have reviewed industrial-scale biomass combustion systems and component hardware suitable for application to rural development.

Industrial wood-burning furnaces are commonly incorporated into boiler systems to produce steam. They range in size from shop-fabricated units with steam capacities up to 100,000 lb/hr (45,000 kg/hr) to large field-erected units that can generate up to 600,000 lb/hr of steam (273,000 kg/hr). Wood-fired boilers that produce 15,000–100,000 lb/hr (6,800–45,000 kg/hr) of steam are the most common type among recent installations.

Furnace design for wood burning falls into two basic classes: grate burners and suspension burners. Dutch ovens, inclined grate furnaces, and spreader stokers burn the fuel on a grate, either in a pile or spread as a thin bed. In suspension burning, the fuel is supported by air during its combustion—for example, suspension-fired burners, cyclonic burners, and fluidized bed burners (see Karchesy and Koch 1979).

The overall conversion efficiency for conventional steam generation of electricity from wood is about 25 percent, including boiler efficiency, steam cycle efficiency, and auxiliary power requirements, all of which are variable (Benemann 1978). Thus if hardwood containing 42 percent moisture were used as a fuel (at a conversion efficiency of 25 percent), 1.8 t/day of biomass would be needed to support a 50-kW electrical generating plant, 5.4 t/day for a 150-kW plant, 18 t/day for a 500-kW plant, and 36 t/day for a 1-MW plant (see Karchesy and Koch 1979).

Pyrolysis

Pyrolysis is a thermochemical process in which a material is thermally decomposed into less complex, mainly volatile, organic compounds in an air- (oxygen-) starved environment. A carbonaceous residue (or char), as well as combustible gas and an oily liquid, are normally produced.

Virtually every form of terrestrial biomass is amenable to thermochemical conversion. Pyrolysis has been applied principally to wood in developing countries, but the process is also applicable to a wide variety of other cellulosic materials: rice husks, peanut shells, cotton gin waste, coconut shells, bagasse, etc. (see Beagle 1978).

The most widely recognized pyrolysis process is "destructive distillation," in which organic or carbonaceous materials are heated to several hundred degrees Celsius in the absence of air and the condensable liquids of decomposition are condensed and collected. A solid, nonvolatile residue–charcoal–is produced. Bituminous coal is used to produce coke, coal tar, and coal gas by destructive distillation. Formerly, wood was used extensively to produce "pyroligneous acid" (methanol, acetone, acetic acid, and other chemicals) by destructive distillation. However, because this method of producing these chemicals could not compete economically with fossil-based systems, the destructive distillation of wood is used almost entirely to produce charcoal, a vital fuel in developing countries.

Established Pyrolysis Technologies. Modern pyrolytic processes and actual commercial applications are well advanced in North America (Bio-Energy Council 1978) and Europe and Japan (Jones, *et al.* 1978). These newer technologies contrast significantly with established methods that are more generally applied in developing countries.

Pyrolytic conversion can occur in batch processes, such as those used historically for making charcoal, or in continuous processes. For batch processes, simple earth-covered kilns and pits are generally used in areas lacking capital and materials; however, kilns constructed of brick, concrete, and metal allow better control of yields and eliminate contamination of the product with earth and stones.

There are generally two types of fabricated kilns: fixed or portable. Some typical kilns are the French Nilmelior system, German DRP kilns, beehive kilns, masonry block kilns, Missouri kilns, CUSAB (charcoal-from-useless-shrubs-and-bushes) kilns, and Mark V kilns (Earl 1974). The first four kilns are cheap and easy to construct and are adaptable to mechanized working methods, but they have not proved strong enough to withstand high pyrolysis (carbonization) temperatures. The Missouri kiln incorporates the advantages of the first four systems and overcomes their disadvantages. Its chief advantage is its strength which allows trucks and tractors to enter the kiln for loading and discharging of products. The Mark V is an adaptation of the CUSAB kiln. Both are portable kilns, relatively inexpensive, and practical to use.

The selection of fixed or portable kilns depends greatly on the location of the available biomass supply and the relative cost of labor.

Retorts overcome the disadvantages of kilns, which do not allow the collection of the volatile oils and gases that are the co-products of pyrolysis, and also permit better control of the process which results in higher charcoal yields. However, retorts are more expensive to manufacture, and they require an external source of energy to maintain the process.

Modern, continuous-flow pyrolysis processes, as developed in the United States and Europe, are high-temperature porcesses that tend to maximize gas production, minimize oil production, and reduce solid residues virtually to ash. This emphasis arises from a concern with municipal solid waste disposal and the assumption of the proximate location of a large, gas-fired industrial or utility-type boiler.

A number of continuous, low-temperature (572°-1112°F or 300°-600°C) processes, applicable to the thermochemical conversion of agricultural and forestry products, have been developed. The simplest, most promising systems appear to be based on vertical retorts through which the feedstock passes. In most cases, because the chemical reactions are exothermic, these processes can be self-sustaining: however, most continuous-flow processes require external heat to dry the as-received biomass feedstock and to maintain acceptable production rates. All these processes are basically similar and differ chiefly in the manner in which the auxiliary heating is introduced.

Shaft furnaces have been designed in a variety of forms to allow for flexibility in handling a wide variety of biomass feedstocks. Although designs differ in the source of auxiliary energy, the following steps are generally employed in all such processes: the biomass feedstock is introduced at the top of the retort; char is removed from the bottom of the retort; the hot volatile gases are passed upward through the downward-moving bed; and the gases are then removed at the top of the retort, and scrubbed, condensed, and collected as liquid fractions (pyroligneous acid or pyrolytic oils). Frequently, the biomass feedstock is dried externally, rather than in the reactor bed itself, by using the noncondensable gas fraction resulting from pyrolysis.

Some examples of this continuous, vertical retort-based conversion system include the Stafford, Mellman (Othmer 1952), Lambiotte, and DeBartolomeis retort designs, and the Barneby and Georgia Tech processes (Knight 1976).

New Pyrolysis Technologies. Pyrolysis of biomass residues is normally accomplished under starved, but controlled, air oxidation conditions. Recently, however, new pyrolysis technologies have resulted from the realization that higher energy efficiencies are obtainable by means of direct combustion or incineration processes. Figure 48 shows a 25-t/day pyrolysis unit developed by the Engineering Experiment Station of the Georgia Institute of Technology. The biomass residues are convertible into more readily usable, more transportable, more compact, and more environmentally acceptable fuels. Figure 49 depicts a flow diagram of a modern pyrolysis process that produces biofuels in three forms: char, with a fuel value of 28-31.5 kJ/kg; oil, at 21-23 kJ/kg; and gas, at 5.6-7.5 MJ/m^3. In a controlled pyrolysis operation, the relative yields of the three products can be varied to suit the needs of the local region.

The variety of products produced, high energy conversion efficiencies (up to 95 percent based on recoverable energy in the dry biomass feedstock), adaptability in developing countries, and consumer acceptance in developing countries of similar fuel products (kerosene and charcoal) make pyrolytic processes highly attractive for energy production in these countries, particularly where large quantities of otherwise unproductive agricultural residues exist. It must be emphasized, however, that all suggestions for the use of pyrolysis depend on recovery of the heat of combustion for reasonable thermodynamic efficiency.

Although pyrolytic oils cannot presently be used directly as fuels in

FIGURE 48 A 25-t/day pyrolysis unit developed by the Engineering Experiment Station of the Georgia Institute of Technology.

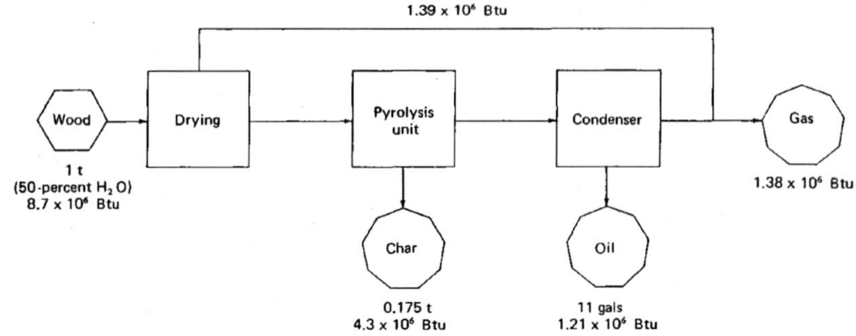

FIGURE 49 Mass-energy flow diagram of a pyrolysis unit.

internal combustion engines, they can be used as fuels for cooking, drying, steam and electrical power generation, and lighting, and they can be substituted for firewood in industrial processes, e.g., the firing of kilns for roof tile or brick manufacture. By mixing pyrolytic oils and charcoal, the fuel can be used to fire boilers for factories, trains, power stations, and steamships (Table 6). Charcoal has long played a prominent role in meeting the energy needs of many developing countries (Table 7).

Since char and oil are storable and more easily transported, they are of greater importance than the low-energy gas produced in the pyrolytic conversion of agricultural wastes. Pyrolytic gases are most practically used on-site for drying feed materials or for firing utility boilers.

Applications for Developing Countries. Developing countries have shown increased interest in the development of pyrolytic conversion systems appro-

TABLE 6 Energy-related applications of pyrolytic oil.

Application	Features/uses
Cooking	Quick starting, clean burning, easy to control
Lighting	Used in lamps (need to modify or control viscosity for wicking)
Heating	Replaces wood or kerosene
Boiler fuel/electrical steam	With charcoal, as slurry
	Rural electrification
	Industrial steam
	District heating
	Transportation (steam engines)
Drying and baking	Kiln fuels
	Structural materials (bricks, roof tiles, etc.)
	Pottery
	Ceramics

TABLE 7 Energy-related applications of charcoal.

Application	Features/uses
Domestic heating and cooking	Smokeless, sulfur-free Adaptable to stoves for cooking and hot water production
Direct drying (agricultural)	Drying, e.g., hops, tobacco, and other crops in same environment
Indirect drying (agricultural)	Central heating system fuel
Internal fuel (industrial)	For lime and cement manufacture Mix with limestone and fire to quicklime Portland cement manufacture using charcoal in rotary kilns (spontaneous ignition)
Metal extraction (industrial)	Charcoal combines with ores of metal oxides and sulfides, reducing the ore to pure metal Copper, iron, steel manufacture
Briquetted charcoal fuel	Cooking (barbecue) fuel
Activated carbon (industrial, agricultural)	Environmental control Sanitation Water purification
Boiler fuel/electrical steam (utility)	Alone or with coal or oil Low sulfur Rural electrification Industrial steam District heating

priate to local needs. The Engineering Experiment Station of the Georgia Institute of Technology collaborated with the Building and Road Research Institute in Kumasi, Ghana, on a feasibility study of the use of pyrolytic conversion of agricultural and forest wastes in Ghana (Chiang, *et al.* 1976). The study concluded that pyrolysis could be feasible economically and presented a preliminary design for an appropriate technology version of a continuous/batch converter based on processing 6 dry tons of sawdust per day (Figure 50). Current work in Ghana involves the construction and demonstration of four 1.5-metric tons-per-day pyrolytic converters, fabricated chiefly from locally available materials and intentionally labor-intensive (Moh and Potts 1976).

Four different conceptual designs were developed for Indonesia by a Georgia Institute of Technology team in Bandung when it became evident that pyrolysis was technically and economically attractive in terms of both domestic energy production and possible export of pyrolysis products (Tatom, *et al.* 1977). Designs were based on processing 1 t/day of biomass (rice hulls, wood waste, etc.) and ranged from a simplified labor-intensive system (two annual crops, sun-drying, three shifts, located at a local rice mill) to a relatively mechanized system (three rice crops, a gas-fired drier, three shifts, located at a local rice mill) to a more capital-intensive system (single-shift

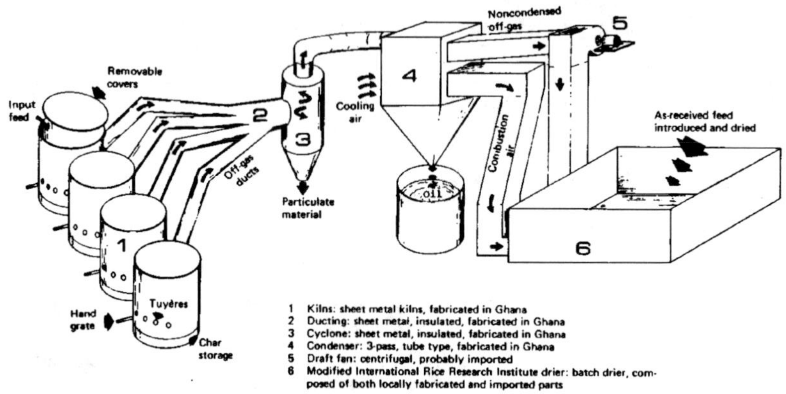

FIGURE 50 A 6-t/day pyrolytic conversion system for Ghana.

operation). All designs essentially use the same kiln design but differ as to the nature of the off-gas systems and inclusion of a drier.

A small-scale pyrolytic converter (Figure 51) has been constructed at the University of the Philippines with the technical support of Georgia Institute of Technology in a U.N. Industrial Development Organization (UNIDO) program (Ministry of Energy, Republic of the Philippines 1979). It employs a design similar to the Georgia Tech converters in Ghana.

Other developments include operation of a 1-t/day pyrolysis unit in New Guinea which is to be followed by construction of a 25-t/day prototype, industrial-scale system in 1980 (Ministry of Minerals and Energy, New Guinea 1979).

The International Development Association has recently approved credits of \$4.5 million and \$4.3 million for wood energy usage surveys in Mali and Burundi, respectively (see National Research Council 1980b). The projects will focus on wood-burning systems and charcoal kilns.

At the regional level, the U.N. Development Program (UNDP) is completing engineering feasibility studies for a large rural electrification project on the island of Savaii in Samoa. The electrical generator will be powered by the combustion of wood wastes and sawdust from a commercial sawmill on the island. It is estimated that electrical power will be provided to 31,000 of the 42,000 inhabitants at a price less than \$.07/kWh.

Gasification

This is not a new technology. In the past, particularly during World War II, "producer gas" was commonly used in Europe, Japan, China, Brazil, New Zealand, and Australia, and this technology is now being reevaluated for use

FIGURE 51 Small-scale pyrolysis unit in the Philippines.

in the developing world.* A forthcoming publication of the National Academy of Sciences, *Producer Gas: An Alternative for Motor Transport* (working title), will be devoted to this subject.

The introduction of reliable, small-scale, portable gasifiers for application in developing countries may be realized in the next 5–10 years. Large-scale gasifier systems will probably be limited in application for a number of years to a few developing countries with the necessary capital and national commitment to production of energy from biomass. For example, Brazil is pursuing the production of ethanol and methanol from biomass for use as a transportation fuel. Davy Powergas, Inc. (Great Britain) is designing an atmospheric, fixed-bed gasifier that would convert eucalyptus trees in Brazil to synthesis gas, with the gas being converted to methanol by the low-pressure

*Producer gas is generated when a limited amount of air passes through a bed of burning, combustible materials—such as wood, charcoal, coal, peat, and compressed agricultural residues—causing self-sustaining but incomplete combustion. This process takes place within a gasifier—a metal tank with a firebox, grate, air inlets, and an outlet for the combustible gases, consisting primarily of carbon monoxide and hydrogen, which can be used in boilers, engines, etc.

loop process. A 2,000-t/day facility is planned. It is projected that 1 t of methanol can be produced from 3 t of eucalyptus containing 35 percent moisture (Webber 1979).

REFERENCES

Abeles, T. P., D. F. Freedman, L. A. DeBaere, and D. A. Ellsworth (1978) Energy and Economic Assessment of Anaerobic Digesters and Biofuels for Rural Waste Management. Report prepared for the U.S. Environmental Protection Agency, Contract R-80-4-457-010, Washington, D.C., USA.

Anonymous (1979) Producing Methane Gas from Solid Waste. Brown and Caldwell Project Highlights, No. 12. Brown and Caldwell Consulting Engineers, Walnut Creek, California, USA.

Beagle, E. C. (1978) Rice-Husk Conversion to Energy. FAO Agricultural Services Bulletin 31, Food and Agriculture Organization of the U.N., Rome, Italy.

Benemann, J. R. (1978) Biofuels: A Survey. Special Report EPRI ER-746-SR, Electric Power Research Institute, Palo Alto, California, USA.

Bio-Energy Council (1978) Bio-Energy Directory. Bio-Energy Council, Washington, D.C., USA.

Brace Research Institute (1977) Annual Report. Report No. M35, McDonald College, McGill University, Ste. Anne de Bellevue, Canada.

Bryant, M. P., V. H. Varel, R. A. Frobish, and H. R. Isaacson (1976) Biological potential of thermophilic methanogenesis from cattle wastes. In: *Microbial Energy Conversion*, edited by H. G. Schlegel and J. Barnes. Pergamon Press, Elmsford, New York, USA.

Bullock, L. D., G. M. Higgins, R. B. Smith, and J. T. Swartzbaugh (1980) Enzymatic enhancement of solid waste bioconversion. Pp. 319–332 in: *Symposium Papers, Energy from Biomass and Wastes IV.* January 21–25, 1980, Lake Buena Vista, Florida. Institute of Gas Technology, Chicago, Illinois, USA.

Chandler, J. A., *et al.* (1979) Predicting Methane Fermentation Biodegradability. Paper presented at the Second Symposium on Biotechnology in Energy Production and Conservation, October 3, 1979, Gatlinburg, Tennessee, USA.

Chen, Y. R. and A. G. Hashimoto (1979) Kinetics of methane fermentation. In: *Biotechnology in Energy Production and Conservation*, edited by C. D. Scott. John Wiley & Sons, New York, New York, USA.

Chiang, T. I., *et al.* (1976) Pyrolytic Conversion of Agricultural and Forest Wastes in Ghana. Final Report, Project A-1823, Engineering Experiment Station, Georgia Institute of Technology, Atlanta, Georgia, USA.

Chou, W. L., R. H. Siddiqi, K. McKeon, and R. E. Speece (1977) Methane Production from the Anaerobic Treatment of Petrochemical Wastewaters. Final Report NSF/GI/43864, AER 74-13977-A0I, prepared for the U.S. National Science Foundation, Washington, D.C., USA.

Chynoweth, D. P., V. J. Srivastava, M. P. Henry, and P. B. Tarman (1980) Biothermal gasification of biomass. Pp. 527–554 in: *Symposium Papers, Energy from Biomass and Wastes IV.* January 21–25, 1980, Lake Buena Vista, Florida. Institute of Gas Technology, Chicago, Illinois, USA.

Cornwell, D. A., J. Zoltec, C. D. Patrinely, T. S. Furman, and J. I. Kim (1977) Nutrient removal by water hyacinths, *Journal of the Water Pollution Control Federation* 70: 57–65.

Dinges, R. (1978) Upgrading stabilization pond effluent by water hyacinth culture, *Journal of the Water Pollution Control Federation* 50:833–845.

Earl, D. E. (1974) A Report on Charcoal. André Mayer Fellowship Report, Food and Agriculture Organization of the U.N., Rome, Italy.

Eckholm, E. (1975) The Other Energy Crisis: Firewood. Worldwatch Paper No. 1. Worldwatch Institute, Washington, D.C., USA.

Forest Products Research Society (1979) Proceedings of the FPRS Conference on Hardware for Energy Generation in the Forest Products Industry, January 30–February 1, 1979, Seattle, Washington. (Available from FPRS, Madison, Wisconsin, USA)

Geertz, C. (1963) *Agricultural Involution: The Process of Ecological Change in Indonesia*. University of California Press, Berkeley, California, USA.

Ghosh, S., J. R. Conrad, and D. L. Klass (1975) Anaerobic acidogenesis of wastewater sludge, *Journal of the Water Pollution Control Federation* 47(1):30–45.

Ghosh, S. and D. L. Klass (1978) Two-phase anaerobic digestion, *Process Biochemistry* 13:15–24.

Ghosh, S. and D. L. Klass (1980) Anaerobic sludge digestion in the presence of lactobacillus additive. Pp. 295–318 in: *Symposium Papers, Energy from Biomass and Wastes IV*. January 21–25, 1980, Lake Buena Vista, Florida. Institute of Gas Technology, Chicago, Illinois, USA.

Gossett, J. M., J. B. Healy, Jr., L. Y. Young, and P. L. McCarty (1975) Heat Treatment of Refuse for Increasing Anaerobic Biodegradability. Semi-Annual Progress Report for the U.S. National Science Foundation, NSF/RANN/SE/GI-43504/75/2, Stanford University, Palo Alto, California, USA.

Gottschall, W. C., J. G. Morse, D. M. Updegraff, and A. Wang (1978) The effect of gamma radiation on yields of methane gas from the anaerobic bacterial digestion of sewage sludge. Pp. 193–214 in: *Symposium Papers, Energy from Biomass and Wastes*. August 14–18, 1978, Washington, D.C. Institute of Gas Technology, Chicago, Illinois, USA.

Harwood, R. R. (1979) *Small Farm Development*. Westview, Boulder, Colorado, USA.

Hashimoto, A. G., Y. R. Chen, and R. P. Prior (1978) Thermophilic anaerobic fermentation of beef cattle residue. Pp. 379–402 in: *Symposium Papers, Energy from Biomass and Wastes*. August 14–18, 1978, Washington, D.C. Institute of Gas Technology, Chicago, Illinois, USA.

IRRI (1977) Annual Report for 1976. International Rice Research Institute, Los Baños, Philippines.

Jewell, W. J. (1980) Development of the Attached Microbial Film Expanded Bed Process for Aerobic and Anaerobic Waste Treatment. Paper presented at the Conference on Biological Fluidized Bed Treatment of Water and Wastewater, April 14–17, 1980, Hertfordshire, England.

Jewell, W. J., *et al.* (1979a) Low cost methane generation on small farms. Pp. 547–580 in: *Proceedings of the Third Annual Biomass Energy Systems Conference*. Solar Energy Research Institute, Golden, Colorado, USA.

Jewell, W. J., M. S. Switzenbaum, and J. W. Morris (1979b) Sewage Treatment with the Anaerobic Attached Microbial Film Expanded Bed Process. Paper presented at the 52nd Water Pollution Control Federation Conference, October 11, 1979, Houston, Texas. WPCF, Washington, D.C., USA.

Jones, J. L., *et al.* (1978) Pyrolysis, Thermal Gasification, and Liquefaction of Solid Wastes and Residues—Worldwide Status of Processes (as of Fall 1977). Paper presented at the Eighth Biennial National Waste Conference, American Society of Mechanical Engineers, May 1978, Chicago, Illinois. (Available from SRI International, Menlo Park, California, USA)

Karchesy, J. and P. Koch (1979) Energy Production from Hardwoods Growing on Southern Pine Sites. General Technical Report 50-24. U.S. Forest Service, U.S.

Department of Agriculture, Washington, D.C. (Available from the U.S. Government Printing Office, Washington, D.C., USA)

Klass, D. L. (1980) Anaerobic Digestion for Methane Production—A Status Report. Paper presented at the Bio-Energy World Congress and Exposition, April 21–24, 1980, Atlanta, Georgia. Bio-Energy Council, Washington, D.C., USA.

Knight, J. A., *et al.* (1976) Pyrolysis–a method for conversion of forestry waste to useful fuels, *Proceedings of the Conference on Energy and Wood Products*. Sponsored by the Forest Products Research Society, November 15–17, 1976. (Available from the Engineering Experiment Station, Georgia Institute of Technology, Atlanta, Georgia, USA)

KVIC (1974) Drawings and Estimates of Vertical Gobar Gas Plants. Khadi and Village Industries Commission, Bombay, India.

Ladisch, M. R., M. C. Flickinger, and G. T. Tsao (1978) Fuels and chemicals from biomass, *Energy* 4:263–275.

Maramba, F. D., Sr. (1979) *Biogas and Waste Recycling: The Philippine Experience*. Maya Farms Division, Liberty Flour Mills, Inc., Metro Manila, Philippines.

McCarty, P. L., L. Young, W. Owen, D. Stuckey, and P. J. Colburg (1979) Heat treatment of biomass for increasing biodegradability. Pp. 411–418 in: *Proceedings of the Third Annual Biomass Energy Systems Conference*. Solar Energy Research Institute, Golden, Colorado, USA.

McConville, T. and W. J. Maier (1978) Use of powdered activated carbon to enhance methane production in sludge digestion. Pp. 345–359 in: *Biotechnology in Energy Production and Conservation*, edited by C. D. Scott. John Wiley & Sons, New York, New York, USA.

McDonald, R. C. and B. C. Wolverton (1980) A comparative study of a domestic wastewater lagoon with and without water hyacinths, *Economic Botany* 34(2):101–110.

McGarry, M. G. and J. Stainforth (1978) *Compost, Fertilizer, and Biogas Production from Human and Farm Wastes in the People's Republic of China*. International Development Research Centre, Ottawa, Canada.

McNabb, C. D., Jr. (1976) The potential of submersed vascular plants for reclamation of wastewater in temperate zone ponds. Pp. 123–132 in: *Biological Control of Water Pollution*, edited by J. Tourbier and R. W. Pierson, Jr. University of Pennsylvania Press, Philadelphia, Pennsylvania, USA.

Ministry of Energy, Republic of the Philippines (1979) The National Nonconventional Energy Resources Development Program, Progress Report.

Ministry of Minerals and Energy, New Guinea (1979) White Paper—Energy Policy and Planning for Papua New Guinea.

Moh, K. C. and P. W. Potts (1976) Demonstrating the Feasibility of Pyrolytic Conversion in Ghana. Project A-2076, Engineering Experiment Station, Georgia Institute of Technology, Atlanta, Georgia, USA.

National Research Council (1976) *Making Aquatic Weeds Useful: Some Perspectives for Developing Countries*. National Academy of Sciences, Washington, D.C., USA.

National Research Council (1977) *Methane Generation from Human, Animal and Agricultural Wastes*. National Academy of Sciences, Washington, D.C., USA.

National Research Council (1979) *Microbial Processes: Promising Technologies for Developing Countries*. National Academy of Sciences, Washington, D.C., USA

National Research Council (1980a) *Firewood Crops: Shrub and Tree Species for Energy Production*. National Academy of Sciences, Washington, D.C., USA.

National Research Council (1980b) *Proceedings, International Workshop on Energy Survey Methodologies for Developing Countries*. National Academy Press, Washington, D.C., USA.

Norrman, J. and B. Frostell (1977) Anaerobic Waste Water Treatment in a Two-Stage Reactor of New Design. Paper presented at the 1977 Industrial Waste Conference, May 10, 1977, Purdue University, Lafayette, Indiana, USA.

Novak, J. T. and M. S. Ramesh (1975) Stimulation in anaerobic digestion, *Water Resources* 9:963–967.

Othmer, D. F. (1952) History and present status of the industry, continuous carbonization, the Mellman Report. P. 49 in: *Charcoal Production and Uses*. Bulletin 37, Northeastern Wood Utilization Council, Inc., Old Forge, New York, USA.

Pitt, W. W. and R. K. Genung (1980) Energy conservation and production in a packed-bed anaerobic bioreactor. Pp. 451–472 in: *Symposium Papers, Energy from Biomass and Wastes IV*. January 21–25, 1980, Lake Buena Vista, Florida. Institute of Gas Technology, Chicago, Illinois, USA.

Sathianathan, M. A. (1975) *Biogas–Achievements and Challenges*. Association of Voluntary Agencies for Rural Development, Sagar Printers and Publishers, New Delhi, India.

Schellenbach, S. (1980) Imperial Valley bio-gas project–operations and methane production from cattle manure, October 1978 to November 1979. Pp. 473–496 in: *Symposium Papers, Energy from Biomass and Wastes IV*. January 21–25, 1980, Lake Buena Vista, Florida. Institute of Gas Technology, Chicago, Illinois, USA.

Seidel, K. (1976) Macrophytes and water purification. Pp. 109–122 in: *Biological Control of Water Pollution*, edited by J. Tourbier and R. W. Pierson, Jr. University of Pennsylvania Press, Philadelphia, Pennsylvania, USA.

Shelton, J. W. and A. B. Shapiro (1978) *The Woodburners Encyclopedia*. Vermont Crossroads Press, Waitsfield, Vermont, USA.

Singh, R. B. (1971) *Bio-gas Plant–Generating Methane from Organic Wastes*. Gobar Gas Research Station, Ajitmal, Etawah, India.

Skrinde, R. T. (1978) Review of international biogas programs, *Proceedings of the Second Annual Symposium on Fuels from Biomass*. Vol. II. U.S. Department of Energy, Washington, D.C., USA.

Soil Conservation Service (1977) Cropland Erosion. U.S. Department of Agriculture, Washington, D.C., USA.

Spano, L. A. (1976) Enzymatic Hydrolysis of Cellulosic Wastes to Fermentable Sugars and the Production of Alcohol. Pollution Abatement Division, Food Sciences Laboratory, U.S. Army Natick Research and Development Command, Natick, Massachusetts, USA.

Spencer, R. R. (1978) Enhancement of methane production in the anaerobic digestion of sewage sludges. Pp. 257–268 in: *Biotechnology in Energy Production and Conservation*, edited by C. D. Scott. John Wiley & Sons, New York, New York, USA.

Subramanian, S. K. (1977) *Biogas Systems in Asia*. Statesman Press, New Delhi, India.

Tatom, J. W., T. I. Chiang, F. Harahap, R. M. Apandi, and H. Wirjosumarto (1977) Pyrolytic Conversion of Agricultural and Forestry Wastes to Alternate Energy Sources in Indonesia, A Feasibility Study. Project A-1914, Engineering Experiment Station, Georgia Institute of Technology, Atlanta, Georgia, USA.

VITA/Intermediate Technology Development Group (1980) *Wood-Conserving Cook Stoves, A Design Study*. Volunteers in Technical Assistance, Mt. Rainier, Maryland, USA.

Walton, J. D., Jr., A. H. Roy, and S. H. Bomar, Jr. (1978) A State of the Art Survey of Solar Powered Irrigation Pumps, Solar Cookers, and Wood-Burning Stoves for Use in Sub-Sahara Africa. Final report, Georgia Tech Project A-2004 for the al Dir'iyyah Institute. (Available from the Engineering Experiment Station, Georgia Institute of Technology, Atlanta, Georgia, USA)

Webber, D. (1979) Methanol: a new wrinkle or two, *Chemical Business*. October 15, p. 57.

Wolverton, B. C. (1979) Engineering Design Data for Vascular Aquatic Plant Wastewater Treatment Systems. Paper presented at an Aquaculture Systems for Wastewater

Treatment Seminar, September 11–13, 1979, University of California at Davis, Davis, California, USA.

Wolverton, B. C. and R. C. McDonald (1976) Water Hyacinths for Upgrading Sewage Lagoons to Meet Advanced Wastewater Treatment Standards: Part II. NASA Technical Memorandum TM-X-72730. U.S. National Aeronautics and Space Administration, Washington, D.C., USA.

Wolverton, B. C. and R. C. McDonald (1978) Nutritional composition of water hyacinths grown on domestic sewage, *Economic Botany* 32(4):363–370.

Wolverton, B. C. and R. C. McDonald (1979a) The water hyacinth: from prolific pest to potential provider, *Ambio* (The Royal Swedish Academy of Sciences) 8(1):2–9.

Wolverton, B. C. and R. C. McDonald (1979b) Upgrading facultative wastewater lagoons with vascular aquatic plants, *Journal of the Water Pollution Control Federation* 51(2):305–313.

Wolverton, B. C. and R. C. McDonald (1979c) Water hyacinth productivity and harvesting studies, *Economic Botany* 33(1):1–10.

Wolverton, B. C. and R. C. McDonald (1980) Energy from Aquatic Plant Wastewater Treatment Systems, NASA Technical Memorandum TM-X-72733. U.S. National Aeronautics and Space Administration, Washington, D.C., USA.

Wujcik, W. J. and W. J. Jewell (1979) Dry Anaerobic Fermentation. Paper presented at the 2nd Symposium on Biotechnology in Energy Production and Conservation, October 3–5, 1979, Gatlinburg, Tennessee, USA.

PART III
OTHER TECHNOLOGIES

Geothermal Energy

Exploration and development of geothermal resources are formidably large tasks for the rural or village economies in developing countries. Although there have been recent developments in both technology and processing, these have not significantly improved the economic, geologic, or engineering feasibility of most geothermal projects. However, these developments could make the use of geothermal energy highly desirable in specific circumstances, depending upon resource availability, energy needs, and the economics of alternatives.

The costs and risks of geothermal exploration remain high. Detailed geophysical and geochemical surveys may cost from $2,000 to $5,000 each per day. Expenditures for detailed geologic investigations of a small prospect several tens of square kilometers in area may amount to $200,000 or more. A single geothermal resource may well cost as little as $300,000 if shallow and small in diameter, or as much as $2 million if more drilling is required or if engineering problems are encountered. While individual wells may yield anywhere from 100 kWe to 10,000 kWe or more when connected to an appropriate conversion facility, even the best geologic exploration program will not guarantee discovery of a reservoir or the output of any one well. (See Appendix C for a list of recent literature on geothermal energy.)

RECENT DEVELOPMENTS IN GEOTHERMAL TECHNOLOGY

Recent developments in the technology for single geothermal wellhead electric generators hold the most promise for use in the rural areas of developing countries. Single wellhead turbogenerating equipment produces electricity in relatively small amounts with a relatively small financial investment. Such devices are capable of generating electricity from a single geothermal well in the range of 20 kW to as much as 10 MW, depending in large part upon the resource capabilities of the well.

Significant effort has been expended recently to develop such wellhead devices for commercial operation, and some wellhead generating systems are now available. Others are still being tested and developed, but may be available in the near future. The following descriptions are offered only as examples of equipment now available or being developed (see Appendix B for a listing of research and development centers). Information on costs, availability, and performance specifications should be obtained from individual manufacturers (see Appendix A).

The helical rotary screw expander system has undergone extensive field tests with hydrothermal fluids in the United States, and a 1-MW packaged power plant was successfully operated for several months. This system is a total flow conversion process, utilizing the entire well flow to turn the turbine. It was designed and built by the Hydrothermal Power Company (USA) with funding from the U.S. National Science Foundation and the U.S. Department of Energy (DOE). The California Institute of Technology's Jet Propulsion Laboratory in Pasadena, California, performed the testing. Units in other sizes may be available in the near future. A similar total flow system is available from Turbines Incorporated (USA), in sizes from 5 to 10 MW. A third total flow turbogenerator system is being made available by the Rotoflow Corporation (USA), which adapted its units from the system it constructs for hydrocarbon pipelines.

A prepackaged wellhead generating unit is available from the Geothermal Power Company, Inc. (USA). The "Mono-Blok" unit utilizes separated steam from the geothermal resource and is available in units from 100-kW to 2.5-MW capacity. A 500-kW Mono-Blok unit has been installed and operated in Indonesia. Mitsubishi (Japan) offers a similar portable geothermal turbine generator with capacities of from 500 kW to 5 MW. Units of this type have been ordered for Japan, the Philippines, and El Salvador. The Philippine government has undertaken a fairly extensive geothermal development program on the island of Luzon. Several wells have been drilled and are being connected to a central steam plant in order to produce electric power.

Biphase Energy Systems (USA) has designed and tested a two-phase rotary turbine for use with geothermal hot water systems. A 20-kW turbine generator system has undergone extensive testing in the Imperial Valley, California, and this company advertises that they are prepared to enter into fixed-price contracts for the manufacture and installation of Biphase turbines for certain applications with appropriate performance warranties. The Ormat turbine, although designed to be used as a solar energy conversion system, appears quite capable of using geothermal heat as its energy source. This closed-cycle, vapor turbogenerator is available from Ormat Turbines, Ltd. (Israel).

Two other wellhead-type, geothermal generating systems are now commencing commercial-size testing operations in the Imperial Valley. Barber–Nichols Engineering Company, under contract to the Lawrence Berkeley Laboratories and funded by DOE, is constructing a 500-kW direct contact, heat exchanger/geothermal power plant at East Mesa. This system will use isobutane as the working fluid for the turbine generator system. The direct contact heat exchanger uses the hot geothermal fluid to heat the isobutane working fluid in a direct contact counterflow system, which eliminates many of the scaling problems inherent in other heat exchangers.

The Sperry Research Center (USA), under contract to DOE, will begin field tests in East Mesa on a 4.3-MW gravity head power plant. In this system, a specially designed heat exchanger is placed in the geothermal well at a depth of approximately 2,000 ft (610 m). The freon working fluid is heated as it is pumped down to the heat exchanger and back up to the turbogenerator system on the surface. A small pump at the bottom of the heat exchanger pumps the geothermal fluid back up to the surface, at which point the fluid has had its heat extracted. Both the Sperry and the Barber–Nichols systems are being designed to increase the efficiency of the geothermal generation system. However, it appears that neither system will be available in the near future.

DIRECT APPLICATIONS OF GEOTHERMAL HEAT ENERGY

Low-temperature (100°-200°F or 38°-93°C) waters are the most abundant accessible geothermal resource and, for the near term, will likely offer the most benefit to developing countries. One potential advantage of utilizing low-temperature geothermal waters for direct applications is that the geothermal fluid can also be used for irrigation, crop drying, or even as a potable water supply, if its salt content is low. Such waters could also be used directly for aquacultural applications. The potential for multiple use of a geothermal resource for space heating or other direct application, and subsequently using the fluid as a traditional water resource, is especially attractive for developing countries with cold climates.

Over the past few years, a number of schemes have been proposed that would successively utilize geothermal waters in a series of increasingly lower-temperature heat applications in an energy cascade. However, most energy cascade systems are capital-intensive, which would limit their suitability in developing countries. In addition, many of these systems are designed to be economical when a medium-temperature (approximately 300°F or 148°C) geothermal resource is available, and it is unlikely that a resource at this temperature would be available at shallow depths except in rare instances. The cascade systems could make economic sense in developing countries if waste heat from geothermal power plants were utilized or if suitable medium-temperature resources were available from otherwise noncommercial exploratory deep wells.

Energy Storage

With current technology, there are ways to store energy in chemical, mechanical, and thermal forms. Significant advances have been made in the chemical and mechanical forms during the past 5 years, particularly in the areas of improved batteries (chemical form) for photovoltaic and wind energy applications and improved flywheel energy storage systems (mechanical form), both of which are discussed here. Another form of mechanical storage, isothermal hydraulic air compression, is discussed in the chapter, "Energy from Falling Water." (See Appendix C for a listing of recent literature on energy storage techniques.) Considerable cost reductions are desirable before energy storage in these forms becomes widely used in developing countries.

STORAGE BATTERIES

Since publication of the original volume of *Energy for Rural Development*, research and development on batteries has, generally, been directed at developing batteries suitable for electric vehicles, stationary energy storage applications, and photovoltaic and wind energy applications (see Appendixes A and B for listings of manufacturers and research and development centers, respectively). As the photovoltaic and wind energy applications seem more suitable for rural areas of developing countries, they are stressed here.

Both photovoltaic arrays and windmills generate direct current suitable for storage in batteries. However, for efficient use of the batteries and to prolong its life, it is desirable to adjust the voltage and control the battery-charging current by means of a power conditioning system. If alternating current of closely regulated voltage is the desired output of the energy source, an inverter will be required within the power conditioning system. Thus a complete photovoltaic or wind energy system will consist of three units: the energy source (windmill or photovoltaic array), a power conditioning system, and a battery energy storage system.

Battery manufacturers will design an entire battery system for photovoltaic or wind applications, including the selection of the power conditioning equipment. For applications in the near term, only two types of batteries should be considered—lead-acid and nickel-cadmium. However, lead-acid batteries should be selected for almost all situations in developing countries.

Several types of lead-acid cells have been developed to meet various applications, and these are summarized in Table 8. The SLI (starting, lighting, and ignition) batteries deliver very high power, but have short cycle life. Such batteries could be considered for photovoltaic or wind energy storage applications only if a low initial cost is an important consideration and if deep discharges occur relatively infrequently. Motive power batteries have thicker plates, and thus lower discharge rate capability, but they have long cycle life. The types made for forklifts and mine vehicles would have sufficient cycle life for many photovoltaic and wind energy installations. Stationary batteries were designed for stand-by and emergency power applications requiring few cycles. They have low self-discharge rates and long calendar life when used under the conditions for which they were designed. For many photovoltaic power systems, the stationary type of battery is satisfactory and is the basis for the design of several photovoltaic batteries. For some small-scale applications, sealed lead-acid batteries that have a relatively poor cycle life, but do not require water addition or ventilation, have been developed.

TABLE 8 Types of lead-acid batteries.

Type	Characteristics	Applications
Automotive (SLI) and diesel starting	High discharge rate, relatively low cost, poor cycle life	Automobile starting, lighting, and ignition; tractor, snowmobile, and other small engine starting; large diesel engine starting
Motive power (traction)	Moderate discharge rate, good cycle life	Forklifts; mine vehicles; golf carts, submarines; other electric vehicles
Stationary	Medium discharge rate, long life (years). Some types have low self-discharge rates, poor cycle life.	Telephone power supplies, uninterruptable power supplies, other standby and emergency power supply applications
Photovoltaic	Low maintenance, low self-discharge, special designs for high and low ambient temperatures, poor deep cycle life	Remote, daily shallow discharge, large reserve (stand-alone) photovoltaic power systems
Sealed	No maintenance, moderate rate, poor cycle life	Lanterns, portable tools, portable electronic equipment

SOURCE: Bechtel National, Inc. (1980).

The positive electrodes of lead–acid batteries frequently contain either antimony or calcium alloying elements. Antimony is added to increase cycle life. Cells containing lead–calcium positive grids have the advantage of low self-discharge rates of a fraction of a percent per week at 70°F (21°C), as compared to self-discharge rates of about 4 percent per week for lead–antimony grids. This self-discharge rate can become as high as 15–20 percent per week for lead–antimony grids near the end of life. If self-discharge is not included, lead–acid batteries will deliver about 70–80 percent of the energy charged.

Nickel–cadmium batteries are very expensive and cost about $340/kWh for large, vented, pocketplate cells to over $900/kWh for sintered-plate, sealed cells (Bechtel National, Inc. 1980). These batteries should be considered for applications requiring long cycle life, high rates of discharge, and relatively small capacity (5 kWh or less).

When selecting a lead–acid battery for a particular application, the following considerations should be kept in mind:

- *Capacity* depends on how the battery is used, and is defined in terms of the number of hours or days of service the battery will provide when fully charged and at the average load. If year-round operation without a back-up system is required, the battery may be designed for as much as one month of operation at the average load without recharging. As solar arrays become less expensive relative to the cost of the battery, larger solar arrays relative to the load may be used. This would permit faster recharging of the battery, which would then be reduced in capacity to as little as 3–5 days of service for stand-alone systems. Wind energy units, even if designed for stand-alone service, are usually provided with only a few days of storage capacity.
- *Cycle life* is inversely related to capacity, as measured in days of storage at normal load conditions. For most photovoltaic and wind energy installations, only 50–500 cycles would be required for 10-year battery life. This can be achieved with batteries having lead–calcium alloy grids and designed for photovoltaic applications. Such batteries also have the important advantages of low gassing rates, low water-addition requirements, low self-discharge rates, and high efficiency. Lead–calcium batteries should be chosen for most photovoltaic applications.
- *Extent of discharge* is best limited to about 60–80 percent of the rated capacity of the battery. Beyond 80 percent discharge, the battery life is shortened considerably.
- *Maintenance* includes (1) proper maintenance of charging conditions by the power conditioning system to achieve a high state of charge in each cycle, acceptable charging efficiency, and long life; and (2) proper temperature control which is important in determining battery capacity, the rate at which a battery will accept a charge, and the lifetime of the battery.

• *Cost* varies considerably with the level of production and service costs, but for many uses it will be in the range of \$60–\$200/kWh (in 1980 U.S. dollars). (See Appendix A for a list of battery manufacturers from whom prices are available.)

Research and Development Programs

The U.S. Department of Energy (DOE) is sponsoring battery development for both electric vehicles and stationary energy storage applications; most of the latter effort has been directed toward load leveling for electric utilities. One program recently initiated by DOE to develop batteries for photovoltaic and wind energy applications was in the evaluation and planning stage at this writing. This effort is being managed by Sandia Laboratories, Albuquerque, New Mexico.

The Electric Power Research Institute (EPRI, USA) is also concerned with load-leveling applications and has funded several battery programs. Table 9 lists those programs supported by DOE and EPRI, among others, that show promise of developing batteries for stationary storage applications and that may be applicable to photovoltaic and wind energy devices.

Programs aimed at developing advanced lead–acid batteries for utility load-leveling applications are striving to develop large cells having very long life (>2,000 cycles) and low cost. For those applications having back-up energy systems and only a few hours of storage, long cycle life will be very desirable. However, for most rural applications, stand-alone service and therefore at least several days of energy storage are required. Under these conditions, the batteries will probably undergo less than the equivalent of 50 full cycles per year; long life is not needed for these applications. The lowest total cost for purchase and operation of the battery system is likely to be achieved with a battery developed specifically for stand-alone service in photovoltaic and wind energy applications. These applications will be addressed in the new DOE program.

All of the battery systems in Table 9 are expected to be available for engineering demonstration for photovoltaic or wind applications by 1984–86. However, none is expected to be an economical alternative to lead–acid batteries until well into the 1990s. When manufacturing experience has been gained, means of producing reliable batteries at a cost below that of lead–acid batteries may evolve. In that case, one or more of these battery systems may become economically attractive for photovoltaic and wind energy systems.

For stand-alone service, which may be needed in developing countries, the large storage requirement relative to the load favors the reduction–oxidation flow systems under development at NASA's (U.S. National Aeronautics and Space Administration) Lewis Research Center. However, it is not certain that any of the battery storage systems will be economical for a week or more of

TABLE 9 Advanced batteries under development in the United States show promise for photovoltaic and wind energy applications.

Battery type	Major developer[a]
Advanced lead–acid	ESB Corp.[b] Eltra Corp.[b]
Flow systems	
Zinc/chlorine	Energy Development Associates
Zinc/bromine	Exxon Enterprises, Inc. Gould, Inc. General Electric Co.
Reduction/oxidation	NASA–Lewis Research Center
High temperature	
Sodium/sulfur	Ford Motor Co. General Electric Co. Dow Chemical Co.
Lithium/metal sulfide	Eagle–Picher Industries[b] Gould Inc.[b] General Motors Corp. Rockwell International Corp.

[a]Most of these efforts are funded by the U.S. Department of Energy or the Electric Power Research Institute, but some are supported wholly or in part by the developer.
[b]Contracts managed by Argonne National Laboratory (USA).

storage when compared to one day of storage and a back-up energy system such as an engine–generator.

Applications of Battery Systems

In many developing countries, little power is available to rural areas from the new, large power projects installed for industrial or domestic use in large urban communities. Many small villages in such countries have no electric energy, except what is provided by dry cells for flashlights and small radios (Doshi 1977). Below are several examples of the type of battery storage systems required for electrification of such small villages with solar or wind-power devices (Bifano, *et al.* 1979; Ratajczak and Bifano 1979; Solarex Corporation 1979).

Schuchuli, a small, remote village on the Papago Indian reservation in southwestern Arizona, utilizes a 3.5-kWp solar photovoltaic electric system provided by DOE (see "Photovoltaics"). All electrical items in the village operate on 12-volt DC in order to avoid the complexity of an inverter system. The Schuchuli system has two batteries—a main battery of 120 volts and 2,380 amp-hr capacity, and a small 12-volt instrumentation and control battery which is charged from the 120-volt system through a DC–DC converter. The batteries are installed in a ventilated room, and they consist of cells that have lead–calcium grids and are designed for deep cycling, low gas evolution, and low self-discharge. The batteries provide approximately 15–20 times the anticipated daily energy requirement; thus interruption of electric power is not expected. However, because of uncertainty in the energy demand of the village, the instrumentation system has been designed to shed the load as battery discharge increases as a result of higher than anticipated demand.

A similar photovoltaic system, with a peak power of 18 kW, was installed in the African village of Tangaye in Upper Volta, and became operational in March 1979 (Ratajczak and Bifano 1979). Like Schuchuli, it is a stand-alone system consisting of a photovoltaic array, batteries, and a control and instrumentation unit (see "Photovoltaics"). This system operates at 120 volts DC to avoid the cost and complexities of DC–AC inverters. The main battery has a capacity of 540 amp-hr, and the 12-volt instrumentation and control battery has a capacity of 200 amp-hr. The latter is charged from a separate 12-volt, 74-W photovoltaic array.

Both the Schuchuli and Tangaye systems have sufficient battery capacity for many days of operation without recharging. Calculations indicate that under the worst conditions the main battery at Schuchuli would discharge to 60 percent and at Tangaye to 30 percent. Because of the high prices of photovoltaic arrays, the battery systems were designed to have large capacities relative to the power of photovoltaic arrays. Combinations of larger photovoltaic arrays and smaller batteries could fulfill the needs just as reliably as the combinations presently used. However, the present systems, both designed by the NASA–Lewis Research Center, provide the maximum service and reliability for the funding available and for the existing price structure.

A photovoltaic system that provides power for irrigation of an 80-a (32-ha) cornfield was installed in Mead, Nebraska, in 1977 (Romaine 1979; Millner 1979c). The power from the system operates a 15-hp pump capable of pumping 1,000 gal/min (3,785 l/min) from an irrigation reservoir. Initially, the pump was driven either by a 120-volt DC motor powered by the photovoltaic system or by a 3-phase 240-volt AC motor driven by the utility power back-up system. In a later revision, a 3-phase inverter was installed, enabling the photovoltaic system to operate the 3-phase pump motor and two 3-phase crop-drying fans, which were added to the load.

The energy storage system for the Mead unit consists of two parallel banks of 6-volt lead–acid batteries, C&D Type 3CBSB-11, rated 375 amp-hr each at an 8-hr discharge rate. These batteries were designed for railroad caboose service and have an expected life of 5 years when deep discharged once a day. The cells have lead–antimony anodes and lead–calcium cathodes. There are 20 6-volt batteries in each bank and a total terminal voltage of 120 volts. The battery bus bar is operated between normal limits of 146 volts maximum during battery charging and a low of 110 volts on discharge. Battery voltage is managed by removing load when the battery reaches a suitable depth of discharge and allowing the photovoltaic array to recharge the battery to near the fully charged state before reapplying load to the photovoltaic system. The battery system at Mead stores enough energy for only about one day of operation at full load. It should be noted that the battery capacity relative to the load is much smaller in the Mead system than in the Schuchuli and Tangaye systems because of the availability of the utility back-up system at Mead.

In addition to the moderately large installations discussed above, many small solar photovoltaic systems with battery storage units have been installed in, for example, New Hampshire, near Washington, D.C., and in Montana (see Solarex Corporation 1976, 1977, 1978). Some additional examples of battery storage systems used in conjunction with photovoltaic arrays are given in Table 10. These examples illustrate that the use of batteries in combination with photovoltaic systems is a well-established commercial technology and that systems are readily available from many companies eager to sell their products.

Wind energy systems that are designed to stand alone usually require a battery system sufficient to supply the load for several days without recharging. If a windmill is designed to supply a full load at very low wind speeds, only a few hours of storage will be required. However, this is impractical, requiring that the windmill be several times larger than if several days of battery storage are provided.

In a design study carried out for installation of a wind energy system at a site near Harlow, Essex, in southeastern England, provision was made for a continuous electrical supply of either 50-volts DC or 240-volts, 50-cycle AC (Hirst and Rees 1976). This unit was envisioned as an energy supply for communications repeaters, navigational aids, or similar applications. In the design study, the battery consisted of 24 lead-acid cells manufactured for stationary storage duty. It was shown that the storage capacity required was inversely related to the power of the windmill. Storage capacity for about one week would be sufficient for a windmill designed for moderate wind speeds of about 7 kmph (4.3 mph).

In another study, required storage capacity was evaluated for wind energy installations designed to supply electricity for homes in a housing project (Littler and Thomas 1978). A vertical-axis unit with a swept area of 24 m^2 (258 ft^2) was investigated. A study of hourly wind speeds at two sites in England considered for the installations showed that the periods of very low wind speed differed considerably from year to year. It was concluded that standby power batteries should be used, because a computer model indicated that the cells are within 80 percent of complete charge for 60 percent of the time and undergo a deep discharge only once or twice a year. Such duty cycles are easily met by lead-calcium batteries.

TABLE 10 Photovoltaic battery systems in operation in the United States.

Location/application	Date installed	Array power (W)	Battery storage: Capacity (amp-hr)	Battery storage: Manufacturer/information	Cost (US$)
Isle Royale, Michigan Refrigerator, 12-volt DC	5/76	220	600	J. C. Penney 29 F, 6 in parallel, 100 amp-hr, 12-volt, sealed automotive battery	$240
Papago Indian Village, Sil Nakya, Arizona Refrigerator, 12-volt DC	7/76	220	600	Globe Union EV 27, 6 in parallel, 100 amp-hr,	$240
	8/77	330	600	12-volt; battery replaced with same type	
Antelope Peak, California Water pump, refrigerator, lights, communications, 12-volt DC	10/76	294	3,015	C&D, KCPA 1005 cells, 6 cells in series, 3 in parallel, 18 total, 1,005-amp-hr cells, lead–calcium photovoltaic type	$2,160
Pilot Peak, California Same as Antelope Peak	10/76	294	3,015	Same as Antelope Peak	$2,160
Stratfort Shoals, New York Remote weather instruments, 24-volt DC	5/77	111	60	Globe Union gel cell 1220, 2 in series, 3 in parallel, 6 total, 20 amp-hr, 12-volt, batteries: sealed, gelled electrolyte	$252
Halfway Rock, Maine	9/77	111	60	Same as Stratfort Shoals	$252
Loggerhead Key, Florida	10/77	74	60	Same as Stratfort Shoals	$252
Clines Corners, New Mexico	4/77	74	60	Same as Stratfort Shoals	$252
South Point, Hawaii	6/77	74	60	Same as Stratfort Shoals	$252
Pt. Retreat, Alaska (near Juneau)	7/77	148	60	Same as Stratfort Shoals plus C&D, KCPA	$252
			1,005	1005 cells, 12 cells in series, lead–calcium photovoltaic cells, 1,005 amp-hr	$1,440
Tucson/Phoenix, Arizona Interstate highway sign, lights, communications, 12-volt DC	4/77	116	200	Globe Union EV 27, 2 in parallel, 100 amp-hr, 12-volt	$80
College Station, Texas	5/77	23	100	Globe Union EV 27, 100 amp-hr, 12-volt	$40
Insect survey traps, 12-volt DC		163	400	Globe Union GC2-8144, 2 in series, 2 in parallel, 4 total, 200 amp-hr, 6-volt golf cart battery	$140
Lone Pine, California Chilled water, 120-volt DC	10/77	446	100	Globe Union EV 27, 9 in series, 100 amp-hr, 12-volt electric vehicle battery	$360

SOURCE: Matlin and Sacco (1978).

FLYWHEEL ENERGY STORAGE SYSTEMS

A flywheel (rotor) is a mechanical device that stores kinetic energy by spinning and then releases the energy as the spinning slows down. Flywheels are found most commonly in every automobile that has an internal combustion engine.

The modern flywheel is emerging from the experimental stages. Several flywheel energy storage systems (FESS) have been tested as experimental prototypes since 1976, and such testing will undoubtedly continue during the 1980s. Technological advances achieved in recent years (Chang and Hirschfeld 1978; Renner-Smith 1979; Millner 1979a, 1979b; Notti, *et al.* 1975) indicate that commercial manufacture of small systems (up to 40 kWh) will most likely commence in the mid-1980s in industrialized nations. Subsequent applications of FESS technology in rural areas of developing nations may become significant after 1985.

A comprehensive, interdisciplinary investigation was conducted during 1976-78 to determine the proper use of energy storage in conjunction with solar photovoltaic and wind energy conversion systems (Johnson, *et al.* 1979). Of the 11 energy storage methods considered (aboveground pumped hydro, underground pumped hydro, thermal-oil, thermal-steam, underground compressed air, small pneumatic systems, lead-acid batteries, advanced batteries, flywheels, hydrogen, and superconducting magnetic systems), the flywheel was found to be the second best choice for residential photovoltaic systems. Advanced batteries were the top choice with attractive projected economics after 1985 (General Electric Co. 1978). The flywheel system considered, however, did not possess such beneficial features as power conditioning, conversion, and magnetic bearings. With the recent advances in flywheel technology, the relative attractiveness of flywheels has improved drastically, and today they are considered competitive with batteries. Even if the projected initial cost of flywheels were several times that of batteries, they would still be competitive, since the flywheel can be designed to last the life of a system, without the need for replacement. In most stationary storage system applications, batteries would have to be replaced several times during the anticipated 20-year system life.

A series of flywheels made of high-strength Kevlar (and glass and graphite) fibers and epoxy resin have been fabricated and tested within the last 4 years by several university laboratories and industrial organizations, resulting in much improved performance. The typical range of energy density of such "composite flywheels" is from 40 to 80 Wh/kg of rotor mass (Chang and Stone 1977), as compared to the typical 7-9 Wh/kg for metal flywheels. Another attractive feature of the composite flywheel is its safety: it disintegrates into something resembling cotton candy or ruptured plywood boards, in contrast to metal flywheels which explode into extremely dangerous jagged projectiles (Chang and Stone 1977; Lustenader, *et al.* 1978). Figure 52 shows the rather safe failure mode of a multi-ring composite rotor capable of storing a few kilowatt-hours of energy. It is made of Kevlar/epoxy resin (with small amounts of s-glass), and is mounted on an aluminum hub. Additional recent technological advances are discussed below.

Recent Advances in Flywheel Technology

Although the basic energy input and output of the FESS is electricity, the flywheel system is capable of accepting, and discharging, mechanical shaft power from various sources such as windmills and hydraulic turbines. Flywheels are particularly suited for the photovoltaic system, in which the FESS serves as an interface between a photovoltaic array and an AC load, providing energy storage as well as regulating the output waveform.

Tenth-Scale FESS Unit for Photovoltaics. A recent (1978–1980) experimental FESS unit designed and tested at the Massachusetts Institute of Technology's (MIT) Lincoln Laboratory includes a low-cost flywheel rotor, an

FIGURE 52 The safe failure mode of a composite flywheel.

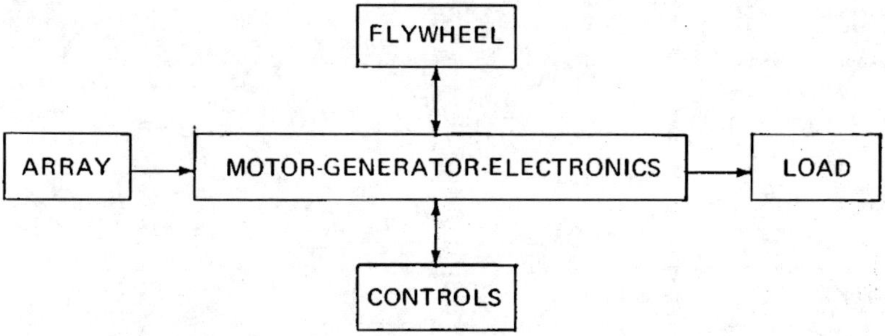

FIGURE 53 Advanced flywheel energy storage and power conditioning.

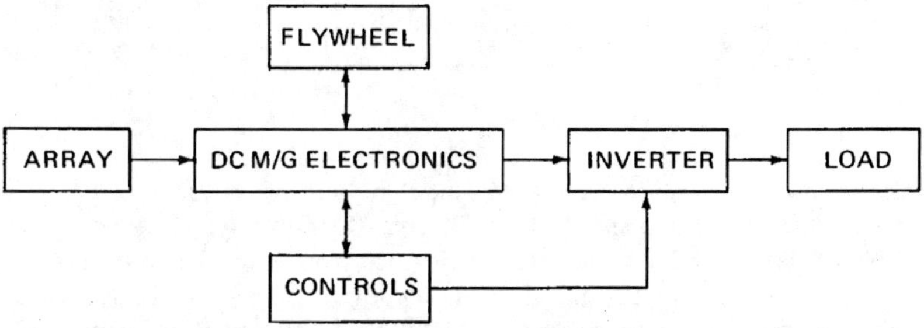

FIGURE 54 Conventional flywheel storage with inverter.

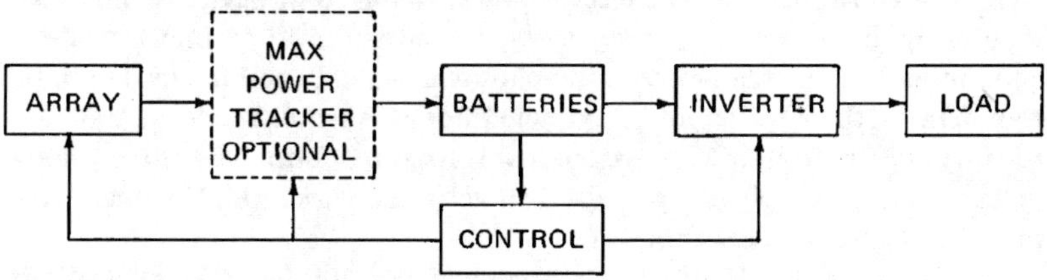

FIGURE 55 Battery storage with inverter.

ironless armature motor-generator, and magnetic bearings (Millner 1979a, 1979b). The schematics of this system are shown in Figure 53.

The MIT system concept contrasts with the more conventional FESS deployment shown in Figure 54, in which an inverter is utilized. In Figure 55, the schematics of a rather common storage/conversion system involving the use of electrochemical batteries are shown for comparison. It can be seen that the advanced MIT system is far simpler than the two competing systems.

The advanced solar photovoltaic/flywheel unit shown in Figures 53 and 56 hinges on the use of an efficient, low-loss motor-generator (Millner 1979a, 1979b). The DC drive motor and a permanent magnetic motor–generator

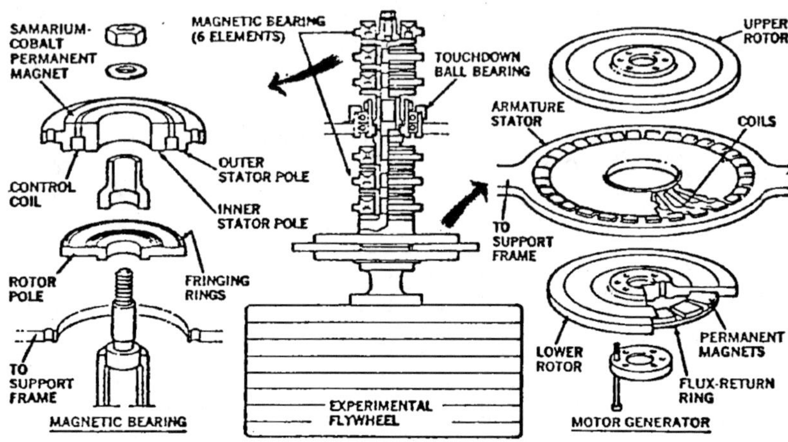

selected for this purpose serve as the input–output mechanisms for the flywheel, which has a top speed of 15,000 rpm. This unit, which stores up to 4 kWh of electricity, has a magnetic bearing that permits low-loss operation.

The flywheel and motor–generator are supported on magnetic bearings in a crude vacuum housing (approximately 10^{-2} Torr). The brushless motor and the generator are in fact the same device with separate input and output electronics. All solar photovoltaic electric power would go through the DC motor to spin up the flywheel. Consequently, the advanced FESS unit provides a good impedance match between the photovoltaic array and the load in order to maximize the electric power extracted from the array. The FESS unit also serves as an inverter to convert the low-voltage DC output from the photovoltaic array to a high-voltage (e.g., 110 volts) AC waveform. Nominally, the motor–generator is rated at 600 W.

As of the end of 1979, this unit contained a 500-lb (227 kg) steel test rotor, which will be replaced by the more desirable low-cost wound rotor and will be ready for prototype testing in 1981. The rotor is expected to coast for days without having to be recharged, and the overall system efficiency is approximately 73 percent.

In the event of foul weather, causing near-zero output from the arrays, the FESS unit will continue to supply electricity for 2–3 days at a satisfactory level.

Full-Size FESS for a Single-Residence Photovoltaic System. Based on the test results of the MIT tenth-scale FESS unit, a full-size unit with a 20-year life is projected to possess the cost characteristics shown in Table 11 (Millner

TABLE 11 Cost characteristics of a full-size FESS for a single-residence photovoltaic system (40 kWh + 8-kW DC + 10-kW AC).

	Cost[a]		
Item	1985 low	1985 high	1978 cost
Rotor[b]	$2,000	$ 4,000	$ 8,000
Motor–generator	750	1,000	1,160
Magnetic bearing	400	800	3,000
Vacuum housing	1,000	1,600	2,680
Generator electronics	300	1,000	1,000
Motor electronics	240	600	800
Enclosure	960	1,280	1,280
Total	$5,650	$10,280	$17,920

[a]All figures are in 1975 U.S. dollars based on American manufacturing costs.
[b]Weight estimates: 4,000 lb (1,814 kg), 2,800 lb (1,270 kg), and 5,000 lb (2,268 kg) for the 1985 low, 1985 high, and 1978 cost, respectively.

1979a, 1979b; Millner and Hay 1979). The low-cost rotor is based on the bare-filament concept developed at the Applied Physics Laboratory of Johns Hopkins University, Baltimore, Maryland (Rabenhorst 1979). The higher rotor cost figure ($4,000) refers to General Electric's (USA) pseudoisotropic fiberglass/epoxy resin design. The vacuum system is based on a tank 1.2 m (4 ft) in diameter, 1.2 m (4 ft) in length, sealed and pumped out every 6 months, and maintaining a pressure of 10 μ maximum. This insures low losses during operation, since no concurrent vacuum pumping is required. This passive vacuum system is only practical with steel wire or steel tape composite rotors. The enclosure is a 3.375 m^3 (119 ft^3) concrete pit. The overall input-output efficiency can be as high as 73 percent, which is some 5 percentage points higher than the competing batteries.

Portable FESS Unit. In 1976, the Garrett Corporation (USA) designed and fabricated a small FESS unit that was capable of receiving 20 Wh of electricity from wall receptacles and used a small AC motor (rated under 76 W). The steel rotor was supported by conventional ball bearings. All components, including a generator, can be neatly packed into a small briefcase with a volume of less than one cubic foot.

Once charging is completed, the unit can be carried around with ease, with the gyroscopic effect felt only when rapid turning motions are applied to the FESS case. The flywheel energy can be converted upon demand to alternating current through conventional power generation and inversion techniques.

Although no attempt was made to optimize the performance of this unit, the rundown time was found to be on the order of nearly an hour. Indeed, the whole unit behaved much like a portable battery, with the self-contained capability for AC–DC–AC conversion.

Integrated Power and Control System. In the mid-1970s, a flywheel unit, known as IPACS, was developed and tested by Rockwell International (USA) for spacecraft electrical power storage as well as attitude control. This special FESS unit has a single titanium rotor rated at 2.5 kW which stores up to 1.5 kWh. Rotor diameter is 454.1 mm (17.9 in.), and it operates with a maximum speed of 35,000 rpm. The Samarium-Cobalt permanent magnet rotors for the dual-purpose motor-generators are integral with the shafts of the flywheel.

Inductor Motor/Alternator/Flywheel Energy Storage System. From 1976 to 1978, General Electric Co. (USA), partially supported by the U.S. Energy Research and Development Administration (ERDA), undertook a FESS development project, the Inductor Motor/Alternator/Flywheel Energy Storage System (Lustenader, *et al.* 1978).

The entire unit, diagrammed in Figure 57, was made into a single package, hermetically sealed, with the rotor assembly operating in a near vacuum with an inert residual gas (helium). Although the unit was primarily developed for electric vehicles, it could be adapted for other applications, such as solar photovoltaic or wind power systems.

The demonstration unit, shown in Figure 58, consists of a 20-kVA, solid-rotor, inductor-type, synchronous motor/alternator coupled directly to a three-disk composite rotor. The entire rotor assembly is hermetically sealed and operates in the range of from 630 to 762 mm (25 to 30 in.) of mercury vacuum. The rotor, 756 mm (30 in.) in diameter, is made of fiberglass layups, operating between 7,500 and 15,000 rpm. The usable energy is 1 kWh, with a rotor weight estimated at 60 kg (132 lb).

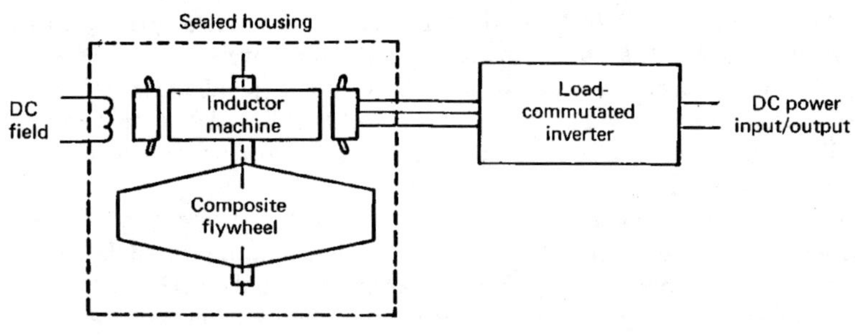

FIGURE 57 Basic concept of the General Electric Inductor Motor/Alternator/Flywheel Energy Storage System.

FIGURE 58 Inductor Motor/Alternator/Flywheel Unit.

This demonstration established the technical feasibility of the underlying concept, but also pointed out the need for better understanding of the composite flywheel.

Residential Flywheel with Wind Turbine Supply. Flywheels can be used to store excess wind energy for a windless day (Place 1979). In late 1978, the Garrett Corporation (USA) was commissioned by the U.S. Department of Energy to develop an engineering design for a mini-residential FESS rated at 5-kW electric output. Then-current low technology was utilized in this design which was completed in late 1979.

After evaluating five candidate systems, the design shown in Figures 59 and 60 was selected for further refinement. Considering cost and safety, the E-glass/epoxy, rim-shaped rotor was selected. The flywheel is housed in a metal container evacuated to less than 10 μ.

If it were produced in quantity (1,000 units per year), the 10-kWh FESS is estimated to cost $3,300 per unit, excluding installation. The unit storage cost, therefore, is $330/kWh, which is slightly lower than the corresponding figure ($448/kWh) for the MIT FESS discussed earlier. However, this unit cost is higher than the average MIT flywheel storage cost based on 1985 technology (i.e., $200/kWh in 1975 dollars).

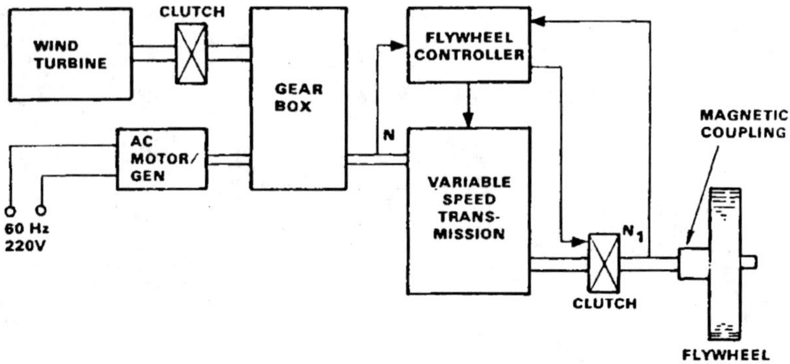

FIGURE 59 Residential FESS with wind turbine supply.

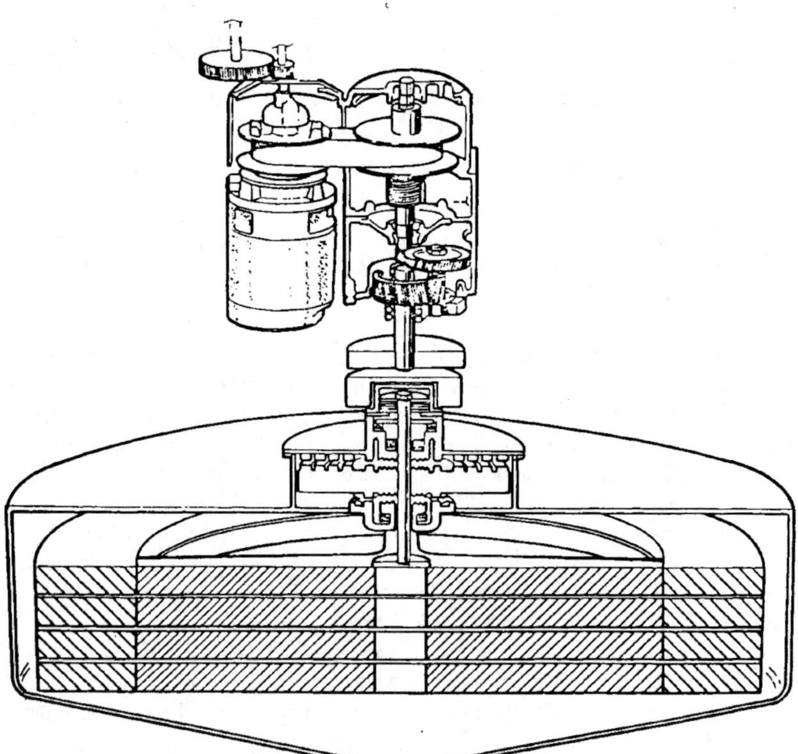

FIGURE 60 A 10-kWh flywheel energy storage unit.

Projected Technological Trends

During the past 4 years, research and development on flywheel energy storage systems has emphasized flywheels used in transportation largely because alternate energy systems (notably solar and wind energy) are still in the early stages of development. Within a few years, the emphasis may shift toward "stationary" flywheel storage systems intended for residential or industrial applications. Such a shift, aided by the natural technology transfer process in which advances associated with transportation flywheels (and other technologies in general) are incorporated into stationary flywheel systems, will greatly accelerate the development of the stationary flywheels under discussion.

Based on the literature to date, and discussions with professionals in the energy and flywheel areas, some aspects of the FESS technological trends can be projected (Table 12). The FESS system envisioned will likely be capable of accepting either direct current from solar photovoltaic arrays or mechanical shaft power from wind turbines. Alternatively, the unit should be capable of receiving power from wall receptacles connected with a central generating power system—possibly a large solar electric system or a conventional electric

TABLE 12 Stationary FESS performance projections.

Factors	Current	Near term (4-5 yr)	Far term (9-10 yr)
Isotropic material rotor energy density (Wh/kg)	7-32	11-31	11-31
Composite rotor energy density (Wh/kg)	22-44	66-88	66-132
Ball bearings			
Life (yr)	3-5	3-5	3-5
Losses (percent of stored energy per hour)	2-4	1-2	1-2
Magnetic bearings			
Life (yr)	A few	6-8	10-20
Losses (percent of ball bearings)	7-10	5-7	<3
Motor-generator efficiency (percent)	60-94	88-97	88-97
System life (yr)	3-5	10-15	20
System energy density (Wh/kg)	6-11	35-44	35-88
System spinning loss (percent of energy per hour)	3-5	1-3	0.5-1.0
Round-trip efficiency (percent)	68-76	73-77	78-83
System cost (US$/kWh)[a]	300-600	200-300	130-170

[a]With complete storage and conversion capabilities; in 1976 dollars.

utility. This residential flywheel unit could have a storage capacity of 20–40 kWh, with a maximum power rating of 5–10 kW, although a much larger growth version of this machine should not be dismissed. Its construction will be dictated by the operating economics as well as reliability (and safety).

Overall efficiency (electrical to electrical) of the FESS unit will probably be more important to the user than the system's energy density. Most likely, the rotor will be made of inexpensive, easy-to-fabricate types of material, and will be supported by magnetic bearings featuring low-cost permanent magnets, operating in a hermetically sealed, crude vacuum of the type discussed (Lustenader, *et al.* 1978; Millner 1979a, 1979b; Place and Rabenhorst 1979). In short, the trend will be for the development of a safe, rugged, highly economical FESS unit.

Toward pursuing these objectives, DOE has taken steps, through its flywheel program at the Lawrence Livermore Laboratory (Livermore, California), to initiate a full-scale home flywheel storage program, and has received proposals from a number of industrial firms. This hardware program, which is expected to run for about 2 years, will emphasize low-cost, fail-safe flywheel technology, low-loss magnetic flywheel suspension, low-loss vacuum systems, and low-cost but highly efficient electrical machines. It will also emphasize commercialization prospects. The program will begin with development of a 10-kWh flywheel energy storage system, and will culminate in a 40-kWh system.

FESS units will likely become more widely utilized when rural areas of the developing world begin using more solar energy—particularly photovoltaic systems, wind turbines, and solar thermal electric-generating systems. FESS units may also be used to store cheap off-peak electricity on a daily basis from a central generating source (such as a conventional utility or a rural alternate energy co-op). Because of the low load factors for central station electricity in many rural areas, this could well be the principal use of the FESS in the next 10 decades.

REFERENCES

Bechtel National, Inc. (1980) Handbook for Battery Energy Storage in Photovoltaic Power Systems. Report SAND-80-7022. U.S. Department of Energy, Washington, D.C., USA.

Bifano, W. J., A. F. Ratajczak, P. M. Bahr, and B. G. Barrett (1979) Social and economic impact of solar electricity at Schuchuli Village: a status report, *Proceedings of Seminar on Solar Technology in Rural Settings; Assessment of Field Experiences*. Sponsored by the United Nations University, June 1–2, 1979, Atlanta, Georgia, USA.

Chang, G. C. and F. Hirschfeld (1978) For the latest in energy storage, try the flywheel, *Mechanical Engineering*. February.

Chang, G. C. and R. G. Stone, eds. (1977) *Proceedings of the 1977 Flywheel Technology Symposium*. U.S. Department of Energy CONF 771053, October 5–7, 1977, San Francisco, California, USA.

Doshi, H. N. (1977) Taking power to rural areas, *Proceedings of the Second National Conference on Power Sources*. November 25–26, 1977, Karaikkudi, India.

General Electric Co. (1978) Applied Research on Energy Storage and Conversion for Photovoltaic and Wind Energy Systems. Report No. HCP/T-22221-01/1. Prepared for the U.S. National Science Foundation and the U.S. Department of Energy, Washington, D.C., USA.

Hirst, P. and D. H. Rees (1976) The regulation, storage, and conversion of wind produced electrical energy at the level of a few hundred watts, *Proceedings of the International Symposium on Wind Energy Systems*. September 7–9, 1976, Cambridge, England.

Johnson, A. W., R. L. McCarthy, and G. C. Chang (1979) Energy storage and conversion in conjunction with photovoltaic energy systems, *Proceedings of the 14th Intersociety Energy Conversion Engineering Conference*. August 5–10, 1979, Boston, Massachusetts, USA.

Littler, J. G. F. and R. B. Thomas (1978) Wind generation of electricity for a novel dwelling independent of servicing networks, *Proceedings of the International Solar Energy Society Congress*. Pergamon Press, Oxford, England.

Lustenader, E. L., *et al.* (1978) Laboratory evaluation of a composite flywheel energy storage system, *Proceedings of the 13th Intersociety Energy Conversion Engineering Conference*. August 21–25, 1978, San Diego, California, USA.

Matlin, R. W. and D. B. Sacco (1978) Photovoltaic projects at MIT Lincoln Laboratory, *Proceedings of the Workshop on Battery Storage for Solar-Photovoltaic Energy Systems*. Argonne National Laboratory Publication ANL/OEPM-78-3, Argonne, Illinois, USA.

Millner, A. R. (1979a) A Flywheel Energy Storage and Conversion System for Photovoltaic Applications. Paper presented at the International Assembly on Energy Storage, May 28–June 1, 1979, Dubrovnik, Yugoslavia.

Millner, A. R. (1979b) Flywheels for energy storage, *MIT Technology Review*. November.

Millner, A. R. (1979c) Mead Photovoltaic System Controller. Lincoln Lab Document No. C00-4094-27. Massachusetts Institute of Technology, Lexington, Massachusetts, USA.

Millner, A. R. and R. D. Hay (1979) Flywheel energy storage interface unit for photovoltaic applications, *Proceedings of the 14th Intersociety Energy Conversion Engineering Conference*. August 5–10, 1979, Boston, Massachusetts, USA.

Notti, J., *et al.* (1975) Economic and Technical Feasibility Study for Energy Storage Flywheels. Publication No. 76-65. U.S. Energy Research and Development Administration, Washington, D.C., USA.

Place, T. W. and D. W. Rabenhorst (1979) Residential flywheel with wind turbine supply, *Proceedings of the 14th Intersociety Energy Conversion Engineering Conference*. August 5–10, 1979, Boston, Massachusetts, USA.

Rabenhorst, D. W. (1979) Low cost flywheel energy storage system demonstration, *Proceedings of the 14th Intersociety Energy Conversion Engineering Conference*, August 5–10, 1979, Boston, Massachusetts, USA.

Ratajczak, A. F. and W. J. Bifano (1979) Description of Photovoltaic Village Power Systems in the United States and Africa. Report No. NASA TM-79149. NASA–Lewis Research Center, Cleveland, Ohio, USA.

Renner–Smith, S. (1979) Basement flywheel, *Popular Science*. October.

Romaine, W. R. (1979) The Mead, Nebraska, 25-kW Photovoltaic Power System. Lincoln Lab Document No. C00-4094-10. Massachusetts Institute of Technology, Lexington, Massachusetts, USA.

Solarex Corporation (1976) Applications Note 6001-2. Rockville, Maryland, USA.

Solarex Corporation (1977) Applications Note 6903. Rockville, Maryland, USA.

Solarex Corporation (1978) Product Bulletin 11-1-78. Rockville, Maryland, USA.

Solarex Corporation (1979) Applications Note 6907-1. Rockville, Maryland, USA.

Pedal Power

INTRODUCTION

One of the most efficient mechanical devices developed by man is the bicycle, and pedaling is the key to the most efficient use of leg power. The thigh muscle (the quadricep) is the largest and most powerful muscle in the human body. Through pedaling, this muscle can be used to perform a variety of useful tasks.

Pedaling enhances the power of the legs, which are then relieved of the effort of standing or supporting the body. The usual pedaling speed, 60–80 rpm, uses leg muscles at their maximum efficiency. Studies have shown that this speed yields somewhat less than 0.1 hp (74.6 W). In some cases, professional cyclists have produced up to 1 hp for short periods of time.

Pedal power is particularly appropriate when small amounts of power are needed at irregular intervals, or when a mobile power source is desirable. Because many pedal power devices are lightweight, they can be carried easily to a worksite to power such devices as winnowers, threshers, and water pumps.

Pedal power units are sometimes referred to as "dynapods" (Greek for power and foot), and usually consist of a stationary bicycle frame, seat, handlebars, and pedals which drive a series of chains or belts. The number of revolutions per minute (rpm) can be changed by adding or subtracting pulleys or gears. Usually a flywheel is necessary to smooth out the high and low power output of a person's natural pedaling rhythm. The use of the flywheel assures constant speed and power to the device being driven by the dynapod (see Weir 1979).

During the past 5 years, new pedal power technologies have evolved that can be used in the transportation, agricultural, and domestic sectors of developing countries. Most of the technologies described in this chapter are discussed in more detail in a book entitled *Pedal Power In Work, Leisure, and Transportation* (McCullagh 1977).

TRANSPORTATION

Oxtrike

A carrier bicycle with a large basket over the front wheel is effective for loads up to 100 lb (45 kg), but heavier weights require three wheels. The Oxtrike (see Figures 61, 62, and 63), designed by Stuart Wilson of Oxford University, is a basic three-wheeled chassis with a choice of bodies to carry a variety of goods or people with a payload up to 330 lb (150 kg).

For the sake of greater strength, the wheels of the Oxtrike are 20 in. (51 cm) in diameter. The rear wheels—Raleigh "Chopper" type (20 × 2.12 in. or 51 × 5.4 cm)—are especially strong against side load, a requirement that

FIGURE 61 Oxtrike chassis. (Reprinted from *Pedal Power* © 1977. Permission granted by Rodale Press, Inc., Emmaus, PA 18049)

does not arise on a bicycle but is a very real problem on a tricycle. They also lower the center of gravity of the load (but give adequate ground clearance), and permit easier access by elderly or infirm passengers and a full-width rear

FIGURE 62 Up-ended view of the Oxtrike chassis, allowing ability to park in a small space, easy inspection of transmission and brakes, and ability to tip out the load. Note Sturmey–Archer three-speed hub gear with double adjustment for primary and secondary chains. (Reprinted from *Pedal Power* © 1977. Permission granted by Rodale Press, Inc., Emmaus, PA 18049)

seat without an increase in the overall vehicle width, which is only 36 in. (91.4 cm). The use of small wheels also reduces the overall length of the tricycle to just over 6.6 ft. (2 m). The front wheel (20 × 1.75 in. or 51 × 4.4 cm) and fork are of the carrier-bicycle type and are designed to carry a large load forward from the frame over the front wheel.

The Oxtrike's three-speed gearbox, a standard Sturmey–Archer hum gear, is used as an intermediate gearbox, as on many motorcycles, with a primary and secondary chain. The overall "gear" of the Oxtrike is 31.5 in. (80 cm) in the bottom gear, 42 in. (1 m) in the middle, and 56 in. (1.4 m) in the top, compared to 66.5 in. (1.7 m) for a normal bicycle with 26-in. (66-cm) wheels, which greatly improves the driver's ability to start with a heavy load and to climb at least a slight gradient, thus improving the range and mobility.

The front brake is the standard pull-rod stirrup type, but the rear brakes are inboard band brakes applied by a foot pedal. This is a powerful and effective method of applying a braking force; each wheel has its own brake drum, mounted at the inboard end of the half-shaft, and the braking effect is applied equally by means of a balance bar. Location of the brakes ensures protection from rain. The brake pedal can be held down by a lever catch to act as a parking brake, an essential feature on a tricycle.

The construction of a normal bicycle frame requires thin-walled tubing brazed into special sockets, which is not really suitable for small-scale local manufacture in developing countries. For this reason, the Oxtrike was des-

FIGURE 63 Oxtrike with temporary seat. (Reprinted from *Pedal Power* ©1977. Permission granted by Rodale Press, Inc., Emmaus, PA 18049)

signed to use a mild steel sheet of a standard thickness, which can be readily cut on a foot-powered guillotine, folded on a hand-operated folding machine, and joined by almost any method of welding, brazing, or riveting.

A variety of bodywork types has been designed, including a two-passenger rickshaw with a hinged tailboard. When the tailboard is lowered, two passengers can sit facing backward; when it is raised, three children can sit facing forward or goods can be transported. Larger loads can be carried with the tailboard horizontal.

The chassis can be fitted with a simple open truck body or an enclosed box body for carrying parcels. A hopper body with sloping ends would be suitable for carrying sand, gravel, or other loose materials (the size of the hopper should be restricted to prevent overloading). Finally, a 40-gal (151-l) drum can be fitted for carrying liquids.

Railbike

The railbike is a regular bike fitted with special attachments so that it can be ridden on train rails (see Figures 64 and 65), and it can take many forms and shapes. It can be fitted with a stationary bar between the bicycle and a third wheel, or it can take the form of a scooter that effectively applies human power.

FIGURE 64 The railbike. (Reprinted from *Pedal Power* © 1977. Permission granted by Rodale Press, Inc., Emmaus, PA 18049)

FIGURE 65 Details of railbike attachments. (Reprinted from *Pedal Power* © 1977. Permission granted by Rodale Press, Inc., Emmaus, PA 18049)

A U.S. inventor, Mark Hansen, of Onamia, Minnesota, has enlarged the concept of the railbike by building a vehicle that will support 2-3 people or 700 lb (315 kg) (see Hansen 1976).

AGRICULTURE

Pumping

Probably one of the oldest applications of pedal power is pumping water. Wood is an excellent material for constructing water pumps. In low salt-flat areas, corrosion of metal pumps would cause maintenance problems. In Taiwan, foot pumps made in the ancient Chinese way use few if any metal parts (see Figure 66). The water being pumped acts as a lubricant, making such a pump quite easy to operate. In large, flat, marshy areas, the water usually needs to be lifted about 2 ft (0.6 m) and the depth of the evaporation ponds is only a few inches. Two people work together to operate the pump as the power rotor has two sets of pedals.

At the International Rice Research Institute (IRRI) at Los Baños, Laguna, the Philippines, engineers have developed a small, lightweight, inexpensive foot-powered pump that will lift large quantities of water several feet, using only moderate amounts of human power. The operator simply stands on two footrests at either end of the pump and rocks back and forth (see Figure 67).

FIGURE 66 Old-style wooden water pumps used in Taiwan for raising sea water into salt-evaporation beds at a coastal town. (Reprinted from *Pedal Power* © 1977. Permission granted by Rodale Press, Inc., Emmaus, PA 18049)

This effort compresses a diaphragm, which forces water from the outlet valve. By operating the pump in a rhythmic manner, a continuous flow of water is pumped. It is estimated that the foot pump will cost only $25–$30 to manufacture. IRRI gives designs for its machines at no charge to fabricators throughout Asia.*

Pedal power can also be applied to existing hand pumps. Figure 68 depicts a pitcher pump; an activator rod from a reciprocating pedal power device can supply the up and down motion to pump water.

A simple, inexpensive (less than $5) centrifugal pump can also be attached to a pedal-driven device (Figure 69). These devices can pump 3,500 gal/hr (13,248 l/hr), sending water streaming out 30 ft (9 m) through a hose nozzle.

*More information on this technology is available from the International Rice Research Institute, Agriculture Engineering Department, Post Office Box 933, Manila, Philippines, and from Volunteers in Technical Assistance, 3706 Rhode Island Avenue, Mt. Rainier, Maryland 20822, USA.

FIGURE 67 Pedal pump designed by the International Rice Research Institute, Los Baños, Laguna, the Phillippines. (Reprinted from *Pedal Power* © 1977. Permission granted by Rodale Press, Inc., Emmaus, PA 18049)

Plowing, Cultivating, and Harrowing

The research and development laboratories of Rodale Press, Inc. (Emmaus, Pennsylvania (USA) have developed a pedal power farming system, the Mechanical Mule™ (see Figure 70). This two-person operation consists of a pedal power winch that is used to reel in a steel cable attached to a mobile tool carrier. The tool carrier can be equipped with a plow, harrow, cultivator, or other tools, all quickly interchangeable.

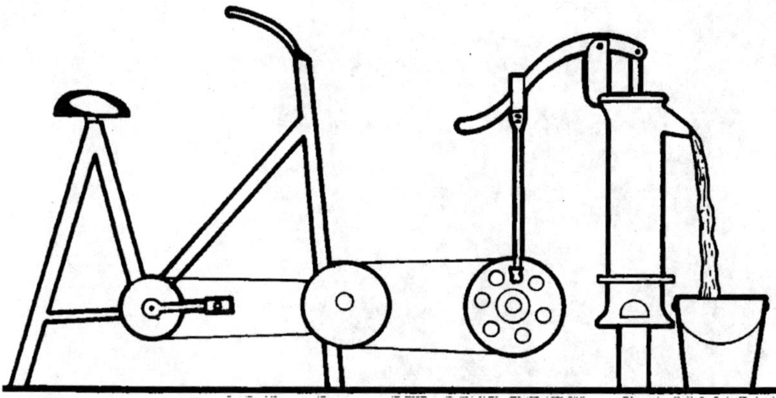

FIGURE 68 Pitcher pump (prototype).

FIGURE 69 Pumping water using a centrifugal pump and a pedal-driven device. (Reprinted from *Pedal Power* © 1977. Permission granted by Rodale Press, Inc., Emmaus, PA 18049)

Seated in the pedal power unit which is positioned at the head of the garden row (spade blades keep it in position), the operator pedals to reel in the cable attached to the tool carrier, which is guided by a helper. The tool carrier, equipped with the right tool for the desired operation (say, a plow), is

FIGURE 70 Plowing operation using the Mechanical Mule™.

positioned at the foot of the same row. When the tool carrier reaches the head of the row, the operator stops pedaling and pushes a lever that slips the reel into "free-reeling" position. The helper then wheels the tool carrier to the foot of the next row, thus unreeling the cable. Simultaneously, the operator moves the pedal power unit to the head of the next row. This sequence is followed until the entire plot of ground is cultivated or plowed.

The pedal power unit or winch (Figure 71) is designed with a three-speed sprocket-and-chain drive (Figure 72) which permits over 1,000 lb (453 kg) of pull with normal pedaling effort. "Low gear" is used for jobs requiring a slow, powerful pull, such as plowing through heavy turf. "Second" or "high" gears are used for easier jobs, such as harrowing or cultivating. The pedal power is much faster than hand spading and raking. Because the unit's design is simple, repairs or adjustments can be handled using simple tools. The unit can be built for less than $150 in quantities of 25 units.

DOMESTIC APPLICATIONS

Most domestic chores require a stationary power source. The use of pedal power in such a capacity necessitates overcoming a number of problems, in-

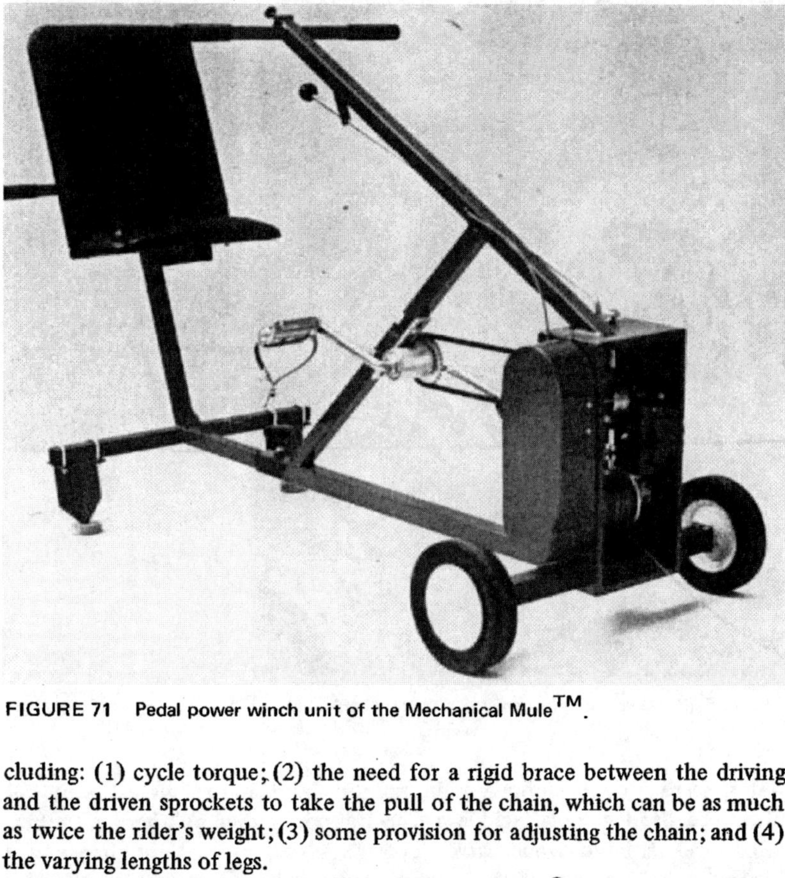

FIGURE 71 Pedal power winch unit of the Mechanical Mule™.

cluding: (1) cycle torque; (2) the need for a rigid brace between the driving and the driven sprockets to take the pull of the chain, which can be as much as twice the rider's weight; (3) some provision for adjusting the chain; and (4) the varying lengths of legs.

The principle behind the stationary Energy Cycle®-powered generator is simple. The energy generated by pedaling is transmitted by a power take-off head, and a conventional chain-and-sprocket is used as the drive chain. The input shaft of the machine or tool to be powered is coupled directly to the power take-off shaft (Figure 73).

If the unit is expected to do a number of jobs, it must be capable of speeds of from 60 to 1,800 rpm. When the higher speed is required, gearing up will be necessary. However, it is best to use a 1:1 ratio whenever possible. As a note of caution, one should never attempt to power a mechanism that requires more than a one-third horsepower motor; one-fourth horsepower or less is feasible.

Stationary pedal-powered units can perform a variety of food preparation tasks, including food grinding, food shredding, dough kneading, batter beating, butter churning, etc. Farm machinery applications include pumping

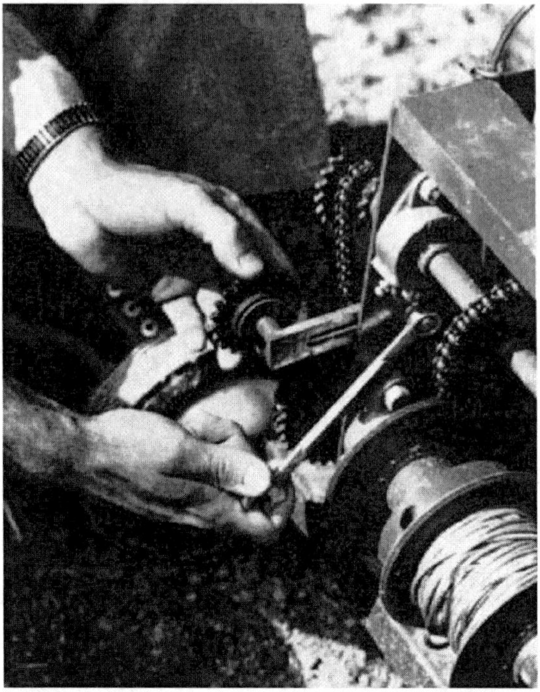

FIGURE 72 Gearshift mechanism of the Mechanical Mule™.

water, shelling corn, cleaning grain, polishing rice, and rolling oatmeal. Still other pedal-tested tools include a wheel grinder, stone polisher and buffer, drill, jeweler's lathe, wood carver, potter's wheel, and battery charger (see Figure 74).* Finally, a stationary pedal-powered unit can provide power for television sets and other electronic equipment. A simple kit—auto generator, 12-volt battery, and inverter—makes it possible to generate and store electricity for use in rural areas. It is possible to use an inverter that converts the

*Plans and more information on pedal-powered devices are available from a number of sources. Plans for a bicycle-operated power take-off unit are available from the American Society of Agricultural Engineers, Box 440, St. Joseph, Michigan 49085, USA. Intermediate Technology Publications, Ltd., 9 King Street, London WC2E 8HN, England, has more information available on a pedal-powered cassava grinding machine. Information on foot-operated machines is available from the American Village Institute, Route 3, Box 3486, Selah, Washington 98942, USA, and information on a foot-powered wood lathe is available from Volunteers in Technical Assistance, 3706 Rhode Island Avenue, Mt. Rainier, Maryland 20822, USA. *Free-wheeling Magazine*, 37 Warrender Park Road, Edinburgh EH1 3JT, England, provides current information on pedal power technologies.

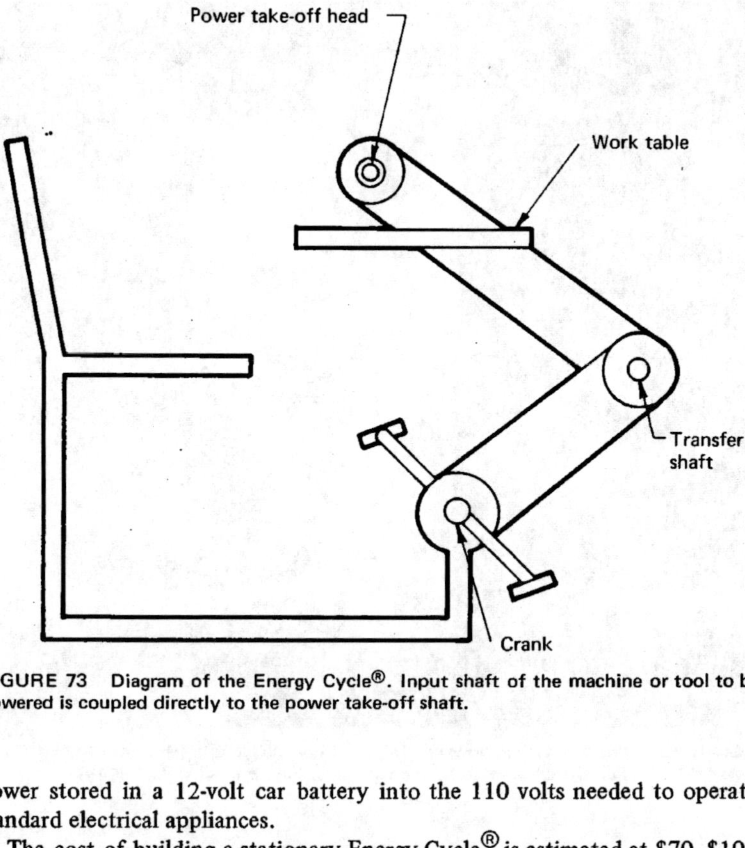

FIGURE 73 Diagram of the Energy Cycle®. Input shaft of the machine or tool to be powered is coupled directly to the power take-off shaft.

power stored in a 12-volt car battery into the 110 volts needed to operate standard electrical appliances.

The cost of building a stationary Energy Cycle® is estimated at \$70–\$100 in new and used parts, plus about 12 hours of labor.*

Other stationary devices include the Bik-O-Generator†, which will grind grain or generate electricity (Figure 75), and a commercially available maize sheller which is driven by a very simple but effective form of a bolted-on pedal drive unit. The latter is sold in several countries in Africa and can shell two cobs at a time. A heavy cast-iron fan serves both as flywheel and winnower to blow away the husks.

*Instructions for building the Energy Cycle® are contained in *Pedal Power in Work, Leisure, and Transportation*, edited by James C. McCullagh, Rodale Press, Emmaus, Pennsylvania, USA, 1977.

†Plans for this device are available from Homestead Industries, Ananda Village, Nevada City, California 95959, USA, and from the North Shore Ecology Center, 3070 Dato, Highland Park, Illinois 60035, USA.

FIGURE 74 Potter's wheel powered by the Energy Cycle®. (Reprinted from *Pedal Power* © 1977. Permission granted by Rodale Press, Inc., Emmaus, PA 18049)

Another important application, shown in Figure 76, is a four-person, pedal-powered grain thresher/mill for sorghum, millet, and wheat (Weir/VITA 1979).* An inexpensive two-person dynapod that can be used to power a variety of mechanical devices or used for limited electrical generation (200 W) has also been developed (Weir/VITA 1979).

*Working diagrams for construction of the four-person, pedal-powered grain thresher/mill are available from Volunteers in Technical Assistance, 3706 Rhode Island Avenue, Mt. Rainier, Maryland 20822, USA. More information on a thresher for rice and other small grains is available from the South Pacific Appropriate Technology Foundation, Post Office Box 6937, Boroko, Papua New Guinea. Complete technical drawings for a treadle-operated peanut thresher are available from Intermediate Technology Publications, Ltd., 9 King Street, London WC2E 8HN, England. One should inquire about the prices of these plans before ordering.

FIGURE 75 The Bik-O-Generator will power a grain mill or generate electricity. A pulley attached to a power take-off wheel could be attached to many other belt-powered tools as well. (Reprinted from *Pedal Power* © 1977. Permission granted by Rodale Press, Inc., Emmaus, PA 18049)

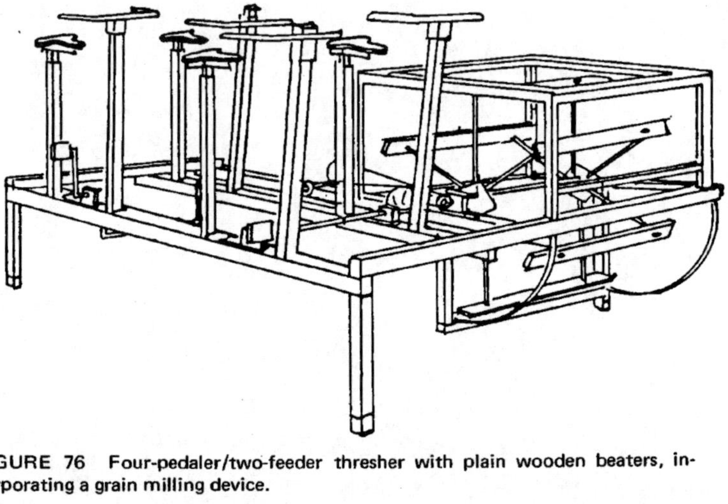

FIGURE 76 Four-pedaler/two-feeder thresher with plain wooden beaters, incorporating a grain milling device.

REFERENCES

Hansen, M. (1976) Rail bike, *Alternative Sources of Energy*. No. 23, December.

McCullagh, J. C. (1977) *Pedal Power in Work, Leisure, and Transportation*. Rodale Press, Emmaus, Pennsylvania, USA.

Weir, A. (1979) The Dynapod: A Pedal-Power Unit. Volunteers in Technical Assistance, Mt. Rainier, Maryland, USA.

Weir, A./VITA (1979) Four-Person Pedal-Powered Grain Thresher/Mill. Volunteers in Technical Assistance, Mt. Rainier, Maryland, USA.

Stirling Engines

Stirling engines are external combustion heat engines that usually utilize air or other gases as working fluid. They are capable of operating on any source of heat such as sun, wood, coal, or field waste. The Stirling cycle lends itself to a wide, and by no means fully explored, variety of physical forms. Some of these forms are sufficiently simple to make and yet sufficiently effective to be strong candidates for power generation in developing countries, using any local fuel or solar energy as their heat source.

There has been a rapid increase in Stirling engine research and development since publication of the first edition of *Energy for Rural Development*. This is evidenced by the number of papers published yearly in the proceedings of the Intersociety Energy Conversion Engineering Conference (IECEC), the recognized forum for dissemination of the latest in Stirling engine work. (See Appendix C for suggested reading.) Of this work, an increasingly larger percentage is being directed toward the free-piston Stirling engine, reflecting the fact that its lack of critical seals and its mechanical simplicity make it the favored form for many applications.

RECENT TECHNOLOGICAL ADVANCES

A large proportion of the work in the United States, the Netherlands, and Sweden is being directed toward automotive applications under the impetus of the heavily funded U.S. Government automotive Stirling engine program. Significant work on solar-electric Stirling engines is also being conducted under the auspices of the U.S. Department of Energy (DOE), targeted toward development of a highly efficient, point-focusing Stirling electric system in the early 1980s. The target price of power from this system is $.05/kWh.

Several small groups in the United States, United Kingdom, the Netherlands, India, and South Africa are directing their efforts toward technology appropriate for developing country use. These groups have made advances in liquid-piston water pumps, free-cylinder water pumps, low-pressure air engines, diaphragm engines, and cooling machines, all potentially applicable in developing countries. (See Appendix B for a listing of research and development centers presently engaged in work on Stirling engines.)

Commercial production and sale of Stirling engines have not yet started but appear to be imminent for several uses, one of which is water pumping. Simple liquid-piston water pumps of reasonable efficiency and reliability are in the late stages of prototype development. Free-cylinder water pumps, in which a free-piston Stirling engine is arranged so that it pumps by direct action of its hermetically sealed cylinder (pressure enclosure) on the water, are now well proven and may attract investment by manufacturers if they see a significant commercial market in the near future. Both the liquid-piston and the free-cylinder Stirling engines can use solar, solid fuel, or biogas heat sources, and both are simple devices easily repaired by any bicycle mechanic.

Modern versions of the crank-drive hot-air (Stirling) engines are far more powerful for their size than earlier versions of this engine, due to better understanding of heat transfer, better materials, and computer optimization procedures. For this reason, the time seems right for the reemergence of the hot-air Stirling engine in its modern form. One new design produces 1 kW of shaft power at 1,200 rpm with a slightly pressurized, dry crankcase using permanently lubricated bearings. This machine is readily adapted to biogas or solid fuels, rice husks, or bagasse, and has universal application as a result of its rotating-shaft power output. It is quite simple mechanically and can be designed for fabrication in motorcycle repair shops, which are found in almost any city in developing countries.

Stirling engines are also being developed for cooling tasks, driving either Rankine-cycle compressors or Stirling-cycle coolers. Both free-piston and rotating-shaft machines are being developed in the United States and Europe for food freezing and domestic heat pumps, for both heating and cooling. Possibly the simplest cooling engine, and the one most likely to be applied to developing countries, is the duplex Stirling, free-piston, heat-driven heat pump. This machine contains both the heat engine and the heat pump within the same hermetically sealed pressure enclosure, which is free of lubricants or any fluid other than the working gas. Applications of the duplex Stirling engine include food coolers or freezers, refrigerators, portable vaccine coolers, and living space air conditioning. These machines could probably best be driven by biogas, alcohol, or other liquid or gaseous fuels, but could be adapted to operate on solids such as coal, wood, rice husks, bagasse, etc.

Another emerging technology of importance to the developing world is the large stationary Stirling engine, ranging from 100 kW to several megawatts of power output. These machines are under development in the United States for stationary power generation and cogeneration using solid fuels and would be useful for those purposes elsewhere. Their first cost per kilowatt is projected to be about 60–80 percent higher than an equivalent diesel engine, but payback times can be quite short if a locally available, inexpensive fuel is used instead of increasingly expensive diesel fuel.

The following sections describe in detail the characteristics of some of the most promising varieties of Stirling engines.

Low-Pressure Air Engine

Figure 77 shows the general arrangement of a modern low-pressure air engine of moderate mechanical complexity. Its mechanism is similar to that of the successful Philips hot-air engine of the 1950s, but it uses solid fuel. A laboratory version of a similar engine operating at an average pressure of 1.5 bar, a rotational speed of 900 rpm, and a piston displacement of about 3 l (.8 gal) was found capable of about 1 kW of shaft power (Senft 1979). A slightly

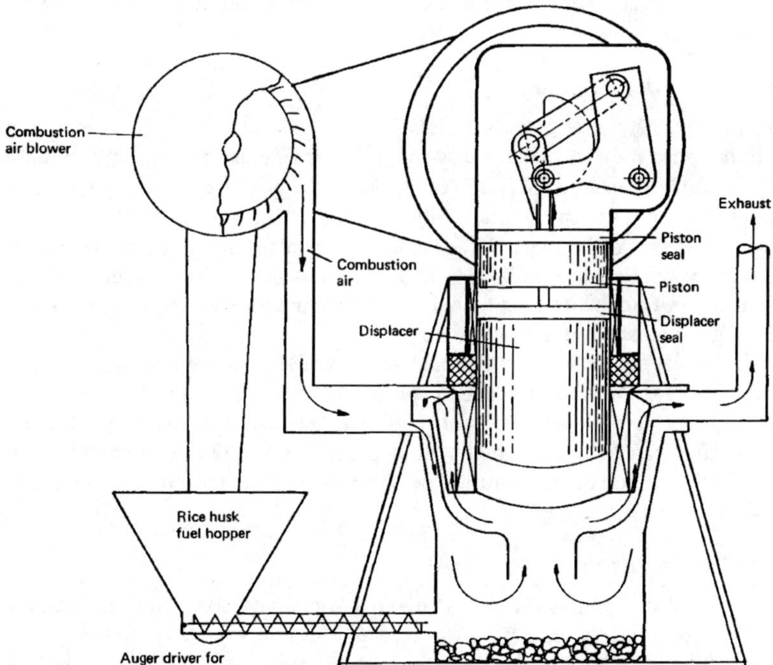

FIGURE 77 Low-pressure air engine.

FIGURE 78 Low-pressure engine designed specifically for pumping water, using a solid fuel.

higher pressure version specifically intended for water pumping using solid fuel is now being developed by Sunpower, Inc. (USA). Its design criteria are simplicity, repairability, low maintenance, and insensitivity to heater fouling (Figure 78).

Free-Cylinder Engine

These extremely simple mechanisms have only a piston, a displacer, and a cylinder that move to produce work. In their simplest form, they use only an annular gap around the displacer to serve the heat exchange and regenerative function. They are hermetically sealed and can use hydrogen or helium without danger of leakage. These engines have an advantageous stroke–force–frequency relationship with very high forces available as cylinder stroke is reduced, and adequate speed stability as external load is removed, thus permitting harmless operation under no load.

Figure 79 shows a small demonstration model arranged as a water pump (Beale 1979). The thermodynamic performance of these machines is in no way inferior to that of more conventional free-piston or crank Stirling engines. They can be used to pump fluids, dig, process grain, and can perform many other tasks including turning a shaft through a tension cable wrapped around a ratchet or sprag clutch.

Free-Piston Alternator Engine

Used today as demonstrators, as research tools, and for other special purposes, these engines are ready for mass production when a market appears. Figure 80 shows the interior arrangement of a typical small alternator engine

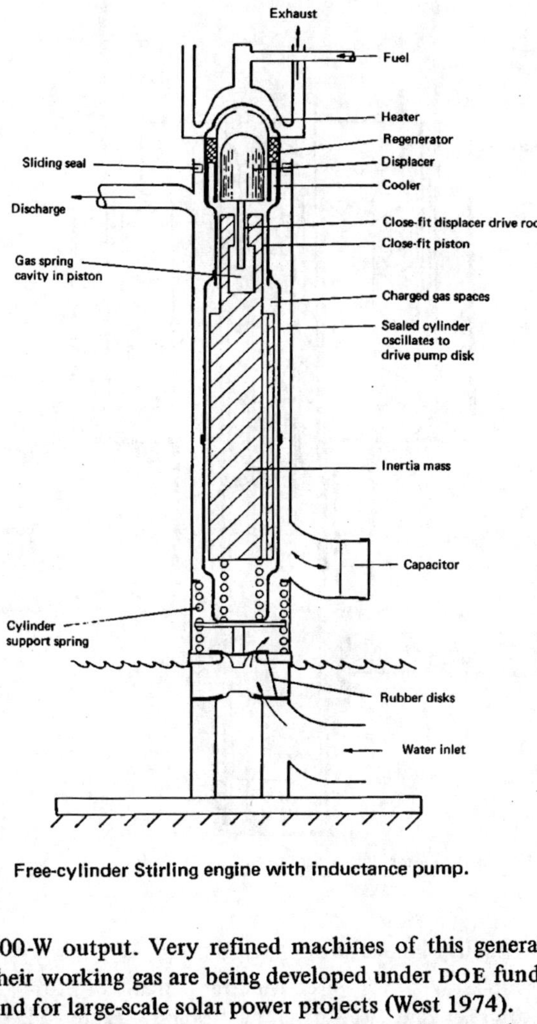

FIGURE 79 Free-cylinder Stirling engine with inductance pump.

of about 100-W output. Very refined machines of this general type using helium as their working gas are being developed under DOE funding for aerospace uses and for large-scale solar power projects (West 1974).

These machines can be scaled up to much larger sizes on the order of 100 kW per cylinder, and used together to form large stationary power plants producing power in the megawatt range. They promise long life, low maintenance, and high thermal efficiency while using solid fuels as their heat source.

In developing countries, simplified designs of the linear alternator engine could be built locally to provide electric power for lighting, communication, and other uses. Its exceptionally low noise and its ability to use solar heat or agricultural wastes make it compatible with village life.

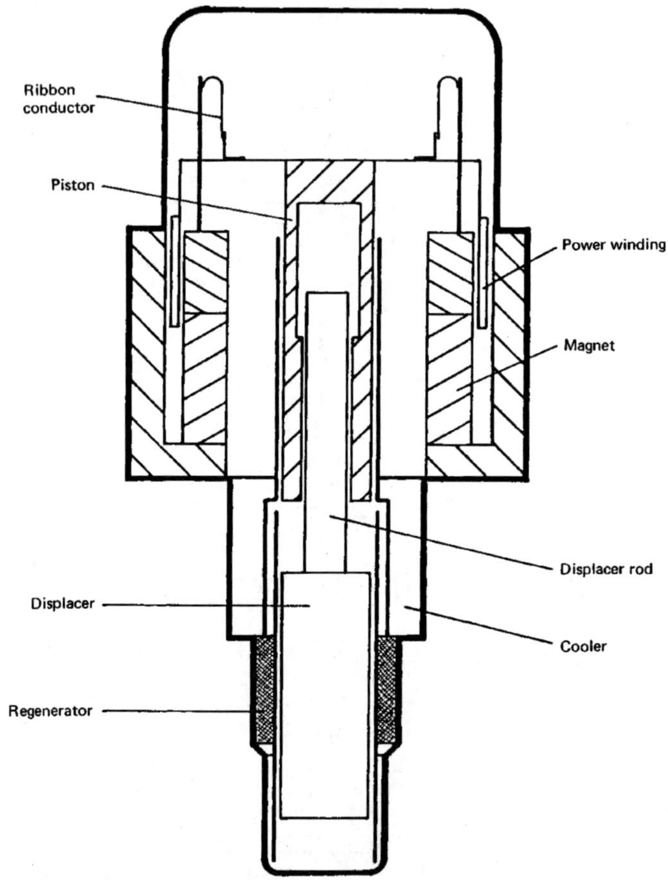

FIGURE 80 Free-piston alternator engine of about 100-W output.

Free-Piston Stirling Heat Pump

One of the simplest yet most effective arrangements of Stirling machines is the duplex Stirling engine, heat-driven heat pump (Figure 81). This machine currently exists only in small sizes and a low-temperature design, but its scaling characteristics are well understood so that it can be made in any size and temperature range. It requires no higher level of manufacturing skill than the free-cylinder engine or alternator engine and is no more complex mechanically than the simple free-cylinder engine. It can be expected to be much more fuel-efficient than absorption refrigeration devices designed for the same purpose, and it requires no lubricant, is hermetically sealed, and is very quiet in

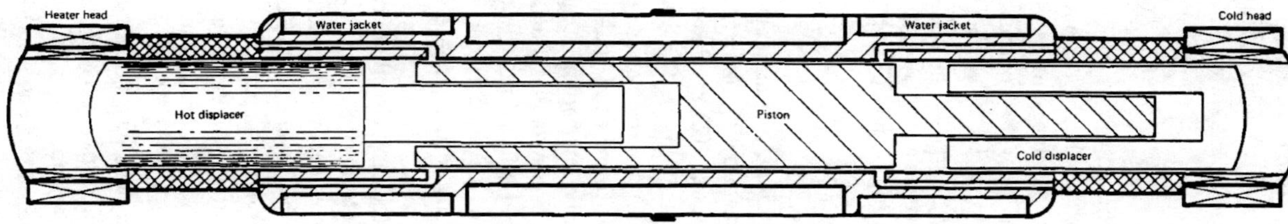

FIGURE 81 Duplex Stirling, heat-driven heat pump.

operation. It would be ideally suited, for example, as a milk cooler heated by biogas from cattle manure, and is readily adaptable to refrigeration, food freezing, space cooling, and similar tasks.

Liquid-Piston Engine

Liquid-piston engines have the advantage of being simple mechanically in their elementary form; however, they have a large volume for the useful power produced and very low efficiency unless modified by the addition of some sort of interface between the water columns and the air working fluid.

Liquid-piston engine development is just beginning in earnest, and it is possible that efforts may soon bear fruit. A practical water pump may result in the future. The question remains as to its competitiveness with the possibly more complex but smaller Stirling engines of the types described here.

FUTURE OUTLOOK

With modest development funding, Stirling engines could be commercially available within 5 years in the following forms: (1) simple hot-air engines up to a few kilowatts in shaft power, (2) specialized hot-air engines for water pumping, (3) free-cylinder water pumps, (4) free-piston electric generators up to 10 kW in output, and (5) heat-driven cooling machines. All of the above machines could be designed to burn solid fuel at about 10 percent overall conversion efficiency using only low-carbon steel, cast iron, and aluminum for their major structure. The free-cylinder and free-piston engines could use solar energy from an inexpensive inflated film collector. If 300-series stainless steel is available for the hot end of the cylinder, significant improvements in life and efficiency would result, with moderate increases in manufacturing cost.

From the viewpoint of the technologist, there is no impediment to the successful design of the entire spectrum of Stirling engines described in this chapter within 10 years. Which ones become economically competitive depends on the capital invested and the nature of the competition, and, probably equally if not more importantly, on the future fuel situation. Increasingly expensive petroleum would lead toward greater popularity of external combustion Stirling engines or those driven by solar heat.

In the far future (10-15 years), it is probable that low-temperature Stirling engines of respectable efficiency will be available, extending the Stirling range of utility into the areas now dominated by Rankine cycles. This can be expected from liquid-piston engines and from those using working fluids other than ideal gases. The other types of engines—water pumps, alternator and heat pump free-piston engines, and crank engines of the low-pressure air type as well as of the high-pressure hydrogen type—will have found their proper best applications in relation to the competition.

DESIGNS FOR LOCAL AVAILABILITY

While the Stirling engine can be very simple mechanically, its design process is not; it requires a deep understanding of complex mechanical, thermodynamic, and dynamic interactions. The problem of introducing Stirling engines in developing countries is essentially one of directing available capital toward that end. Likely end users have little capital, and groups with the design knowledge have no economic incentive to invest their effort in the expensive job of engineering a practical design.

Is it possible to make an effective Stirling engine from material already found in developing countries or from materials easily acquired? Some materials already available that can be used to fabricate these devices include low-carbon steel, cast iron, and aluminum; containers of various sizes (cans, drums, barrels); automotive and bicycle parts (tires, inner tubes); indigenous materials (brick, clay, glass, wood, and fibers); and, by no means least, human ingenuity, determination, and effort.

As an example, any of the engines described in this chapter could be used to pump water, as could the 1900 Erickson engine, although less efficiently

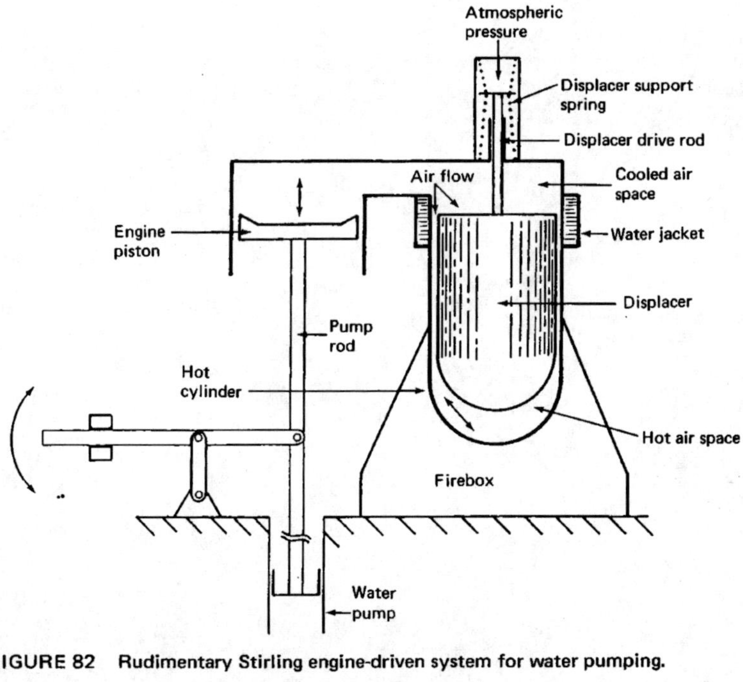

FIGURE 82 Rudimentary Stirling engine-driven system for water pumping.

and with much more metal weight. However, if an otherwise unused agricultural by-product is used as fuel, and abundant low-quality cast iron, perhaps with food containers modified as displacers, is used as the metal, then the Erickson engine has the primary quality of availability, if designs and examples are made available for local artisans to copy.

Even simpler configurations are possible. For example, Figure 82 shows a conceptual layout that could well be used to pump water. It could be constructed of sheet steel, and could use a leather cup seal around a wooden disk for the working piston. It would have no cranks or bearings, and would be self-starting. The firebox could be made of brick or clay and the hot cylinder of either steel, cast iron, or possibly even reinforced, fired clay.

REFERENCES

Beale, W. T. (1979) A Free Cylinder Stirling Engine Water Pump. Paper presented at the 1979 International Solar Energy Society Meeting, August 1979, Atlanta, Georgia, USA. ISES, Victoria, Australia.

Senft, J. R. (1979) *IECEC Proceedings*. Intersociety Energy Conversion Engineering Conference, Boston, Massachusetts, USA.

West, C. D. (1974) Fluidyne Heat Engine. Paper No. 5, Record of the International Solar Energy Society (U.K. Section).

Solid-State Electric Inverters

Renewable energy sources generally produce direct current that is unregulated and/or intermittent. However, most devices that use electricity are designed to operate over a reasonably narrow range of voltages and frequencies, and thus cannot be attached directly to most renewable energy sources. For that reason, suitable power conditioning equipment is essential for use of renewable energy sources, and that equipment will almost certainly be in the form of solid-state power converters, sometimes used in conjunction with battery, or perhaps flywheel, storage.

The most difficult problem connected with electric power conditioning is converting an intermittent source of electricity, continuously varying in voltage, to a continuous source of relatively constant voltage and frequency, which is required for electronic equipment, television sets, standard refrigeration units, etc. Sometimes the cost of such a power conditioning system, when battery storage is included, is more than the cost of the original electrical energy source.

Most electrical appliances are designed to be operated from a stable electrical supply from a grid system (or at least a motor-driven electric generator) capable of handling many times the size of the load of any individual appliance. Thus electric motors, which may draw many times as much starting current as they do normal running load current, present particularly difficult problems. A power source with a 3- or 4-kW instantaneous capacity or more is required to line start a motor that draws only a kilowatt at full load. A solid-state inverter capable of delivering such power surges must necessarily be significantly overdesigned for its average load and therefore rather expensive on a per-kilowatt basis. Further, whatever source is used to power the solid-state inverter must be stable and capable of the surge currents required.

Some low-cost inverters on the market have very poor waveform and surge capability for their average wattage rating. As a result, their performance and reliability in practical field applications have been very disappointing. Consequently, solid-state inverters in general have been considered unreliable. It is quite possible to design solid-state inverters with adequate surge capability and excellent reliability, but at somewhat higher overall cost, of course.

It is obviously uneconomical to buy or build electrical generation systems operated from renewable energy sources that have the surge capability, and therefore capacity, of several times the average load requirement. Thus, in remote regions methods must be developed for delivering suitable energy to difficult types of loads when that energy is derived from essentially unstable sources. There are two approaches to dealing with this problem:

- Build one central power conditioning and storage system capable of handling the most extreme load conditions. This method allows the use of standard appliances, motors, and other commercially available units. For small power sources, however, the cost per installed kilowatt is generally prohibitively high.
- Accept the electrical energy sources as they are and find ways to use the intermittent and variable energy as effectively as possible. This generally would not allow the use of unmodified commercial appliances. However, extensive modification sometimes is not required and the overall cost is often less than for the first approach.

METHODS FOR POWER CONDITIONING WITH INTERMEDIATE STORAGE

If the power source is intermittent to the extent that for some period of time there is no power, or if the type of load calls for short but heavy surge currents, then an intermediate storage system must be used. Most likely, this storage device might be a lead–acid battery, the size of which is determined by the surge peaks required and the time duration of the outages expected. Power conditioning equipment requirements under these conditions are essentially the same as those of a system operating directly from batteries, since batteries (particularly lead–acid batteries) are superb power smoothing devices and have excellent surge capabilities, at least for 10–20 sec. Thus if the solid-state inversion system can withstand the surge currents required, the battery system can provide them even though the original power source cannot. The battery is often necessary for that reason alone, irrespective of long-term storage needs.

Higher-quality solid-state inverters have become available in recent years, while, at the same time, the cost of inverter systems has been decreasing slightly, even in the face of rapid inflation. The surge-current capabilities of silicon control rectifiers, for example, have been increasing, and the cost per kilowatt of energy-handling capability has been decreasing. Thus power inverters in the 25–100-kW range have decreased in price from about $2,000/kW to about $500/kW.

Inverters (DC–AC) operating at power levels under 25 kW are more expensive on a per-unit basis, but there are also encouraging developments in this area. High-current transistors that can operate at up to several hundred volts are becoming available, and the price of such devices is dropping. Recently, field effect transistors (FETs) that operate at 10 amp and 600 volts have been developed. The price is currently high, but as production proceeds, the price should drop an order of magnitude or more.

In short, while there are not now many reasonably priced, reliable and efficient DC–AC inversion systems on the market, it appears that they may become available within the next 5 years. For those applications where well-regulated AC power with controlled frequency and good sine waveform is necessary, suitable units will be available.

METHODS FOR POWER CONDITIONING WITHOUT INTERMEDIATE STORAGE

A number of available generation schemes can be applied in situations in which stable utility-grade power is to be generated from a variable-speed mechanical energy source, but battery storage is not used. Such situations might include: windmills operated in the variable-speed (constant tip-speed ratio) mode; steam turbines fed from geothermal wells; and point-focusing, distributed-receiver, solar–thermal–electric systems employing high-speed (Brayton cycle) turbines operating in the variable-speed mode.

Variable-Speed, Constant-Frequency (VSCF) Power Generation

Several VSCF generation schemes have been developed by judiciously combining rotating electrical machines with mechanical (commutator) and/or electronic (thyristors and diodes) switching devices. Some promising schemes are the AC–DC–AC link and the special generator/AC link or DC–AC link, which could take the form of a field-modulated generator system, double-output induction generator, or an AC commutator generator.

AC–DC–AC Link. In this case, a conventional AC alternator and rectifiers are used to generate DC power to be fed to a DC–AC inverter. Naturally commutated (also known as line-commutated) inverters are used for converting direct current into alternating current. They utilize a timing device (either electronic, or perhaps an existing power line) that periodically reverses polarity and causes commutation to occur naturally. Such converters are controlled by building controllable phase delay into the switching control signal generator. They can be used either to feed power directly to an existing power grid, or to provide alternating current independently, depending on variations in design.

Special Generator/AC Link or DC-AC Link. A *field-modulated generator-system* employs a three-phase, high-frequency alternator to generate single-phase power at the required low frequency. The rotating field coil of the alternator is supplied with alternating current at the desired low frequency instead of with conventional direct current.

Power delivered by a field-modulated generator system to a power grid can be controlled by regulating the magnitude and phase of excitation voltage applied to the rotor terminals from the power grid. Alternatively, a separate field excitation AC source will permit the system to stand alone. This generation scheme has many unique features, such as operation in a single-, two-, or three-phase mode and self-excitation, which allow versatility in operation and control.

Conventional induction generators, by their very nature, must operate at essentially constant speeds. One approach to overcoming this restriction and allowing the induction generator to become a variable-speed, constant-frequency device is the use of rotor power regeneration. The squirrel cage rotor is replaced by a wound rotor with leads brought out through slip rings. Rotor power output at slip frequency is converted to line-frequency power by an AC-DC-AC link consisting of a rectifier and inverter.

Because output power is obtained both from the stator and rotor, this device is called a *double-output induction generator*. Rotor output power has the electrical equivalence of an additional impedance in a rotor circuit, and thus increasing rotor outputs leads to increasing slips and higher speeds. Such an operation broadens the operating speed range from synchronous to twice the synchronous speed. This corresponds to a slip range per unit of 0-1.0.

The AC commutator generator, also known as a Scherbius machine, has two polyphase windings in the stator. The exciting winding provides the basic excitation to the magnetic circuit and sets up the rotating field, while the compensating winding is connected in series with the brushes in such a way as to neutralize the armature reaction field set up by the armature (rotor) currents. The rotor winding is of the commutator type and is connected to the commutator.

For a given rotating field in the air gap, the frequency of the voltage and current collected by the polyphase brush gear resting on the commutator is the same as in the stator, irrespective of the rotational speed. In essence, the commutator acts as a mechanical frequency changer. Therefore, the brush gear can be connected in series with the stator (compensating) winding, and the frequency of the output is independent of the shaft speed. The principle of operation is similar to that of a compensated DC generator except that instead of direct current there is constant (line) frequency alternating current and three leads instead of two.

Additional windings, such as interpole and damping windings, may be provided to improve commutation. Basic problems in employing this device are the cost and the additional maintenance and care required by the commutator and the brush gear.

DIRECT USE OF INTERMITTENT ELECTRICITY

Water pumping, space heating, crop drying, grain grinding, small lathes, hydraulic logsplitters, small plows and tillers similar to those described in the chapter on pedal power, and other small machines, can use intermittent (as available) electricity. If the intermittent sources provide energy on at least a daily basis, perhaps even small refrigeration units could be operated. However, some equipment may require modification so that electric sources that vary widely in voltage, frequency, and energy availability can be used.

For any application requiring low-grade heat (e.g., water heating, space heating, grain drying), the problem is relatively simple. A resistive (heat-producing) load must be designed that fully absorbs the output of the electric generator. For example, if a simple fixed-pitch wind turbine is to deliver full power for any wind speed, it must operate at an approximately constant tip-speed ratio, which means that the turbine's revolutions per minute must vary proportionately to the wind velocity. However, the available power output varies as the third power of the wind velocity. Thus a nonlinear resistive load or some method of automatic generator voltage adjustment is required. Considerable progress is being made in the development of such loads and automatic controls.

One problem with using sources of intermittent electricity such as geothermal wells, windmills, or solar thermal collectors (driving free-piston, sealed Stirling engines or small, simple turbines) is that the electric generators they drive sometimes operate over wide speed, voltage, frequency, and power ranges. It must be kept in mind that most small mechanical devices and machines run at more or less constant speed because the electric motor driving them is generally a single-phase induction motor operating on a power line of more or less constant frequency and voltage.

Typical mechanical loads do not necessarily have to operate over a narrow range of speeds. Further, induction motors (particularly polyphase motors) will run successfully over a wide range of speeds if the input voltage and frequency vary over a similar range. The primary concern with these motors is that the maximum load, applied voltage, and power line frequency all remain proportional to each other. To provide continuous rotational torque, it is desirable that the power source and motors both be polyphase. Otherwise, there is no simple way to start the motor.

One of the simplest generation schemes that can be used to deal with these problems is a permanent magnet alternator. The stator of the alternator is wound for polyphase (two- or three-phase) output, and the rotor consists of permanent magnets of alternating polarity mechanically embedded around the rotor periphery. The output frequency is strictly proportional to the rotational speed. Under open-circuit (no-load) conditions, output voltage is also proportional to the rotational speed. However, under load, armature reaction effects and internal impedance drops contribute to a departure from this proportionality.

If the prime mover is operated at a constant speed, the output under load will be at a constant frequency, but the voltage will be variable, depending on the load. If operated at variable speed, the output will be variable both in voltage and in frequency. Either of these outputs can be converted to constant-voltage, constant-frequency alternating current using a solid-state rectifier-inverter combination with appropriate controls, but that requirement can sometimes be avoided.

It is well known that a conventional squirrel cage induction motor can be operated in the variable-speed mode by supplying its stator with variable-voltage, variable-frequency power. To avoid saturating the magnetic circuit, the applied voltage and the load demand should vary proportionally with the frequency. The air-gap flux will then be approximately constant and satisfactory operation will result. Hence, the output of a variable-speed, permanent magnet alternator is ideally suited for connection to an induction motor.

Such an induction motor can be used to drive a centrifugal water pump, a small shop lathe, a grain grinder, or even a small plow (see "Pedal Power"). The result will be simply an electrical transmission between the prime mover and the electric motor. In all of the cases discussed above, a set of capacitors connected across the machine terminals improves performance significantly because of the leading currents drawn by them and the resulting self-excitation of the alternator caused by the leading current.

A small, three-phase alternator (driven by a variable-speed mechanical source), which could, in turn, drive a small, three-phase electrical motor, can be constructed relatively simply. First, the rotor from a standard three-phase induction electric motor is slotted in a milling machine. Then rare earth (samarium-cobalt) magnets are glued (with a high-temperature epoxy glue) into the slots and the rotor is reground to the proper diameter and rebalanced. Finally, the rotor is reinserted in the original stator assembly. The result is a polyphase permanent magnet alternator. Alternators built this way have been used successfully to drive deep-well water pumps and for other motor-drive applications. Such a system can transmit the mechanical power of a windmill (or other intermittent mechanical source) to some suitable remote location easily and without any inverters, batteries, or other high-technology equipment. Systems of this type, up to several horsepower, have been built and successfully operated in the field for several years. (See Appendix A for a list of manufacturers of solid-state electric inverters and Appendix C for a list of suggested reading.)

APPENDIXES

Appendix A

MANUFACTURERS OF ENERGY-RELATED DEVICES

This appendix represents the best information available to the NAS staff at the time of publication, but is not necessarily complete. It was not possible to verify many of the following addresses.

Country	Company/address	Product
TERRESTRIAL PHOTOVOLTAIC SYSTEM AND COMPONENT MANUFACTURERS[a]		
Australia	Amalgamated Wireless Australasia, Ltd. 47 York Street Sydney, New South Wales 2000	
	Lucas Industries Australia Batteries Division 1156 Nepean Highway Cheltenham, Victoria 3192	Solar cell power generation systems
	Philips 161 Starts Street South Melbourne, Victoria	
	Spectrolab 34–40 Clayton Road North Clayton, Victoria 3168	
Federal Republic of Germany	AEG-Telefunken Raumfahrttechnik und neue Technologien Industriestrasse 29 2000 Wedel, Holstein	Solar cell module, solar cell array field
	Christeva Sonnenenergietechnik GmbH Schramminger Weg 74a 8000 Munich	Storage tanks, solar cell array field, flat plate collectors
	Contronic Peer-G, Cruse Postfach 605, 344 2000 Hamburg 60	Solar cell array field
	Contronic Solar–Electronic Bornstrasse 32 2000 Hamburg 13	Solar cell array field

ELMEG GmbH, Abt. ESOTRON Rudigerstrasse 12 5300 Bonn 2	Novelty items
Ferranti GmbH Widenmayerstrasse 5 8000 Munich 22	Solar cell module, solar cells, solar cell array field
Herrmann Helfried Postfach 1450 8998 Lindenberg	Solar cell power generation systems
Herwi Solar GmbH Am Stein 7 8755 Alzenau	Solar cells
Leybold-Heraeus GmbH & Co. KG Postfach 549, Wilhelm-Rohn-Strasse 25 6450 Hanau 1	
Philips GmbH, Unternehmensbereich Licht und Anlagen-Energie-Systeme Monckebergstrasse 7 2000 Hamburg 1	Heat pumps, evacuated tube concentrating collectors, solar cell array field
Rucker GmbH, UB Solartechnik Postfach 1232 5210 Troisdorf 1	Solar cell power generation systems
Siemans AG Postfach 103 8000 Munich 1	Solar cell module, silicon solar cells
Valvo GmbH Burchardstrasse 19 Postfach 106.323 2000 Hamburg 1	Solar cell array field, electric batteries

Country	Company/address	Product
TERRESTRIAL PHOTOVOLTAIC SYSTEM AND COMPONENT MANUFACTURERS[a]		
	Wacker–Chemitronic Gesellschaft fur Elektronik-Grundstoffe MbH Johannes-Hess-Strasse 24 Postfach 1140 8263 Burghausen	Silicon solar cells
France	Aérospatiale 37, Boulevard de Montmorency 75781 Paris Cedex 16	Flywheels/inertial systems, heliostats, solar cell array, tower focus power plant components
	Centre National d'Etudes des Télécommunications (CNET) 38–40, rue du General Leclerc 92131 Issy les Moulineaux	Solar cell array field, wind turbine generator systems
	Compagnie Industrielle des Piles Electriques (CIPEL) Division des Produits Industriels 125, rue du President Wilson 92302 Levallois Perret	Solar cell array field, electric batteries
	France Photon BP 119 Usine des Agriers 16004 Angoulême Cedex	Solar cell array, solar cells
	(R.T.C.) La Radiotechnique–Compelec 130, Avenue Ledru-Rollin 75540 Paris Cedex 11	Solar cells, space power satellite systems
	Leroy W. Somer Boulevard Marcellin Leroy 16004 Angoulême	Heat pumps, educational aids, vertical-axis wind machines, small hydropower systems, solar cell array field

	Motorola Semi-Conducteur S.A. 126, Chemin de Canto Laouzetto 31000 Toulouse	Solar cell module, solar cells, solar cell array field
	Pompes Gutnard Etablissements 179, Boulevard Saint Denis 92402 Courbevoie	Solar cell system components
	Société des Applications de Hélioénergie (SAHEL) 50, rue J. P. Timbaud BP 301 92402 Courbevoie Cedex Paris	
	Thomson–CSF S.A. 173, Boulevard Haussmann 75008 Paris	Solar cell system components
Great Britain	Ferranti Engineering, Ltd. Electronic Components Division Post Office Box 20 Hollinwood Avenue Chadderton Oldham OL9 8EL	Solar cell array field, solar cells, solar cell protective coating
	Grumann International, Inc. 64–65 Grosvenor Street London W1X DB	Solar cell power generation systems
	International Research and Development Fossway Newcastle upon Tyne	Solar cells
	Lucas Electrical, Ltd. Pes Division Great Hampton Street Birmingham B18 6AH	Solar cell power generation systems

Country	Company/address	Product
TERRESTRIAL PHOTOVOLTAIC SYSTEM AND COMPONENT MANUFACTURERS[a]		
	Solar Power, Ltd. 101–111 Strand London WC2R 0AA	Solar cell power generation systems
Italy	Aerimpianti S.p.A. Via Bergano, 21 20135 Milano	Solar cell power generation systems
	Ansaldo Electric Systems Division Via A. Pacinotti, 20 16151 Genova–San Pier d'Arena	Tower focus power plants, dispersed solar electric power plants, domestic hot water systems, solar cell array fields
	Officine Galileo Meccanotessile ESO (Solar Energy Dept.) Via Carlo Bini, 44 50134 Firenze	Tower focus plants, solar cell array field
	Termomeccanica Italiana Via del Molo, 1 19100 La Spezia	Solar cell power generation systems
Mexico	Philips Mexicana, S.A. de C.V., Division Energia Solar Durango 167 Mexico 1, D.F. 07	
The Netherlands	N. V. Philips' Gloeilampenfabrieken Bldg. Tam Pieter Zeemanstraat 6 Eindhoven, Nb 5621 CT	Solar cell panels, solar cell module, solar cell array field

United States	Acurex Corporation Alternate Energy Division 485 Clyde Avenue Mountain View, CA 94042	Domestic hot water/space heating/cooling systems, steam supply systems, decentralized power plants, process hot water systems, irrigation systems, parabolic trough collectors, concentrator solar cell assembly
	Aidco Maine Corporation Orr's Island, ME 04066	Domestic hot water systems, space heating systems, solar cell array fields, decentralized power plants, thermal collector subsystems, process heat systems, greenhouses, wind turbine generator systems, solar ponds
	Applied Solar Energy Corporation (formerly Optical Coating Lab, Inc.) 15251 East Don Julian Road City of Industry, CA 91746	Solar cell module, silicon solar cells
	ARCO Solar, Inc. 20554 Plummer Street Chatsworth, CA 91311	Solar cells, solar cell protective coating, solar cell array,solar cell module
	Arthur D. Little, Inc. 25 Acorn Park Cambridge, MA 02140	Photoelectric solar energy satellite systems, solar cell array field
	Automatic Power, Inc. Post Office Box 18738 Houston, TX 77023	Solar cell array, solar cell module
	Columbia Chase Corporation Solar Energy Division 55 High Street Holbrook, MA 02243	Solar cells; solar cell array field; do-it-yourself domestic hot water systems; domestic hot water systems, liquid; absorbers, copper; flat plate collectors, liquid; housing/framing
	Crystal Systems, Inc. Shetland Industrial Park 35 Congress Street Salem, MA 01970	Solar cells
	Dow Chemical USA FP&S/TS&D, Lab 3, 2020 Dow Center Midland, MI 48640	Latent heat storage modules, pipe insulation, collector insulation, vapor barriers, neoprene sealants and adhesives, plastic sheets and films, protective coatings, solar cell materials, solar cell protective coating, silicone sealants and adhesives, antifreeze, eutectic salts

Country	Company/address	Product
TERRESTRIAL PHOTOVOLTAIC SYSTEM AND COMPONENT MANUFACTURERS[a]		
	Dow Corning Corporation Solar Energy 2200 West Salzburg Road, Box 1767 Midland, MI 48640	Solar cell protective coating, collector insulation, solar cell materials, special insulating units, silicone sealants and adhesives, silicone heat transfer fluids, protective coatings, plastic sheets and films
	Ecotronics, Inc. 8502 E. Cactus Wren Road Scottsdale, AZ 85253	Microprocessors, proportional controls, power conditioning equipment, solar cells
	Free Energy Systems, Inc. Price & Pine Streets Holmes, PA 19043	Solar cell module, solar cell protective coating, solar cell array field
	General Electric Company Advanced Energy Programs Post Office Box 13601 Philadelphia, PA 19101	Solar cell panels, solar cell module, solar cell array field, heat pumps, v-trough concentrating collectors, evacuated tube collectors, large wind turbine generator systems
	Gould, Inc. Gould Laboratories 540 East 105th Street Cleveland, OH 44108	Electric batteries, solar cell system components, heat pipe heat exchangers
	International Rectifier Semiconductor Division 233 Kansas Street El Segundo, CA 90245	Solar cell module, silicon solar cells
	International Solar Leasing Company 7010 Convoy Court San Diego, CA 92111	

InterTechnology/Solar Corporation 100 Main Street Warrenton, VA 22186	Heat pump systems, water-to-air; compound parabolic concentrators; domestic hot water systems, liquid; flat plate collectors, liquid; parabolic trough collectors; support systems; sealants, adhesives, and gaskets; space heating systems, liquid; solar-assisted heat pump systems; domestic hot water/space heating/cooling systems; process hot water systems; energy management systems; air-to-water heat pumps; water-to-water heat pumps; special insulating units; solar cell array field; solar cell module; concentrator solar cell assembly; wood-burning tools/fireplace accessories; gasification systems; organic decomposition systems; liquefaction systems; conversion plants; educational aids
Lockheed Missiles & Space Company, Inc. P.O. Box 504 Sunnyvale, CA 94088	Solar cells, ocean thermal gradient systems, electric batteries, domestic hot water/space heating/cooling systems
Martin–Marietta Aerospace 12250 S. Highway 75, Box 179 Denver, CO 80201	Tower focus power plants, concentrator solar cell assembly
McGraw–Edison Company Power Systems Division Post Office Box 28-T Bloomfield, NJ 07003	Solar cell module
Megatech Corporation 29 Cook Street Billerica, MA 01821	Educational aids, solar cell array field
Mobil Tyco Solar Energy Corporation 16 Hickory Drive Waltham, MA 02154	Concentrator solar cell assembly, solar cell module, solar cells, solar cell array field, silicon solar cells
Monsanto Industrial Chemicals Company 800 North Lindbergh Boulevard St. Louis, MO 63166	Greenhouses, special insulating units, plastic sheets and films, silicon solar cells, coolants

Country	Company/address	Product
TERRESTRIAL PHOTOVOLTAIC SYSTEM AND COMPONENT MANUFACTURERS[a]		
	Motorola, Inc. Semiconductor Group 1303 Algonquin Road Schaumburg, IL 60196	Collector support subsystems, power conditioning equipment, solar cell array, solar cell module, silicon solar cells
	National Semiconductors, Ltd. 331 Cornelia Street Plattsburgh, NY 12901	Silicon solar cells
	Semicon, Inc. 10 North Avenue Burlington, MA 01803	Silicon solar cells
	SES, Inc. (Subsidiary of Shell Oil Company) Tralee Industrial Park Newark, DE 19711	Solar cell array field, solar cell protective coating, solar cell panels, solar cell module, cadmium sulfide solar cells
	Silicon Material, Inc. 341 Moffett Boulevard Mountain View, CA 94043	Solar cell module, silicon solar cells
	Silicon Sensors, Inc. Solar Systems Division Highway 18 East Dodgeville, WI 53533	Cadmium sulfide solar cells, selenium solar cells, solar cell array field, silicon solar cells, solar cell module, radiation measurement devices
	Siltec Corporation 3717 Haven Avenue Menlo Park, CA 94025	Solar cells

Solar Energy Products, Inc. 1208 N.W. 8th Avenue Gainesville, FL 32601	Roll bond heat exchangers; steel tanks; back-up water heaters; pumps; differential controls; flat plate collectors, liquid; support systems; solar cell module; absorption cooling systems; domestic hot water systems, liquid; heat pump systems, water-to-air; domestic hot water/space heating/cooling systems
Solar Power Corporation 5-T North Executive Park Drive North Billerica, MA 01862	Power conditioning equipment, solar cell panels, solar cell module, silicon solar cells, solar cell array field
Solarex Corporation 1335 Piccard Drive Rockville, MD 20850	Blowers, educational aids, tracking devices, housing/framing, antireflective glazing coatings, solar cell panels, solar cell module, silicon solar cells, solar cell array field
Solec International, Inc. 12533 Chadron Avenue Hawthorne, CA 90250	Concentrator solar cell assembly, solar cell module, solar cells, parabolic trough collectors, solar cell array field, active–passive hybrid systems
Solectro Thermo, Inc. 1934 Lakeview Avenue Dracut, MA 01826	Point-focusing collectors, concentrator solar cell assembly, active–passive hybrid systems, nontracking concentrating collectors
Solenergy Corporation 23 North Avenue Wakefield, MA 01880	Solar cell protective coating, solar cell panels, solar cell module, power conditioning equipment
Sollos, Inc. 2231 Carmelina Avenue Los Angeles, CA 90064	Solar cell array field, solar cells, solar cell module
Spectrolab 12500 Gladstone Avenue Sylmar, CA 91342	Solar simulators, solar cell panels, concentrator solar cell assembly, solar cell array, solar cell module, silicon solar cells, solar cell array field
Spire Corporation Patriots Park Bedford, MA 01730	Solar cells, gallium arsenide solar cells, silicon solar cells, solar cell protective coating, solar cell array, solar cell module

Country	Company/address	Product
TERRESTRIAL PHOTOVOLTAIC SYSTEM AND COMPONENT MANUFACTURERS[a]		
	Sun Tap, Inc. Post Office Box 754 Arlington Heights, IL 60006	Power conditioning equipment, solar cell array field
	Sun Trac Corporation 540 Zenith Drive Glenview, IL 60025	Solar cell array field, solar cell module, concentrator solar cell assembly
	Technidyne Associates 58–06 69th Place Maspeth, NY 11378	Linear-focusing collectors, decentralized power plants, solar cell array field, window unit collectors
	Thermo Electron Corporation, Energy Systems Department T 101 First Avenue Waltham, MA 02154	Irrigation systems, Rankine cycle engines, total energy systems, solar cell array field
	Tideland Signal Corporation 4310 Directors Row Houston, TX 77052	Solar cell array, solar cell protective coating, power conditioning equipment, solar cell array field, solar cell module
	Uce, Inc. 20 North Main East Norwalk, CT 06854	Solar cell array field
	Vactec, Inc. 2423 Northline Ind. Boulevard Maryland Heights, MO 63043	Radiation measurement devices, solar cell module, selenium solar cells, silicon solar cells
	Varian Associates 611 Hansen Way Palo Alto, CA 94303	Concentrator solar cell assembly, gallium arsenide solar cells

	Westinghouse Electric Corporation Advanced Energy Systems Division Post Office Box 10864 Pittsburgh, PA 15236	Tower focus power plants; concrete tanks; large wind turbine generator systems; heat pump systems, air-to-air; cooling systems; space heating systems; solar cell systems components; solar cell array field; total energy systems; heliostats
ELECTRICAL- AND MECHANICAL-OUTPUT WIND MACHINES[b]		
Australia	Davey-Dunlite (Division of Pye Industries) 2 Hargreaves Street Huntingdale, Victoria 3166	Electrical-output
Canada	Bristol Aerospace, Ltd. Post Office Box 874 Winnipeg R3C 2S4	Electrical-output
	Crowfield Associates, Ltd. Crapaud RR1 Prince Edward Island C0A 1J0	Electrical- and mechanical-output
	Dominion Aluminum Fabricating, Ltd. 3570 Hawkstone Road Mississauga, Ontario L5C 2V8	Electrical-output
	Winflo Power, Ltd. 90 Esna Drive, Unit 15 Markham, Ontario L3R 2R7	Electrical-output (Wingen machines)
Denmark	Dansk Vindkraft Industri ApS Vendevej 6 Bureso 3350 Slangerup	Electrical- and mechanical-output
	Riisager El-Vindmolle Skaerbaek 7400 Herning	Electrical-output

Country	Company/address	Product
ELECTRICAL- AND MECHANICAL-OUTPUT WIND MACHINES[b]		
	S-J Wind-Power ApS Suderborej 4 990 Frederikshavn	Electrical-output
	Sparco c/o Enertech, Inc. P.O. Box 420 Norwich, VT 05055 USA	Mechanical-output
Federal Republic of Germany	Dornier-System GmbH Friedrichshafen 7559 Immenstadt	Electrical-output (Darrieus-type systems)
	Lubing Maschinenfabrik Postfach 110 2847 Barnstorf	Electrical- and mechanical-output
	Noah Energie Systeme GmbH Wippenhohnerstrasse 31 5202 Hennef 1	Electrical-output
France	Aerowatt Company 37, rue Chanzy 75011 Paris	Electrical-output
	ENAG S.A. Rue de Pont-l'Abbé 29000 Quimper Finistère	Electrical-output
Great Britain	Industrial Institute, Ltd. Stanley Road Bromley BR2 9JF Kent	Electrical-output

	P.I. Enquiries, Ltd. The Dean Alresford Hants	Electrical- and mechanical-output
Italy	Domenico Sperandio & Ager Via Cimarosa 13-21 58022 Folloncia (GR)	Electrical-output
Japan	Japan Wind Power Generator Co., Ltd. Sudo Building 1-4-2 Nakameguro Meguro-Ku Tokyo	Electrical-output
The Netherlands	Fokker-VFW Post Office Box 7600 Schiphol-cost	Electrical-output
South Africa	P. Andrag and Sons (Pty.), Ltd. Post Office Box 364 Belleville, Cape Province	Mechanical-output (Aermotor)
Sweden	Ingenjorsfirma Ture Ekenburg AB Sleipnevagen 15 136 42 Handen	Electrical-output
	Saab-Scania AB Linköping, Ostergtlnd 581 88	Electrical-output
	Sikob AB Engineering Centre S-19178 Sollentuna	Electrical-output
Switzerland	Elektro GmbH St. Gallerstrasse 27 8400 Winterhur	Electrical-output

Country	Company/address	Product
ELECTRICAL- AND MECHANICAL-OUTPUT WIND MACHINES[b]		
Union of Soviet Socialist Republics	USSR Enercgomashexport 35 Mosfilmovskaya UI Moscow V-330	Electrical-output
	USSR Institute for Farm Electrification I-U Veshniakovski Prc. Dom Moscow J-456	Electrical-output
United States	Aero Lectric 13517 Winter Lane Cresaptown, MD 21502	Electrical-output (Wind Wizard)
	Aermotor Post Office Box 1364 Conway, AZ 72032	Mechanical-output
	Aero Power 2398 Fourth Street Berkeley, CA 94710	Electrical-output
	Aerospace Systems 1620 S. Broadmoor Wichita, KS 67207	
	Alcoa Corporation Alcoa Building Pittsburgh, PA 15219	Electrical-output
	All-American Engineering Co. Post Office Box 1247 Wilmington, DE 19899	
	Altos: The Alternate Current Post Office Box 905 Boulder, CO 80302	Electrical-output

Astral/Wilcon Post Office Box 291 Millbury, MA 01527	Electrical-output
Automatic Power, Inc. Post Office Box 18738 Houston, TX 77023	Wind machine generators, dual blade wind machines, small wind turbine generator systems, towers, wind turbine blades, controllers
Bertoia Studio 644 Main Street Bally, PA 19503	Electrical-output
Bowjon 2829 Burton Avenue Burbank, CA 91504	Mechanical-output
Jay Carter Enterprises Post Office Box 684 Burkburnett, TX 76354	Electrical-output
Dakota Wind & Sun, Ltd. Post Office Box 178 Aberdeen, SD 57401	Electrical-output
Dempster Industries, Inc. Post Office Box 848 Beatrice, NB 68310	Mechanical-output
Dragonfly Wind Electric Post Office Box 155 Elk, CA 95432	Electrical-output
Dynergy Corporation Post Office Box 428 1269 Union Avenue Laconia, NH 03246	Electrical- and mechanical-output
Enertech Corporation Post Office Box 420 Norwich, VT 05055	Electrical-output

Country	Company/address	Product
ELECTRICAL- AND MECHANICAL-OUTPUT WIND MACHINES[b]		
	Grumman Energy Systems 4175 Veterans Memorial Highway Ronkonkoma, NY 11779	Electrical-output
	Heller–Aller Company Post Office Box 29 Napoleon, OH 43545	Mechanical-output (Baker windmills)
	Hinton Research 417 Kensington Salt Lake City, UT 84115	Electrical-output
	Independent Energy Systems, Inc. 6043 Sterrettania Road Fairview, PA 16415	Electrical-output (Skyhawk)
	Jacobs Wind Electric Co., Inc. Route 13, Box 722 Fort Myers, FL 33908	Electrical-output
	Kaman Aerospace Corporation Old Windsor Road Bloomfield, CT 06002	
	Kedco, Inc. 9016 Aviation Boulevard Inglewood, CA 90301	Electrical-output
	McDonnell Douglas Corporation P.O. Box 516 St. Louis, MO 63166	
	Megatech Corporation 29 Cook Street Billerica, MA 01866	Electrical-output

Millville Windmills & Solar Equipment Company Post Office Box 32 Millville, CA 96062	Electrical-output
North Wind Power Company, Inc. Post Office Box 315 Warren, VT 05674	Electrical-output
Pinson Energy Corporation Post Office Box 7 Marston Mills, MA 02648	Electrical-output
Power Group International Corp. Suite 106 13315 Stuebener-Airline Road Houston, TX 77014	Electrical-output (Hummingbird)
Product Development Institute 508 South Byrne Road Toledo, OH 43609	Electrical-output (Wind Jennie)
Sencenbaugh Wind Electric Post Office Box 11174 Palo Alto, CA 94306	Electrical-output
Tumac Industries 650 Fort Street Colorado Springs, CO 80915	Electrical-output
TWR Enterprises Sun-Wind-Home Concepts 72 West Meadow Lane Sandy, UT 84070	Electrical-output (Wind Titan)
United Technologies Research Center Silver Lane East Hartford, CT 06118	

Country	Company/address	Product
ELECTRICAL- AND MECHANICAL-OUTPUT WIND MACHINES[b]		
	Wadler Manufacturing Co., Inc. Route 2, Box 76 Galena, KS 66739	Mechanical-output
	Whirlwind Power Company Box 18530 Denver, CO 80211	Electrical-output
	Windco Dyna Technology, Inc. Post Office Box 3263 Sioux City, IA 51102	Electrical-output (Wincharger)
	Wind Engineering Corporation Box 5926 Lubbock, TX 79417	
	Wind Power Systems, Inc. Post Office Box 17323 San Diego, CA 92117	Electrical-output (Storm Master)
	Windworks Post Office Box 329, Route 3 Mukwonago, WI 53149	Electrical-output
	WTG Energy Systems, Inc. Post Office Box 87, 1 La Salle Street Angola, NY 14006	Electrical-output
	Zephyr Wind Dynamo Company Post Office Box 241 Brunswick, ME 09011	Electrical-output

SMALL-SCALE HYDROELECTRIC SYSTEMS AND COMPONENTS

Country	Company	Products
Austria	Maschinenfabrik Kössler GmbH A-3151 St. Pölten	Two types of small-scale hydroelectric systems, with outputs varying from 12 to 1,250 kVA
Canada	Niagara Waterwheels, Ltd. 706 East Main Street Welland, Ontario L3B 3Y4	Six models of horizontal-axis, axial flow propeller turbine packages, with outputs in the 20–300-kW range
Federal Republic of Germany	Drees & Co. GmbH 4760 Werl/Schützenstrasse 36	Small micro-hydro plants to full-size hydroelectric stations
France	Briau S.A. BP 43 37009 Tours Cedex	Packaged low-head turbine–generator systems with outputs from 4 to 50 kVA
Great Britain	Evans Engineering Land & Leisure (Services), Ltd. Priory Lane, St. Thomas Launceston PL15 8DQ, Cornwall	Simple bare-shaft turbines, packaged reaction and impulse turbines, electronic control units, instrument and control panels, simple and duplex hydraulic rams
	Gilbert Gilkes & Gordon, Ltd. Canal Iron Works Kendal Cumbria LA9 7BZ	Wide range of turbines—"Hydec," spiral-cased Francis, open-type Francis, Pelton wheels, and water motors
	Westward Mouldings, Ltd. Greenhill Works Delaware Road Gunnislake, Cornwall	Made-to-order fiberglass water wheels
India	Jyoti Limited Industrial Area P.O. Chemical Industries Baroda, Gujarat 390 003	Range of turbines with outputs from 25 to 100 kW

Country	Company/address	Product
SMALL-SCALE HYDROELECTRIC SYSTEMS AND COMPONENTS		
Sweden	Karlstads Mekaniska Werkstad Fack. S-681 01 Kristinehamn	Turbines from micro to large multi-megawatt units
Switzerland	Elektro GmbH St. Gallerstrasse 27 8400 Winterthur	Various models of Pelton and Francis turbines
United States	Allis-Chalmers Hydro-Turbine Division Post Office Box 712 York, PA 17405	Range of standardized hydroelectric generating units, starting at an output of 50 kW
	Independent Power Developers, Inc. Route 3, Box 174H Sand Point, ID 83864	Range of small turbines
	James Leffel & Company Springfield, OH 45501	Water turbine units ranging in output from 5,000 hp down
	Northern Waterpower, Inc. Post Office Box 49 Harrisville, NH 03450	Six models of horizontal-axis, axial flow propeller turbine packages, with outputs in the 20–300 kW range
	Small Hydroelectric Systems Post Office Box 124 Custer, WA 98240	Small impulse (Pelton) hydroelectric systems producing 5–25 kW
METHANE DIGESTER UNITS AND DESIGNS		
Australia	Sanamatic Tanks Pty., Ltd. Post Office Box 119 Bendigo 3550	Range of biogas plants for utilizing 500 lb to 120 t of effluent per day

Great Britain	Farm Gas, Ltd. Heath Workshop Lydham Bishops Castle Salop SY9 5HB	Two chief types of methane digesters–2,500-gal (11.25-m^3) capacity for farms and a 400-gal (1.8-m^3) capacity for domestic wastes, etc.
India	Khadi & Village Industries Commission Gramodaya 3 Irla Road Vile Parle (West) Bombay 400 056	Biogas plant designs for construction by approved makers
Kenya	Tunnel Co., Ltd. Tunnel Estate P.O. Fort Ternan	Hutchinson methane plants and others
United States	Aidco Maine Corporation Orr's Island, ME 04066	Biomass conversion systems, ethanol plant equipment
	Bio Gas of Colorado, Inc. 5620 Kendall Court, Unit G Arvada, CO 80002	Biogas plant designs, consultancy services
GEOTHERMAL EQUIPMENT		
Israel	Ormat Turbines, Ltd. Post Office Box 68 Yavne	Low-maintenance, closed-cycle, vapor turbogenerators proven in fossil fuel and solar applications
Japan	Mitsubishi Heavy Industries, Ltd. 5-1, Marunouchi 2-chome Chiyoda-Ku, Tokyo 100 Also: Suite 3000, Union Bank Building 50 California Street San Francisco, CA 94111	Portable geothermal turbine generators with capacities of 500–5,000 kW (back pressure type) and 5,000–10,000 kW (condensing type)

Country	Company/address	Product
GEOTHERMAL EQUIPMENT		
United States	Barber–Nichols Engineering Co. Arvada, CO 80002	Now testing a 500-kW power plant that uses a direct contact heat exchanger
	Biphase Energy Systems (Subsidiary of Research-Cottrell, Inc.) 2800 Airport Avenue Santa Monica, CA 90405	Two-phase turbine–separator system that uses steam and hot liquid is now available for certain applications
	Geothermal Power Co., Inc. 1460 West Water Street Elmira, NY 14905	Geothermal Power Mono-Blok, a self-contained power-generating system of from 100-kW to 2,500-kW capacity
	Hydrothermal Power Co. Post Office Box 2794 Mission Viejo, CA 92790	Packaged geothermal power plants up to 5,000 kW using the helical screw expander
	Rotoflow Corporation 2235 Carmelina Avenue Los Angeles, CA 90064	Total flow geothermal turbogenerators
	Sperry Rand Research Center 100 North Road Sudbury, MA 01776	Now testing a 4,300-kW gravity-head power plant that uses a down-hole heat exchanger system
	Turbines Incorporated 950 East Katella Avenue, Suite 9 Orange, CA 92667	Total flow geothermal turbogenerators up to 10,000-kW capacity
STORAGE BATTERIES AND INVERTERS		
Federal Republic of Germany	Accumulatorenfabrik Sonnenschein GmbH Thiergarten 6470 Buedingen/Hessen	Maintenance-free, portable lead–acid batteries, power supply units, battery chargers

Great Britain	Chloride Alcad Post Office Box 4 Union Street Redditch Worcs. B98 7BW	Nickel–cadmium batteries
	Varley Dry Accumulators, Ltd. Alfred's Way Barking Essex 1G11 OTB	Lead–acid batteries that have the electrolyte absorbed in a porous separating medium
United States	Allied Electronics 2400 W. Washington Boulevard Chicago, IL 60612	Inverters: 350–1,000 W; 12-volt input
	ATR Electronics, Inc. 300 East 4th Street St. Paul, MN 55101	Inverters: 275–1,100 W; 12-volt input
	Best Energy Systems for Tomorrow, Route 1 Post Office Box 280 Necedah, WI 54646	Inverters: 1,000–5,000 W; 12–48-volt input
	Carter Motor Company 2711 West George Street Chicago, IL 60618	Inverters: 210–740 W; variable voltage input
	C&D Batteries Division 3404 Walton Road Plymouth Meeting, PA 19462	Almost every type of lead–acid battery
	Delatron Systems Corporation 553 Lively Boulevard Elk Grove Village, IL 60007	Lead–acid deep discharge batteries
	Dynamote Corporation 1130 N.W. 85th Street Seattle, WA 98117	Inverters: automatic DC to 120-volt AC; 60 Hz static inverters

Country	Company/address	Product
STORAGE BATTERIES AND INVERTERS		
	Eico Electronic Instrument Co. 283 Malta Street Brooklyn, NY 11207	Inverters: 110–220 W; 12-volt input
	Electro Sales Co., Inc. 100 Fellsway West Somerville, MA 02145	Inverters: 20–2,000 W; 12–200-volt input
	Elgar Corporation 8225 Mercury Court San Diego, CA 92111	Inverters: 600–1,000 W; 24–48-volt DC input
	ESB Inc. Five Penn Center Plaza Philadelphia, PA 19103	
	Gemini Synchronous Inverters Windworks Box 329, Route 3 Mukwonago, WI 53149	Inverters: 4–1,000 kW; variable voltage input
	Globe Battery Globe-Union Inc. 5757 N. Green Bay Avenue Milwaukee, WI 53201	Storage batteries
	Gould Inc. Industrial Battery Division 2050 Cabot Boulevard West Langhorne, PA 19047	
	Heath Company Benton Harbor, MI 49022	Inverters: 175 W; 12-volt input
	Independent Power Developers, Inc. Route 3, Box 174H Sand Point, ID 83864	Inverters in 7 models–1–12 kW–for use with wind-electric and micro-hydro systems

INEN Corporation 1016 East Airport Road Stillwater, OK 74078	Custom-manufactured inverters up to 500 kW
La Marche Manufacturing Co. 106 Bradrock Drive Des Plaines, IL 60018	Inverters: 100–10,000 W; 24–120-volt input
Newark Electronics 500 N. Pulaski Road Chicago, IL 60624	Inverters: 100–250 W; 12-volt input
North Wind Power Company, Inc. Post Office Box 315 Warren, VT 05674	Synchronous inverters
Ratelco, Inc. 610 Pontius Avenue, N. Seattle, WA 98109	Inverters: 250–2,500 W; 24–100-volt input
Real Gas & Electric, Inc. Box A Guerneville, CA 95446	Inverters: 5-kVA maximum AC output; 75–200-volt DC input
Soleq Corporation 5969 Elston Avenue Chicago, IL 60646	Inverters: 1,500–6,000 W; 12–112-volt input
Westinghouse Electric Corporation Gateway Center Westinghouse Boulevard Pittsburgh, PA 15222	Storage batteries

[a] This list was prepared by the Solar Energy Information Center, Solar Energy Research Institute, 1617 Cole Boulevard, Golden, CO 80401, USA.

[b] This list was compiled by the Energy Systems Group, Rocky Flats Plant, Rockwell International, Post Office Box 464, Golden, CO 80401, USA.

APPENDIX B

Centers of Research and Development

This appendix represents the best information available to the NAS staff at the time of publication, but is not necessarily complete. It was not possible to verify many of the following addresses.

Country	Company/location	Activity
PHOTOVOLTAICS		
France	Centre de Recherches Nucléaires de Strasbourg BP 20 CR Cronenbourg 67037 Strasbourg Cedex	
	Centre National de la Recherche Scientifique 15 Quai Anatole France 75700 Paris	
	Commissariat à l'Energie Solaire (COMES) Ministère de l'Industrie 208, rue Raymond Losserand 75014 Paris	
	Laboratoires d'Electronique et de Physique Appliquée BP 15 3, Avenue Descartes 94450 Limeil–Brezannes	
United States	Institute of Energy Conversion University of Delaware One Pike Creek Center Wilmington, DE 19808	
	Jet Propulsion Laboratory California Institute of Technology Pasadena, CA 91106	
	NASA–Lewis Research Center Technology Utilization Office 21000 Brookpark Road Cleveland, OH 44135	
	Solar Energy Research Institute 1617 Cole Boulevard Golden, CO 80401	

Country	Company/location	Activity
WIND ENERGY		
Canada	Brace Research Institute McDonald College of McGill University Ste. Anne de Bellevue Quebec HOA ICO	
	Institute of Research of Hydro-Quebec 1800 Montec Ste. Julie Varrenes, Quebec JOL 2 PO	
	National Research Council of Canada Research Office Building M-2, Montreal Road Ottawa, Ontario KLA ORS	
Denmark	Aalborg Universitets Center Institute of Industrial Structures & Engineering Langagervej, Post Office Box 159 9100 Aalborg	
	Danish Wind Power Projects Government of Denmark Copenhagen	
	DEFU 100 Lundtoftevej 2800 Lyngby	
Federal Republic of Germany	Center for Wind Energy Research Forschungsprojekt Windenergie University of Regensburg 8440 Regensburg	
Great Britain	Energy Technology Support Unit AERE, Harwell Oxford OX11 ORA	

The Netherlands	Netherlands Energy Research Foundation Westerduinweg 3 Petten N.H.
New Zealand	Department of Mechanical Engineering University of Canterbury Private Bag Christchurch
Sweden	National Swedish Board for Energy Source Development 77 Karlsbergsvagen Box 21048 S-100 31 Stockholm
	Namnden For Energiproduktionforskning Box 1103 S16312 Spanga
United States	Agricultural Research Service Rural and Remote Wind Energy Applications U.S. Department of Agriculture Beltsville, MD 20705
	Battelle–Pacific Northwest Laboratories Wind Characteristics Program Post Office Box 999 Richland, WA 99352
	NASA–Lewis Research Center Large Wind Turbine Program 21000 Brookpark Road Cleveland, OH 44135

Country	Company/location	Activity
WIND ENERGY		
	Rockwell International Energy Systems Group Small Wind Systems Program Post Office Box 464 Golden, CO 80401	
	Sandia Laboratories Darrieus Wind Systems Program Division 4715 Albuquerque, NM 87185	
	Solar Energy Research Institute Wind Systems Research and Analysis 1617 Cole Boulevard Golden, CO 80401	
	United Technologies Research Center Silver Lane East Hartford, CT 06108	
	West Wind Co. Box 1465 Farmington, NM 87401	
GEOTHERMAL		
United States	EG&G Idaho, Inc. Idaho National Engineering Laboratory P.O. Box 1625 Idaho Falls, ID 83401	Major clearinghouse for information on direct heat applications of geothermal energy and a major research and development contractor to the U.S. Department of Energy
	Geo-Heat Utilization Center Oregon Institute of Technology Klamath Falls, OR 97601	Major center for information on direct heat applications of geothermal energy

	U.S. Department of Energy Division of Geothermal Energy Washington, D.C. 20545	Primary information and funding source for technological developments in geothermal energy production and utilization
	Lawrence Livermore Laboratories University of California P.O. Box 808 Livermore, CA 94550	

STORAGE BATTERIES

United States	Dow Chemical Co. 2030 Dow Center Midland, MI 48640	Sodium/sulphur high-temperature batteries
	Eagle-Picher Industries, Inc. 580 Walnut Street Cincinnati, OH 45201	Lithium/metal sulfide high-temperature batteries
	Eltra Corporation 2 Pennsylvania Plaza New York, NY 10001	Advanced lead–acid batteries
	Energy Development Associates 1100 W. Whitcomb Avenue Madison Heights, MI 48071	Zinc/chlorine flow systems
	ESB Corporation Five Penn Center Plaza Philadelphia, PA 19103	Advanced lead–acid batteries
	Exxon Research and Engineering Company P.O. Box 101 Florham, NJ 07932	Zinc/bromine flow systems
	Ford Motor Company American Road Dearborn, MI 48121	Sodium/sulfur high-temperature batteries

Country	Company/location	Activity
STORAGE BATTERIES		
	General Electric Company 1 River Road Schenectady, NY 12345	Sodium/sulfur high-temperature batteries
	General Motors Corporation 767 Fifth Avenue New York, NY 10022	Lithium/metal sulfide high-temperature batteries
	Gould Inc. 8550 Bryn Mawr Avenue Chicago, IL 60631	Lithium/metal sulfide high-temperature batteries
	NASA–Lewis Research Center 21000 Brookpark Road Cleveland, OH 44135	Reduction/oxidation flow systems
	Rockwell International Corporation 600 Grant Street Pittsburgh, PA 15219	Lithium/metal sulfide high-temperature batteries
STIRLING ENGINES		
Canada	University of Calgary Department of Mechanical Engineering Calgary, Alberta T2N 1N4	Student projects on all aspects of Stirling engine design and experimental work
Federal Republic of Germany	Maschinenfabrik-Augsburg-Nurnburg Stadtbachstrasse 1 8900 Augsburg 1	Large automotive and stationary power plants
Great Britain	Harwell Laboratory Instrumentation and Applied Physics Division AERE, Harwell Oxford OX11 ORA	Electric generators, diaphragm type; liquid-piston

	University of Reading Department of Engineering and Cybernetics Whiteknights Reading, Berks RG6 ZAY	Analysis and experimental work on fluidized bed-heated Stirling engines, liquid-piston
India	Metal Box India Limited B2 Hide Road 700 043 Calcutta	Liquid-piston engines
The Netherlands	Philips Nederland BV Stirling Engine Division Boschdiji 525 Eindhoven	Free-piston, linear-alternator cooling engines; hot air, crank-type engines
South Africa	University of Witwatersrand Department of Mechanical Engineering 1 Jan Smuts Avenue Johannesburg 2001	Analytical studies of a wide variety of Stirling engines; experimental work on liquid-piston, free-piston, and others
Sweden	United Stirling (Sweden) AB and Co. Fack 201 10 Malmo 1	Automotive Stirlings, solar-electric
United States	Energy Research and Generation Inc. Lowell and 57th Streets Oakland, CA 94608	Solar-electric, cooling, cryogenic coolers, automotive Stirlings
	Mechanical Technology, Inc. 968 Albany-Shaker Road Latham, NY 12110	Advanced designs for aerospace, military and U.S. domestic service, automotive Stirlings, solar-electric generators

Country	Company/location	Activity
STIRLING ENGINES		
	Stirling Power Systems Corp. 7101 Jackson Road Ann Arbor, MI 48103	10-kW electric generator-crank engine
	Sunpower, Inc. 6 Byard Street Athens, OH 45701	Analysis, design, and construction of free-piston engines; alternators; heat-driven coolers; water pumps; hot-air crank engines; solar-powered Stirling engines of all types

APPENDIX C

Suggested Reading

Where indicated, some of the following publications are available from the National Technical Information Service, (NTIS), 5285 Port Royal Road, Springfield, VA 22161, USA.

GENERAL: ENERGY FOR RURAL DEVELOPMENT

Anonymous (1978) An evaluation of new forms of energy for African development: a study published by the Ministry of Cooperation in the "Technology and Development" series, *Bulletin de l'Afrique Noire* 960:18747–18750.

Bach, W., W. Manshard, W. H. Matthews, and H. Brown, eds. (1980) *Renewable Energy Prospects*. Pergamon Press, Oxford, England.

Bhatia, R. (1977) Energy and rural development in India: some issues. Pp. 559–579 in: *Agriculture and Energy*, edited by W. Lockeretz. Academic Press, New York, New York, USA.

Bhatia, R. (1978) Energy Alternatives for Irrigation Pumping: Some Results for Small Farms in North Bihar. Institute of Economic Growth, Delhi, India.

Brown, N. L., ed. (1978) *Renewable Energy Resources and Rural Applications in the Developing World*. Westview Press, Boulder, Colorado, USA.

Cecelski, E., J. Dunkerley, and W. Ramsey (1978) Household Energy and the Poor in the Third World. Resource Paper R-15. Resources for the Future, Washington, D.C., USA.

Chatterjee, R. (1978) Energy and Rural Development: The Program of the Institute for Energy Research. Seminar, Resource Systems Institute, East–West Center, Honolulu, Hawaii, USA.

Comité Permanent Interétats de Lutte Contre la Secheresse dans le Sahel, Club du Sahel (1978) Energy in the Development Strategy of the Sahel. CILSS, Ouagadougou, Upper Volta.

Conant, M. A. (1978) Commercial energy and agriculture in the developing world, *Common Ground* 4:19–25.

deLucia, R. J. and P. Rogers (1978) Report of the Brookhaven/USAID Workshop on Rural/Domestic Energy Issues, January 5 and 6, 1978. Meta Systems Inc., Cambridge, Massachusetts, USA.

Eckholm, E. (1976) *Losing Ground: Environmental Stress and World Food Prospects*. Norton, New York, New York, USA.

Energy Technology Department, Electricity Generating Authority of Thailand (EGAT) (1979) EGAT's Activities on Non-Conventional/Renewable Energy Resource [*sic*]. Paper presented at a Workshop on Energy and Rural Development, March 1979, Bangkok, Thailand.

Ernst, E. (1977) Fuel Consumption Among Rural Families in Upper Volta, West Africa. Peace Corps, Washington, D.C., USA.

Floor, W. M. (1978) Energy Options in Rural Areas of the Third World. Paper prepared for the Eighth World Forestry Congress, October 16–28, 1978, Jakarta. Food and Agriculture Organization of the United Nations, Rome, Italy.

French, D. (1978) Energy for Africa's future, *Africa Report* 23(3):9–14.

Hayes, D. (1977) Energy for Development: Third World Options. Worldwatch Paper 15. Worldwatch Institute, Washington, D.C., USA.

Howe, J. W., J. A. Bever, and J. Tarrant (1978) Prospects for the Use of Small-Scale Renewable Energy Technologies for the Developing World. Background paper prepared for the World Bank by the Overseas Development Council, Washington, D.C., USA.

Howe, J. W., R. Nathans, and J. Goldenberg (1978) *Energy Problems of Developing Countries: Implications for U.S. Policy*. Overseas Development Council, Washington, D.C., USA.

Hughart, D. (1979) *Prospects for Traditional and Nonconventional Energy Sources in Developing Countries*. World Bank, Washington, D.C., USA.

Institute for Energy Research (1978) Energy and Rural Development Research Materials. State University of New York at Stony Brook, Stony Brook, New York, USA.

International Bank for Reconstruction and Development (1978) An Approach to Planning and Implementing Rural Energy Projects (Report on a Pilot Exercise in Colombia). Central Projects Staff, Energy, Water and Telecommunications Department, IBRD, Washington, D.C., USA.

Islam, M. N. (1979) Issues in the Implementation of Research on Energy and Rural Development. Paper presented at the Energy and Rural Development Research Planning Workshop, May 6–19, 1979, Resource Systems Institute, East-West Center, Honolulu, Hawaii, USA.

LeSieur, H. A. (1979) Appropriate Technology and Scaling Considerations for Developing Countries. Paper presented at the Fifth Chilean Congress of Chemical Engineering, October 18–20, 1979, Valparaiso, Chile.

Levinson, A., C. Rosenberg, and A. Yansane (1977) The political economy of energy and agriculture in the Third World. Pp. 639–655 in: *Agriculture and Energy*, edited by W. Lockeretz. Academic Press, New York, New York, USA.

Lillywhite, M. (1978) Lesotho Renewable Energy Technology Program. Domestic Technology Institute, Evergreen, Colorado, USA.

Lillywhite, M. (1978) Swaziland–Small Scale Renewable Energy Technology Program. Domestic Technology Institute, Project No. 645-0207, Evergreen, Colorado, USA.

Makhijani, A. (1976) *Energy Policy for the Rural Third World*. International Institute for Environment and Development, London, England.

Meta Systems Inc. (1978) Traditional Energy and Rural Development, Issues and Recommendations for Peru. Report prepared for Brookhaven National Laboratory and the U.S. Department of Energy. Meta Systems Inc., Cambridge, Massachusetts, USA.

National Research Council (1977) *Energy and Climate*. National Academy of Sciences, Washington, D.C., USA.

National Research Council (Thailand) (1979) *Report of Workshop on Energy and Rural Development*. March 1979, Bangkok, Thailand.

Palmedo, P. F., R. Nathans, E. Beardsworth, and S. Hale (1978) *Energy Needs, Uses and Resources in Developing Countries*. Brookhaven National Laboratory, Upton, New York. Available from NTIS, Order No. BNL 50784.

Perera, A. F. and S. Warkow (1979) Energy for Development in Semi-arid Areas of Northeastern Brazil. Brazilian School of Public Administration, Getulia Vargas Foundation, Rio de Janeiro, Brazil.

Pimentel, D. (1977) Energy Needs, Uses and Resources in the Food Systems of Developing Countries. Workshop Report. College of Agriculture and Life Sciences, Cornell University, Ithaca, New York, USA.

Policy Analysis Division, National Center for the Analysis of Energy Systems, Brookhaven National Laboratory (1978) Analytical Framework for the Assessment of Energy Resource and Technology Options for Developing Countries. Report prepared for the U.S. Department of Energy International Energy Program, Washington, D.C., USA.

Ramachandran, A. and J. Gururaja (1977) Perspectives on energy in India. Pp. 365–386 in: *Annual Review of Energy*, edited by J. M. Hollander. Vol. 2. Annual Reviews, Inc., Palo Alto, California, USA.

Resource Systems Institute, East–West Center (1979) Energy and Rural Development Research Planning Workshop: Indicative Research Proposals and Implementation Strategies. May 6–19, 1979, East–West Center, Honolulu, Hawaii, USA.

Revelle, R. (1976) Energy use in rural India, *Science* 192:969–975.

Revelle, R. (1977) Energy and rural development, *Habitat* 2(½):133–142.

Revelle, R. (1978) Requirements for energy in the rural areas of developing countries. Pp. 11–26 in: *Renewable Energy Resources and Rural Applications in the Developing World*, edited by N. L. Brown. Westview Press, Boulder, Colorado, USA.

Revelle, R. (1979) Energy Alternatives for Rural Areas. Paper presented at the Conference on Energy Alternatives, January 9–12, 1980, East–West Center, Honolulu, Hawaii, USA.

Stout, B. A. (1979) Energy for World Agriculture. FAO Agriculture Series No. 7. Food and Agriculture Organization of the United Nations, Rome, Italy.

Tillman, D. A., K. Sarkanen, and L. L. Anderson, eds. (1977) *Fuels and Energy from Renewable Resources*. Academic Press, New York, New York, USA.

United Nations Economic and Social Commission for Asia and the Pacific (1976) *Proceedings of the Meeting of the Expert Working Group on the Use of Solar and Wind Energy*. Energy Resources Development Series No. 16. ESCAP, Manila, Philippines.

United Nations Industrial Development Organization (n.d.) Appropriate technology for rural energy. Energy for rural requirements. In: *International Forum on Appropriate Industrial Technology*. Discussion Paper ID/WG 282/100. UNIDO, New York, New York, USA.

U.S. Agency for International Development (1978) Philippines–Nonconventional Energy Development. AID/DLC/P-2288. U.S. Agency for International Development, Washington, D.C., USA.

Weingart, J. (1979) The Role of Renewable Energy Resources in the Developing Countries. Seminar, Resource Systems and Environment and Policy Institutes, East–West Center, Honolulu, Hawaii, USA.

Wilson, C. L. (1977) *Energy: Global Prospects 1985-2000*. Workshop on Alternative Energy Strategies. McGraw–Hill, New York, New York, USA.

World Bank (1976) *Critical Factors in Economic Evaluation of Small Decentralized Energy Projects*. Science and Technology Report Series No. 25. World Bank, Washington, D.C., USA.

PHOTOVOLTAICS

ARCO Solar, Inc. (1979) Vacuum Die Casting of Silicon Sheet for Photovoltaic Applications. Second Quarterly Report, 1 July–30 September 1979. ARCO Solar, Inc., Chatsworth, California. Available from NTIS, Order No. DOE/JPL/955325–2.

Auh, C. (1978) Solar Energy: Its Technologies and Applications. Brookhaven National Laboratory, Upton, New York. Available from NTIS, Order No. BNL-24832.

Backus, C. E. (1978) Can Solar Energy Contribute Significantly to the Solution of the World's Energy Famine? Paper presented at the 13th Intersociety Energy Conversion Engineering Conference, August 20–25, 1978, San Diego, California, USA.

Bell, C. R. (1978) Solar Energy Task Progress Report for 1975/76: Evaluation of Solar Energy Options. RM-77-20. International Institute for Applied Systems Analysis, Laxenburg, Austria.

Bernard, D. M. and M. J. Pheline (1978) Report on the Development of Solar Energy in France. Paper presented at the conference, Mankind's Future Source of Energy, January 1978, New Delhi, India. International Solar Energy Society, Victoria, Australia.

Bifano, W., A. Ratajczak, D. Bahr, and B. Garrett (1979) Social and Economic Impact of Solar Electricity at Schuchuli Village: A Status Report. Report No. NASA TM-79194. NASA–Lewis Research Center, Cleveland, Ohio, USA.

Boldt, J. (1978) A Solar Powered Water Pump for the Rural Third World. Technical University of Denmark, Lyngby, Denmark.

Bolton, J. R., ed. (1977) *Solar Power and Fuels*. Academic Press, New York, New York, USA.

Brown, N. L. and J. W. Howe (1978) Solar energy for village development, *Science* 199 (4329):651–657.

Bucciarelli, L. L., J. D. Cremin, H. A. Fenton, R. F. Hopkinson, and E. F. Lyon (1978) Mead 25 Kilowatt Photovoltaic System. Lincoln Laboratory, Massachusetts Institute of Technology, Lexington, Massachusetts. Available from NTIS, Order No. COO-4094-19.

Burgess, E. L. (1979) Photovoltaic Concentrator System Technology and Applications Experiments. Sandia Laboratories, Albuquerque, New Mexico. Available from NTIS, Order No. SAND-78-2184C.

Burgess, E. L. (1980) Photovoltaic System Cost Experience for Intermediate-Sized Applications. Sandia Laboratories, Albuquerque, New Mexico. Available from NTIS, Order No. SAND-80-0141C.

Burgess, E. L. and E. A. Walker (1979) Summary of Photovoltaic Application Experiments Designs. Sandia Laboratories, Albuquerque, New Mexico. Available from NTIS, Order No. ALO-71.

Calaghan, W. (1979) Department of Energy Low-Cost Solar Array Project, *Proceedings of the Photovoltaic Advanced Research and Development Conference*. September 17–19, 1979, Denver, Colorado. Solar Energy Research Institute, Golden, Colorado, USA.

Cape, J. A., J. S. Harris, Jr., and J. J. Wiczer (1980) Gallium Arsenide Photovoltaic Dense Array for Concentrator Applications. Rockwell International, Thousand Oaks, California. Available from NTIS, Order No. SAND-79-2270C.

Carlson, D. E., I. Balberg, R. S. Crandall, B. C. Goldstein, and J. J. Hanak (1980) Amorphous Thin Films for Solar-Cell Applications. Final Report, September 11, 1978–September 10, 1979. RCA Laboratories, Princeton, New Jersey. Available from NTIS , Order No. DOE/ET/21074-4.

Chu, T. L. (1977) Development of Low Cost Thin Film Polycrystalline Silicon Solar Cells for Terrestrial Applications: Final Report. Southern Methodist University, Dallas, Texas. Available from NTIS, Order No. PB-266 057/9ST.

Cremin, J. D. and H. A. Fenton (1978) Design, Development and Implementation of Data Acquisition Systems for Photovoltaic Tests and Applications Project. Lincoln Laboratory, Massachusetts Institute of Technology, Lexington, Massachusetts. Available from NTIS, Order No. COO-4094-24.

Durand, J. P., M. G. Clemot, J. P. Girardier, and M. Y. Vergnet (1978) Utilization of Solar Energy in the Development of Arid Zones: Solar Water Pumps. Development and Transfer Technology Series No. 5. U.N. Industrial Development Organization, New York, New York, USA.

Forest Products Research Society (1977) Practical Application of Solar Energy to Wood Processing. Proceedings of a workshop at the Virginia Polytechnic Institute, January 6–7, 1977, Blacksburg, Virginia, USA.

Fossum, J. F. and E. L. Burgess (1976) Silicon Solar Cell Development for Concentrated-Sunlight, High-Temperature Applications. Sandia Laboratories, Albuquerque, New Mexico. Available from NTIS, Order No. SAND-76-5311.

Hayes, D. (1978) Entering the solar era, *Science Forum* 11(2):11–14.

Hayes, D. (1978) The Solar Energy Timetable. Worldwatch Paper 19. Worldwatch Institute, Washington, D.C., USA.

Hayes, D. (1979) Solar possibilities, *Energy* 4(5):761–768.

Heeger, A. J. and A. G. MacDiarmid (1980) Polyacetylene as an Emerging Material for Solar Cell Applications. Technical Progress Report, October, November, December 1979. University of Pennsylvania, Philadelphia, Pennsylvania. Available from NTIS, Order No. DOE/ET/23002-T7.

Hovel, H. (1977) Review paper, solar cells for terrestrial applications, *Solar Energy* 19: 605–615.

International Solar Energy Society (1979) *International Congress Proceedings*. May 28–June 1, 1979, Atlanta, Georgia. International Solar Energy Society, Victoria, Australia.

Javetski, J. (1979) A burst of energy in photovoltaics; special report, *Electronics* 52(15): 105–122.

Jet Propulsion Laboratory (1979) *Proceedings of the 13th Project Integration Meeting, DOE Low-Cost Solar Array Project*. August 22–23, 1979, Pasadena, California, USA.

Jet Propulsion Laboratory (1979) *Proceedings of the US DOE Photovoltaics Technology Development and Applications Program Review*. May 16, 1979, Gatlinburg, Tennessee, Jet Propulsion Laboratory, Pasadena, California, USA.

Jones, G. J. (1980) Photovoltaic Systems and Applications Perspective. Sandia Laboratories, Albuquerque, New Mexico. Available from NTIS, Order No. SAND-80-0926C.

Jones, G. J. and J. L. Watkins (1979) Summary of Application Analysis for Photovoltaic Systems. Sandia Laboratories, Albuquerque, New Mexico. Available from NTIS, Order No. SAND-78-0980C.

Kaplan, S. I. (1980) Photovoltaic Applications: Past and Future. Oak Ridge National Laboratory, Oak Ridge, Tennessee. Available from NTIS, Order No. CONF-800106-4.

Kelly, H. (1978) Photovoltaic power systems: a tour through the alternatives, *Science* 199:634.

Langerhorst, J., G. Prast, T. Thalhammer, and W. M. Floor (1977) Solar Energy. Report on a Study of Difficulties Involved in Applying Solar Energy in LDC's. Ministry for Development Cooperation, The Hague, Netherlands. February 11.

Mengel, R. W., T. P. Nadolski, D. C. Sparks, S. K. Young, and A. Yingst (1980) Photovoltaic Applications Definition and Photovoltaic System Definition Study in the Agricultural Sector. 3 vols. BDM Corporation, McLean, Virginia. Available from NTIS Order No. SAND-79-7018/1.

Merchant, M. C. (1978) Photovoltaic review, *Sunworld* 2(1):2–10.

Ministry of Energy, Republic of the Philippines (1978) *Proceedings of the Solar Drying Workshop*. October 18–21, 1978, Manila, Philippines.

Neville, R. (1978) *Solar Energy Conversion: The Solar Cell*. Elsevier, New York, New York, USA.

Newkirk, H. W. (1978) Solar Technology Applications: A Survey of Solar Powered Irrigation Systems. Lawrence Livermore Laboratory, University of California, Livermore, California. Available from NTIS, Order No. UCID-17510 (Rev. 1).

Palz, W. (1977) Solar electricity, *Proceedings of the Unesco/WMO Symposium on Solar Energy*. August 30–September 3, 1976, Geneva, Switzerland. WMO–No. 477, World Meteorological Organization, Geneva, Switzerland.

Palz, W. (1979) Photovoltaics for Development. Paper presented at the Commission of European Communities 2nd Photovoltaic Solar Energy Conference, April 23–26, 1979, West Berlin, FRG.

Prince, M. B. (1978) *Photovoltaic Technology: Renewable Energy Resources and Rural Applications in the Developing World*. Westview Press, Boulder, Colorado, USA.

Pulfrey, D. (1978) *Photovoltaic Power Generation*. Van Nostrand Reinhold Company, New York, New York, USA.

Rauh, R. D., T. L. Rose, T. G. Hoover, and D. L. Natwig (1980) Electrodeposition of Polycrystalline and Amorphous Silicon for Photovoltaic Applications. Quarterly Technical Progress Report No. 02, September 11, 1979–November 30, 1979. EIC Corporation, Newton, Massachusetts. Available from NTIS, Order No. DOE/ET/23046-2.

Rios, M., Jr. (1980) Simple Economic Evaluation and Applications Experiments for Photovoltaic Systems for Remote Sites. Sandia Laboratories, Albuquerque, New Mexico. Available from NTIS, Order No. SAND-80-0749C.

Romaine, W. R. (1979) Mead, Nebraska, 25-KW Photovoltaic Power System. Lincoln Laboratory, Massachusetts Institute of Technology, Lexington, Massachusetts. Available from NTIS, Order No. COO-4094-10.

Rosenblum, L., W. J. Bifano, W. A. Poley, and L. R. Scudder (1977) Photovoltaic Refrigeration Application: Assessment of the Near-Term Market. NASA–Lewis Research Center, Cleveland, Ohio. Available from NTIS, Order No. DOE/NASA/1022-77/23.

Rosenblum, L., L. R. Scudder, W. J. Bifano, and W. A. Poley (1977) Photovoltaic Highway Applications: Assessment of the Near-Term Market. NASA–Lewis Research Center, Cleveland, Ohio. Available from NTIS, Order No. DOE/NASA/1022-77/22.

Rosenblum, L., L. R. Scudder, W. A. Poley, and W. J. Bifano (1977) Photovoltaic Remote Instrument Applications: Assessment of the Near-Term Market. NASA–Lewis Research Center, Cleveland, Ohio. Available from NTIS, Order No. N78-19710/OST.

Savitz, M., J. Deutch, R. Thorne, and D. Beattie (1978) DOE Role in Support of Small-Scale Appropriately Distributed Technology. Report CONF-780132. U.S. Department of Energy, Washington, D.C., USA.

Seager, C. H. and D. S. Ginley (1980) Fundamental Studies of Grain Boundary Passivation in Polycrystalline Silicon with Application to Improved Photovoltaic Devices. A Final Research Report Covering Work Completed from February–December 1979. Sandia Laboratories, Albuquerque, New Mexico. Available from NTIS, Order No. SAND-80-0879.

Smith, D. V. (1977) Photovoltaic Power in Less Developed Countries. Lincoln Lab Document No. C00-4094-1. Massachusetts Institute of Technology, Lexington, Massachusetts, USA.

Smith, D. V. and S. V. Allison (1978) Micro Irrigation with Photovoltaics. MIT Energy Laboratory Report No. MIT-EL-78-006. Massachusetts Institute of Technology, Cambridge, Massachusetts, USA.

Solar Energy Research Institute (n.d.) Analysis Methods for Photovoltaic Applications. SERI, Golden, Colorado. Available from NTIS, Order No. SERI/SP-35-230.

Solar Energy Research Institute (1979) *Proceedings of the Photovoltaic Advanced Research and Development Conference*. September 17–19, 1979, Denver, Colorado. SERI, Golden, Colorado, USA.

Tabors, R. D. (1978) The Economics of Water Lifting for Small Scale Irrigation in the Third World: Traditional and Photovoltaic Technologies. MIT Energy Laboratory Working Paper No. MIT-EL-78-015. Massachusetts Institute of Technology, Cambridge, Massachusetts, USA.

Tabors, R. D., *et al*. (1978) Planning and Analysis for Development and Commercialization of Photovoltaic Energy Conversion Systems. Final Report MIT-EL-78-033TR. Massachusetts Institute of Technology, Cambridge, Massachusetts, USA.

Tanzania National Scientific Research Council (1978) *Workshop on Solar Energy for the Villages of Tanzania*. August 11–19, 1977, Dar es Salaam, Tanzania.

Tata Energy Research Institute (1978) Directory of Solar Energy Research Projects in India. Tata Energy Research Institute, Bombay, India. Available from NTIS, Order No. PB-293 760/5ST.

Tleimat, B. W., E. D. Howe, and A. D. K. Laird (1977) Solar assisted still with vapor compression option, *Desalination* 22:41–54.

Tobak, J. P. (1979) The Application of DC–DC Energy Conversion in a Solar System; Final Report. U.S. Army Armament Research and Development Command, Dover, New Jersey. Available from NTIS, Order No. AD-A077 112/1.

United Nations Industrial Development Organization (1978) *Technology for Solar Energy Utilization*. Development and Transfer of Technology Series. UNIDO, New York, New York, USA.

U.S. Agency for International Development (1978) Solar Technology Applications: A Survey of Solar Powered Irrigation Systems. USAID, Washington, D.C., USA.

U.S. Agency for International Development (1978) Utilization of Solar Energy, in Developing Countries: Identifying Some Potential Markets. Available from NTIS, Order No. N78-29578/9.

U.S. Office of Technology Assessment (1978) Application of Solar Technology to Today's Energy Needs. OTA, Washington, D.C. Available from NTIS, Order No. PB-289, 762/7ST.

Van Overstraeten, R. (1979) European progress in photovoltaic development, *Proceedings, 1979 International Congress of the International Solar Energy Society*. May 28–June 1, 1979, Atlanta, Georgia, USA, ISES, Victoria, Australia.

Vojdani, S. (1978) Research in Photovoltaic Cells. Paper presented at the Bahrain Ministry of Development and Industry Middle East Solar Technology Conference, April 24–27, 1978, Bahrain.

SUN/EARTH-TEMPERED BUILDING DESIGNS

Anderson, B. (1976) *The Solar Home Book: Heating, Cooling, and Designing with the Sun*. Brick House Publishing, Andover, Massachusetts, USA.

Anderson, B. (1977) *Solar Energy: Fundamentals in Building Design*. McGraw-Hill, New York, New York, USA.

Bahadori, M. N. (1978) Passive cooling systems in Iranian architecture, *Scientific American* 238:144.

Bainbridge, D., J. Corbett, and J. Hofacre (1979) *Village Homes' Solar House Designs*. Rodale Press, Emmaus, Pennsylvania, USA.

Bourgeois, J-L and C. Pelos (1980) Welcoming the wind, *Soft Energy Notes* 3(5):12–14.

Butti, K. and J. Perlin (1980) *A Golden Thread: 2000 Years of Solar Architecture and Technology*. Van Nostrand Reinhold, New York, New York, USA.

Fathy, H. (1973) *Architecture for the Poor*. University of Chicago Press, Chicago, Illinois, USA.

Franta, G. W., ed. (1979) *Proceedings of the 4th National Passive Solar Conference*. October 4–5, 1979, Kansas City, Missouri. American Section, International Solar Energy Society, University of Delaware, Newark, Delaware, USA.

Mazria, E. (1979) *The Passive Solar Energy Book*. Rodale Press, Emmaus, Pennsylvania, USA.

Naumann, H., F. Quivik, and T. Riley (1979) *Natural Cooling for Homes*. National Center for Appropriate Technology, Butte, Montana, USA.

Newburgh, S. A. (1980) Rediscovering energy-conscious architecture, *Technology Review* 82(8):68–78.

Olgyay, V. (1963) *Design with Climate: A Bioclimatic Approach to Architectural Regionalism*. Princeton University Press, Princeton, New Jersey, USA.

Stein, R. G. (1978) *Architecture and Energy*. Anchor Press/Doubleday, Garden City, New York, USA.

Sterling, R. (1978) Structural systems for earth sheltered housing, *Underground Space* 3(2):75–81.

WIND ENERGY

Banas, J. F., E. C. Kadlec, and W. N. Sullivan (1975) Application of the Darrieus Vertical-Axis Wind Turbine to Synchronous Electrical Power Generation. Sandia Laboratories, Albuquerque, New Mexico. Available from NTIS, Order No. SAND-75-0165.

Blake, S. (1978) *Wind Driven Water Pumps–Economics, Technology, Current Activities*. Available from Sunflower Power, Route 1, Box 93-A, Oskaloosa, KS 66066, USA.

Braasch, R. H. (1978) Design, Construction, Testing and Manufacturing of Vertical Axis Wind Turbines. Sandia Laboratories, Albuquerque, New Mexico. Available from NTIS, Order No. SAND 78-1253.

Dodge, D. M. and J. V. Stafford, eds. (1979) *Proceedings, Small Wind Turbine Systems 1979, A Workshop on R&D Requirements and Utility Interface/Institutional Issues*. Rockwell International, Rocky Flats Wind Systems Program. Available from NTIS, Order No. RFP-3014/3533/79-8.

Eldridge, F. R. (1975) *Wind Machines*. Mitre Corporation, McLean, Virginia. Available from the U.S. Government Printing Office, Stock No. 038-000-00272-4, Washington, D.C., USA.

Enertech Corporation (1977) *Planning a Wind-Powered Generating System*. Available from Enertech Corporation, Post Office Box 420, Norwick, VT 05055, USA.

Hackleman, M. A. (1975) *The Homebuilt Wind-generated Electricity Handbook*. Available from Earthmind, 5246 Boyer Road, Mariposa, CA 95338, USA.

Hackleman, M. A. (1975) *Wind and Windspinners: A "Nuts 'N Bolts" Approach to Wind-Electric Systems*. Available from Earthmind, 5246 Boyer Road, Mariposa, CA 95338, USA.

Hughes, W. L., H. J. Allison, and R. G. Ramakumar (1975) Development of an Electrical Generator and Electrolysis Cell for a Wind Energy Conversion System. Oklahoma State University, Stillwater, Oklahoma. Available from NTIS, Order No. NSF/RAN/75-043.

Inglis, D. R. (1978) *Wind Power and Other Energy Options*. University of Michigan Press, Ann Arbor, Michigan, USA.

Jagadev, T. S. (1976) Induction Generators for Wind Energy Conversion Systems. University of Wisconsin, Madison, Wisconsin. Available from NTIS, Order No. AER-75-00653.

Kagin, S. (1978) *Buyers Guide to Wind Power*. Available from the Real Gas & Electric Company, Inc., Post Office Box F, Santa Rosa, CA 95402, USA.

Kovarik, T., C. Pipher, and J. Hurst (1979) *Wind Energy: The Generation, Storage and Conversion of Wind Power for Practical Use Today*. Domus Books, Chicago, Illinois, USA.

LeBoff, J. P. (1976) Wind power feasibility, *Energy Resource* 2(4):361–376.

Maughmer, M. D. (1976) Optimization and Characteristics of a Sailwing Windmill Rotor. Princeton University, Princeton, New Jersey. Available from NTIS, Order No. NSF/RANN/GI-41891/PR/75/4.

Meroney, R. and B. L. Burke (n.d.) *Energy from the Wind: Annotated Bibliography*. Available from The Engineering Research Center, Colorado State University, Fort Collins, CO 80523, USA.

McGuigan, D. (1978) *Harnessing the Wind for Home Energy*. Available from Garden Way Publishing, Charlotte, VT 05445, USA.

Mother Earth News (n.d.) *Mother Earth News Handbook of Homemade Power*. Available from the Mother Earth News, Post Office Box 70, Hendersonville, NC 28739, USA.

National Swedish Board for Energy Source Development (1978) *Small Wind Power Units, Technology–Cost Competitiveness–Potential Market*. Available from Forslagsorder, 162 89 Vallingby, Sweden.

Pangborn, J. B. (1976) Wind-Powered Hydrogen-Electric Systems for Farm and Rural Use. Institute of Gas Technology, Chicago, Illinois, USA.

Park, J. (1980) *Wind Power Book*. Cheshire Books, Palo Alto, California, USA.

Power Technologies, Inc. (1979) Induction and Synchronous Machines for Vertical Axis Wind Turbines. Available from NTIS, Order No. SAND 79-7017.

Ramakumar, R. G. (1975) *Harnessing Wind Power in Developing Countries*. Oklahoma State University, Stillwater, Oklahoma. Available from the Publications Department, Institute of Electrical and Electronics Engineers, 345 E. 47th Street, New York, NY 10017, USA.

Ramakumar, R. G., *et al.* (1976) Development and Adaptation of Field Modulated Generator System for Wind Energy Applications. Oklahoma State University, Stillwater, Oklahoma. Available from NTIS, Order No. ERDA/NSF/AER/00647-76/2.

Rockwell International (1978) A Guide to Commercially Available Wind Machines. Available from NTIS, Order No. RFP-2836/3533/78/3.

Rockwell International, Rocky Flats Wind Systems Program (1978) First Semiannual Report, Rocky Flats Small Wind Systems Test Center Activities. Available from NTIS, Order No. RFP-2920/3533/78/6-2.

Rockwell International, Rocky Flats Wind Systems Program (1979) Technical and Management Support for the Development of Small Wind Systems, Annual Report for the Period October 1, 1977–September 30, 1978. Available from NTIS, Order No. RFP-2974/3533/79-2.

Rockwell International, Rocky Flats Wind Systems Program (1980) Second Interim Report, Rocky Flats Small Wind Systems Test Center Activities. Available from NTIS, Order Nos. RFP-3004/3533/797-1 and 2.

Sandia Laboratories (1979) *Proceedings of the Workshop on Mechanical Storage of Wind Energy*. Available from NTIS, Order No. SAND 79-0001.

Schierholz, P. M., W. L. Somerville, Jr., W. Babcock, R. Hartel, and K. Timbre (1976) Wind-Powered Aeration for Remote Locations. Colorado State University, Fort Collins, Colorado. Available from NTIS, Order No. ERDA/NSF/00833-75/1.

Sweeney, T. E. (1976) The Bishop's Windmill. Available from the Forrestal Campus Library, Princeton University, Princeton, NJ 08540, USA.

Tewari, S. K. (1978) Economics of wind energy use for irrigation in India, *Science* 202 (4367):481–486.

Traci, R. M., G. T. Phillips, P. C. Patnaik and B. E. Freeman (1977) Development of a Wind Energy Site Selection Methodology. Available from NTIS, Order No. RLO-2440-11.

U.S. Department of Energy (1977) *Proceedings of the Third Biennial Conference on Wind Energy Conversion Systems*. October 1977, Washington, D.C. Available from NTIS, Order No. CONF-770921.

U.S. Department of Energy (1978) Wind Energy Systems–Program Summary. Available from NTIS, Order No. DOE/ET-0093.

U.S. Department of Energy (1979) *Wind Energy Information Sources*. Available from the Document Distribution Service, Solar Energy Research Institute, 1617 Cole Boulevard, Golden, CO 80401, USA.

Veneruso, A. F. (1974) Electrical System for Extracting Maximum Power from the Wind. Sandia Laboratories, Albuquerque, New Mexico. Available from NTIS, Order No. SAND-74-0105.

Weatherbolt, L., ed. (1976) *Vertical Axis Wind Turbine Technology Workshop, Proceedings*. Sandia Laboratories, Albuquerque, New Mexico. Available from NTIS, Order No. SAND-76-5586.

Wells, W. D. and J. G. McGowan (1976) Design and Installation of Heating System for UMass Solar Habitat. University of Massachusetts, Amherst, Massachusetts. Available from NTIS, Order No. UM-WF-TR-76-10.

Wilson, R. E., P. B. Lissaman, and S. N. Walker (1976) Aerodynamic Performance of Wind Turbines. Oregon State University, Corvallis, Oregon. Available from NTIS, Order No. PB-259-089.

Wind Power Digest (1979) *1980 Wind Energy Directory*. Available from Wind Power Digest, 109 E. Lexington, Elkhart, IN 46514, USA.

ENERGY FROM FALLING WATER

Ahrens, F. W. and J. A. Berghmans (1978) Preliminary Evaluation of the Use of Hydraulic Air Compressors in Water-Compensated Reservoir Compressed Air Storage Power Plants. Argonne National Laboratory, Argonne, Illinois, USA.

Armstrong-Evans, R. (1979) Micro-Hydro as an Appropriate Technology in Developing Countries. Paper presented at Waterpower '79, International Conference on Small Scale Hydro, October 1–3, 1979. U.S. Army Corps of Engineers, Washington, D.C., USA.

Berghmans, J. A. and F. W. Ahrens (1978) Performance of a hydraulic air compressor for use in compressed air energy storage power systems. In: *Fluids Engineering in Advanced Energy Systems*, edited by C. H. Marston. American Society of Mechanical Engineers, New York, New York, USA.

Broome, K. R. and H. A. Mayo (1979) The Optimization of Civil Works Design for Small-Scale Hydropower Projects. Paper presented at Waterpower '79, International Conference on Small Scale Hydro, October 1–3, 1979. U. S. Army Corps of Engineers, Washington, D.C., USA.

Davis, D. W. (1979) Planning Small Scale Hydropower Additions. Paper presented at Waterpower '79, International Conference on Small Scale Hydro, October 1–3, 1979. U.S. Army Corps of Engineers, Washington, D.C., USA.

Davis, D. W. (1979) Technical Factors in Small Hydropower Planning. Paper presented at the Water Systems Specialty Conference of the American Society of Civil Engineers, February 1979, Houston, Texas. Published as Hydrologic Engineering Center Technical Paper No. 61, U.S. Army Corps of Engineers, Washington, D.C., USA.

E. L. Armstrong Associates (1978) Report on Hydropower–Peru. Salt Lake City, Utah, USA.

Feller, E. W. F. (1944) *Air Compressors*. McGraw-Hill, New York, New York, USA.

Ghate, P. (1977) The scope for irrigation through hydraulic rama [*sic*] in the hills of India, *Agriculture Engineering Today* November–December.

Gladwell, J. S. (1980) Low-head hydro–a state-of-the-art discussion, *Proceedings, American Nuclear Society Meeting*. February 27–29, 1980, Los Angeles, California, USA.

Gladwell, J. S. (1980) Small Hydro. Idaho Water Resources Research Institute, University of Idaho, Moscow, Idaho, USA.

Gladwell, J. S. (1980) Small Hydroelectric Energy Generation–Some Technical Considerations. Paper prepared for the Costa Rican Workshop on Energy and Development co-sponsored by the National Academy of Sciences (USA) and the National Research Council of Costa Rica, March 17–21, 1980, San Jose, Costa Rica.

Gladwell, J. S. and C. C. Warnick (1978) Low-Head Hydro. Idaho Water Resources Research Institute, University of Idaho, Moscow, Idaho, USA.

Gladwell, J. S. and C. C. Warnick (1979) Small Hydro–A Viable "Alternative" Now. Paper presented at Waterpower '79, International Conference on Small Scale Hydro. October 1–3, 1979. U.S. Army Corps of Engineers, Washington, D.C., USA.

Gordon, J. L. (1978) Small hydro sets can yield competitive energy, *Energy International* 15(8).

Langborne, P. L. (1979) Hydraulic air compression: old invention–new energy source? *Chartered Mechanical Engineer* 26(10):76–81.

Liboff, R. L. (1978) Hydroelectric solar energy conversion, *Energy Resources* 3(3/4): 255–271.

Mayo, H. A., Jr. (1979) Turbine Selection. Paper presented at a Short Course on Small Scale Hydro Power Feasibility, Planning and Design, University of Idaho, Moscow, Idaho, USA.

Mayo, H. A., Jr., and C. W. Smith (1979) Small Hydro Control and Operation. Paper presented at a Short Course on Small Scale Hydro Power Feasibility, Planning and Design, University of Idaho, Moscow, Idaho, USA.

Moore, B. and J. S. Gladwell (n.d.) Micro-Hydro, A Bibliography. Idaho Water Resources Institute, University of Idaho, Moscow, Idaho, USA.

Norton, J. R. (1978) "Innovating" old concept for hydroelectric power generation, *Power Transmission Design* 48(5):31–32.

Norton, J. R. and C. R. Huddleston (1979) Exploiting dam potential, *International Water Power and Dam Construction* 31(1):36–38.

Plüss, A. (1977) Laos hydro scheme survives political storm, *Energy International* 14(8).

Rice, W. (1976) Performance of hydraulic gas compressors, *Journal of Fluids Engineering* 98(4):645–653.

Rice, W. and B. D. Wood (1980) Characteristics of the hydraulic air compressor as a small hydropower device, *Small Hydro-Power Fluid Machinery*. Proceedings of the Winter Annual Meeting of the American Society of Mechanical Engineers, November 16–21, 1980, Chicago, Illinois, USA.

Schulze, L. E. (1954) Hydraulic Air Compressors. Information Circular 7683. U.S. Department of the Interior, Bureau of Mines, Department of Energy, Washington, D.C., USA.

Smith, D. and R. Revelle (1977) Development of Small-Scale Hydroelectric Power and Fertilizer Production in Nepal. Report of a seminar, Asia Society, New York, New York, USA.

U.S. Army Corps of Engineers, North Pacific Division (1979) Hydropower Cost Estimating Manual. Paper prepared for the Institute of Water Resources National Hydropower Study. May.

Volunteers in Technical Assistance (1980) Micro-Hydro Power Projects. Energy Fact Sheet Number 4. VITA, Mt. Rainier, Maryland, USA.

ENERGY FROM BIOMASS

Ahlich, J. A. and J. G. Witwer (1977) Agricultural and forestry wastes as an energy resource, *Solar Energy* 19:625–629.

American Petroleum Institute (1979) *Net Energy Analysis of Alcohol Fuels*. API Publication No. 4312. API, Washington, D.C., USA.

Anderson, L. L. and D. A. Tillman (1977) *Fuels from Waste*. Academic Press, New York, New York, USA.

Anonymous (n.d.) *The Lorena Cookstove*. Available from Estacion Experimental Choqui, Apartado Postal 159, Quezaltenango, Guatemala.

Anonymous (1977) *The Haybox (A Fireless Cooker)*. Available from Low Energy Systems, Larkfield Gardens, Dublin 6, Ireland.

Appell, H. R., Y. C. Fu, E. G. Illig, F. W. Steffgen, and R. O. Miller (1975) *Conversion of Cellulosic Wastes to Oil*. U.S. Bureau of Mines, Washington, D.C., USA.

Applied Scientific Research Corporation of Thailand (1979) Pre-Feasibility Study of the Biogas Technology Application in Rural Areas of Thailand. Energy-WP-04, ASRCT, Bangkok, Thailand.

Argal, P. (1978) Role of Wood Energy in the Rural Economy in India. Paper prepared for the Eighth World Forestry Congress, October 16–28, 1978, Jakarta. Food and Agriculture Organization of the United Nations, Rome, Italy.

Arnold, J. E. M. (1978) Wood Energy and Rural Communities. Paper prepared for the Eighth World Forestry Congress, October 16–28, 1978, Jakarta. Food and Agriculture Organization of the United Nations, Rome, Italy.

Arnold, J. E. M. and J. Jongma (1978) Fuelwood and charcoal in developing countries, *Unasylva* 29(118).

Avery, D. (1978) Firewood in the Less Developed Countries. Paper prepared for the Eighth World Forestry Congress, October 16–28, 1978, Jakarta. Food and Agriculture Organization of the United Nations, Rome, Italy.

Badger, D. M., M. J. Bogue, and D. J. Stewart (1979) Biogas production from crops and organic wastes 1. Results of batch digestions, *New Zealand Journal of Science* 22:11–20.

Barnett, A., L. Pyle, and S. K. Subramanian (1978) *Biogas Technology in the Third World: A Multidisciplinary Review*. International Development Research Centre, Ottawa, Canada.

Bene, J. G., H. W. Beall, and H. B. Marshall (1979) *Energy from Biomass for Developing Countries: a State of the Art Report*. International Development Research Centre, Ottawa, Canada.

Berry, R. I. (1980) Wood: old fuel provides energy for modern times, *Chemical Engineering* 87(8):73–76.

Bhatia, R. and M. Niamir (1979) *Renewable Energy Sources: The Community Biogas Plant*. Harvard University Press, Cambridge, Massachusetts, USA.

Bio-Energy Council (1976) *Proceedings of the Conference on Capturing the Sun Through Bioconversion*. March 10–12, 1976, Washington, D.C. Available from the Bio-Energy Council, 1625 Eye Street, N.W., Washington, D.C. 20006, USA.

Biogas Newsletter. Published quarterly by the Gobar Gas Development Committee, Box 1309, Kathmandu, Nepal.

Blankenhorn, P. R., *et al.* (1978) Evaluation Procedure for Consideration of Forest Biomass as a Fuel Source for a 100 Megawatt Electric Generating Facility. Bulletin 820. Pennsylvania State University College of Agriculture, University Park, Pennsylvania, USA.

Boer, K. W. and G. W. Franta, eds. (1978) *Solar Diversification*. International Solar Energy Society, American Section, University of Delaware, Newark, Delaware, USA.

Boersma, L. L., E. Gaspin, J. R. Miner, J. E. Oldfield, H. K. Phinney, and P. R. Cheeke (1978) Utilization of Waste Heat in a System for Management of Animal Residuals to Recover and Recycle Nutrients. Department of Soil Science, Oregon State University, Corvallis, Oregon, USA.

Brace Research Institute and the Canadian Hunger Foundation (1976) Small scale biogas plant in India. Case Study #6 in: *Appropriate Technology Handbook*. Canadian Hunger Foundation, Ottawa, Canada.

Breag, G. R. and A. E. Chittenden (1979) *Producer Gas: Its Potential and Application in Developing Countries*. Publication G-130. Available from the Tropical Products Institute, 56–62 Gray's Inn Road, London WC1X 8LU, England.

Breag, G. R. and A. P. Harker (1979) The Utilisation of Waste Heat Produced During the Manufacture of Coconut Shell Charcoal for the Centralised Production of Copra. Report G127. Available from the Tropical Products Institute, 56–62 Gray's Inn Road, London WC1X 8LU, England.

Brokensha, D. and B. Riley (1978) Forest, Foraging, Fences and Fuel in a Marginal Area in Kenya. Paper presented at the U.S. Agency for International Development/Africa Bureau Firewood Workshop, June 12–14, 1978, Washington, D.C., USA.

Brown L. R. (1980) Food or Fuel: New Competition for the World's Cropland. Worldwatch Paper 35. Worldwatch Institute, Washington, D.C., USA.

Campbell-Platt, G. (1979) Improvement of the traditional Ghanaian baking oven, *Appropriate Technology* 6:28–30.

Carioca, J. and J. Scares (1978) Production of ethyl alcohol from babassu, *Biotechnology and Bioengineering* 20:443–445.

Chanco, M. P. (1978) The rice husk stove, *Appropriate Technology* 5(3):18–19.

Chandler J. A., *et al.* (1976) Development of Waste Utilization-Energy Systems, Final Report for the National Science Foundation, SMI-76-08061. California State Polytechnic University, Pomona, California, USA.

Chittenden, A. E. and G. R. Breag (1978) Wood for Energy in Developing Countries. Paper prepared for the Eighth World Forestry Congress, October 16–28, 1978, Jakarta. Food and Agriculture Organization of the United Nations, Rome, Italy.

Cohen, A., *et al.* (1979) Anaerobic digestion of glucose with separated acid production and methane formation, *Water Resources* 13:571–580.

Cruz, I. E. (1980) Studies on the practical application of producer gas from agricultural residues as supplementary fuel for diesel engines. Pp. 649–670 in: *Thermal Conversion of Solid Wastes and Biomass*, edited by J. L. Jones and S. B. Radding. ACS Symposium Series, Vol. 130. American Chemical Society, Washington, D.C., USA.

Cysewski, G. R. and C. R. Wilke (1978) Process design and economic studies of alternative fermentation methods for the production of ethanol, *Biotechnology and Bioengineering* 20:1421–1444.

DaSilva, E. J. (1979) Biogas generation: developments, problems, and tasks—an overview. Pp. 84–98 in: *Bioconversion of Organic Residues for Rural Communities*. Food and Nutrition Bulletin Supplement 2. United Nations University, Tokyo, Japan.

Dennetiere, A. A., F. Leorat, and G. F. Bonnici (1980) Power generation from biomass residues using the gasifier/dual-fuel engine technique. Pp. 635–648 in: *Thermal Conversion of Solid Wastes and Biomass*, edited by J. L. Jones and S. B. Radding. ACS Symposium Series, Vol. 130. American Chemical Society, Washington, D.C., USA.

Digernes, T. H. (1977) Wood for Fuel: Energy Crisis Implying Desertification—The Case of Bara, The Sudan. Doctoral thesis, University of Bergen, Norway.

Draper, S. A. (1977) Wood Processing and Utilization at the Village Level. Third FAO/SIDA Expert Consultation on Forestry for Local Community Development. Food and Agriculture Organization of the United Nations, Rome, Italy.

Duke, J. A. (n.d.) Energy-Producing Crop Diversification Potential in Lowland Panama. Plant Taxonomy Laboratory, Plant Genetics and Germ Plasm Institute, U.S. Department of Agriculture, Beltsville, Maryland, USA.

Electric Power Research Institute (1978) *Biofuels: A Survey*. EPRI, Palo Alto, California, USA.

Eoff, K. M. and F. M. Post (1980) How to Power a Gasoline Engine with Wood. Institute of Food and Agricultural Sciences, University of Florida, Gainesville, Florida, USA.

Evans, I. (1978) Using Firewood More Efficiently. Paper prepared for the Eighth World Forestry Congress, October 16–28, 1978, Jakarta. Food and Agriculture Organization of the United Nations, Rome, Italy.

Evans, I., with J. Kalin and K. Darrow. (1979) *Lorena Owner-Built Stoves*. Volunteers in Asia, Stanford, California, USA.

Evans, I. and D. Wharton (1977) The Lorena mudstove—a wood-conserving cookstove, *Appropriate Technology* 4(2):8–10.

Feber, R. C. and M. J. Antal (1977) *Synthetic Fuel Production from Solid Wastes.* Los Alamos Scientific Laboratory, Los Alamos, New Mexico. Available from NTIS, Order No. PB-272423.

Feldman, H. R., S. P. Chauhan, K. T. Liu, B. C. Kim, P. S. Choi, and H. N. Conkle (1979) Conversion of forest residue to a methane-rich gas, *Proceedings of the Third Annual Biomass Energy Systems Conference.* Solar Energy Research Institute, Golden, Colorado. Available from NTIS, Order No. SERI/TP 33-285.

Fischer, J. R., N. F. Meador, D. M. Sievers, C. D. Fulhage, and E. L. Iannotti (1979) Design and Operation of a Farm Anaerobic Digester for Swine Manure. Agricultural Engineering Department, Missouri University, Columbia, Missouri, USA.

Fleuret, P. and A. Fleuret (1978) Fuelwood use in a peasant community: a Tanzanian case study, *Journal of Developing Areas* 12(3):315–322.

Food and Agriculture Organization of the United Nations (1977) *China: Recycling of Organic Wastes in Agriculture.* FAO Soils Bulletin No. 40. FAO, Rome, Italy.

Food and Agriculture Organization of the United Nations (1978) *China: Azolla Propagation and Small-Scale Biogas Technology.* FAO Soils Bulletin No. 41. FAO, Rome, Italy.

Food and Agriculture Organization of the United Nations (1978) *Rice-Husk Conversion to Energy.* Agricultural Services Bulletin No. 34. FAO, Rome, Italy.

Ford, D. J. (1976) *Microbe Power.* Stein and Day, New York, New York, USA.

Ghate, P. (1979) Biogas–A Pilot Project to Investigate a Decentralized Energy System. Paper delivered at the International Seminar on Energy, January 3–7, 1979, Administrative Staff College of India, Hyderabad, India.

Ghate, P. and K. K. Singh (1978) Action Research in Biogas Development. Paper presented at the India FAO/Norad Seminar on the Complimentary [*sic*] Use of Mineral Fertilizer and Organic Material in India, September 14–19, 1978, Vijyan Bhavan, India.

Ghosh, S. and D. L. Klass (1979) The Production of Substitute Natural Gas and Recyclables from Municipal Solid Wastes. Paper presented at the Second International Recycling Congress CRE/MER, sponsored by the World Health Organization, October 1–3, 1979, Berlin, FRG.

Goddard, K. (1979) Liquid fuels from biomass, *The Chartered Mechanical Engineer* 26(11):33–35.

Golueke, C. G. (1980) Principles of alcohol production from waste. Pp. 43–49 in: *Biogas and Alcohol Fuels Production: Proceedings of a Seminar on Biomass Energy for City, Farm, and Industry.* October 25–26, 1979, Chicago, Illinois. JG Press, Emmaus, Pennsylvania, USA.

Goss, J. R. and R. H. Coppick (1980) Producing gas from crop residues, *California Agriculture* 34(5):4–6.

Groeneveld, M. J. and K. R. Westerterp (1980) Social and economic aspects of the introduction of gasification technology in rural areas of developing countries (Tanzania). Pp. 705–719 in: *Thermal Conversion of Solid Wastes and Biomass,* edited by J. L. Jones and S. B. Radding. ACS Symposium Series, Vol. 130. American Chemical Society, Washington, D.C., USA.

Hanisak, M. D., L. D. Williams, and J. H. Ryther (1980) Recycling the nutrients in residues from methane digesters of aquatic macrophytes for new biomass production, *Resource Recovery and Conservation* 4:313–323.

Hills, D. J. and R. S. Dykstra (1980) Anaerobic digestion of cannery tomato solid wastes, *Journal of the Environmental Engineering Division* (American Society of Chemical Engineers) 106(EE2):257–266.

Islam, M. N. (1979) A Report on Biogas Programme of China. Department of Chemical Engineering, Bangladesh University of Engineering and Technology, Dacca, Bangladesh.

Islam, M. N. (1980) Study of the Problems and Prospects of Biogas Technology as a Mechanism for Rural Development: Study in a Pilot Area of Bangladesh. Bangladesh University of Engineering and Technology, Dacca, Bangladesh.

Jones, J. L. and A. K. Chatterjee (1980) A preliminary analysis of the potential for developing nations to use biomass fuels in vehicle engines. Pp. 689–704 in: *Thermal Conversion of Solid Wastes and Biomass*, edited by J. L. Jones and S. B. Radding. ACS Symposium Series, Vol. 130. American Chemical Society, Washington, D.C., USA.

Jones, J. L. and S. B. Radding, eds. (1980) *Thermal Conversion of Solid Wastes and Biomass*. ACS Symposium Series, Vol. 130. American Chemical Society, Washington, D.C., USA.

Klass, D. L. (1980) Energy from biomass and wastes: 1979 update. Pp. 1–42 in: *Symposium Papers, Energy from Biomass and Wastes IV*. January 21–25, 1980, Lake Buena Vista, Florida. Institute of Gas Technology, Chicago, Illinois, USA.

Klass, D. L. (1980) The anaerobic digestion of *Macrocystis Pyrifera* under mesophilic conditions. Pp. 323–351 in: *Symposium Papers, Clean Fuels from Biomass and Wastes*. January 25–28, 1977, Orlando, Florida. Institute of Gas Technology, Chicago, Illinois, USA.

Klass, D. L., *et al.* (1978) Brazilian BIOGAS®. Final Report, Project 8896. Institute of Gas Technology, Chicago, Illinois, USA.

Klass, D. L. and S. Ghosh (1980) Methane Production by Anaerobic Digestion of Water Hyacinth (*Eichhornia Crassipes*). Paper presented at the American Chemical Society National Meeting, Fuel Chemistry Division, Fuels From Biomass Symposium, August 24–29, 1980, San Francisco, California, USA.

Klass, D. L., S. Ghosh, and D. P. Chynoweth (1979) Methane production from aquatic biomass by anaerobic digestion of giant brown kelp, *Process Biochemistry* 14(4): 18–22.

Klass, D. L., S. Ghosh, and J. R. Conrad (1976) The conversion of grass to fuel gas for captive use. Pp. 229–252 in: *Symposium Papers, Clean Fuels from Biomass, Sewage, Urban Refuse, and Agricultural Wastes*. January 27–30, 1976, Orlando, Florida. Institute of Gas Technology, Chicago, Illinois, USA.

Knight, J. A. (1976) Pyrolysis of pine sawdust. Pp. 159–173 in: *Thermal Uses and Properties of Carbohydrates and Lignins*. Academic Press, New York, New York, USA.

Kuester, J. L. and L. Lutes (1976) Fuel and feedstock from refuse, *Environmental Science and Technology* 10(4):339–344.

Ladisch, M. R. (1979) Fermentable sugars from cellulosic residues, *Process Biochemistry* 14(1):21–25.

Lane, A. G. (1979) Methane from anaerobic digestion of fruit and vegetable processing wastes, *Food Technology in Australia* 31(5):201–210.

Lindemuth, P. E. (1978) Biomass liquefaction program. Pp. 301–319 in: *Proceedings of the Second Annual Symposium on Fuels from Biomass*. Rensselaer Polytechnic Institute, Troy, New York, USA.

Little, E. C. S. (1978) The Mini-Cusab kiln for rapid small-scale manufacture of charcoal from scrub, coconut wood and coconut shells, *Appropriate Technology* 5(1):12–14.

Mabonga-Mwisaka, J. (1978) Wood Energy and Rural Communities in Zambia. Paper prepared for the Eighth World Forestry Congress, October 16–28, 1978, Jakarta. Food and Agriculture Organization of the United Nations, Rome, Italy.

Mandels, M. (1979) Enzymatic saccharification of waste cellulose. Pp. 281–289 in: *Proceedings of the Third Annual Biomass Energy Systems Conference*. Solar Energy Research Institute, Golden, Colorado. Available from NTIS, Order No. SERI/TP 33-285.

Maynall, P. J. (1976) *Methane: Planning a Digester*. Schocken Books, New York, New York, USA.

Mubayi, V., J. Lee, and R. Chatterjee (1980) The potential of biomass conversion in meeting the energy needs of the rural populations of developing countries—an overview. Pp. 617–634 in: *Thermal Conversion of Solid Wastes and Biomass*, edited by J. L. Jones and S. B. Radding. ACS Symposium Series, Vol. 130. American Chemical Society, Washington, D.C., USA.

National Center for Appropriate Technology (1979) *Methane from Anaerobic Fermentation of Manure: An Annotated Bibliography*. NCAT, Butte, Montana, USA.

National Research Council (1977) *Supporting Papers: World Food and Nutrition Study*. Vol. II. National Academy of Sciences, Washington, D.C., USA.

Norman, C. (1978) Soft Technologies, Hard Choices. Worldwatch Paper 21. Worldwatch Institute, Washington, D.C., USA.

Onaji, P. B., S. S. Adefila, and A. Beenackers (1980) Economic feasibility of gasification in Nigeria, *Chemical Age of India* 31:194–197.

O'Neil, D., *et al.* (1979) Alcohol Fermentation Process Development Unit. Engineering Experiment Station, Georgia Institute of Technology, Atlanta, Georgia, USA.

Paddon, A. R. and A. P. Harker (1979) *The Production of Charcoal in a Portable Metal Kiln*. Report G119. Available from the Tropical Products Institute, 56–62 Gray's Inn Road, London WC1X 8LU, England.

Perron, C. H. (1980) Refuse conversion to methane. Pp. 509–526 in: *Symposium Papers, Energy from Biomass and Wastes*. January 21–25, 1980, Lake Buena Vista, Florida. Institute of Gas Technology, Chicago, Illinois, USA.

Polprasert, C. (1979) A low-cost biogas digester, *Appropriate Technology* 6(3):22–24.

Poole, A. D. and J. R. Moreira (1979) A Working Paper on Ethanol and Methanol in Alternatives for Petroleum Substitution in Brazil. Institute de Fisica de Universidade de São Paulo, Brazil.

Prakasam, T. B. S. (1979) Application of biogas technology in India. Pp. 202–211 in: *Biogas and Alcohol Fuels Production: Proceedings of a Seminar on Biomass Energy for City, Farm, and Industry*. October 25–26, 1979, Chicago, Illinois. JG Press, Emmaus, Pennsylvania, USA.

Reed, T. B. (1978) Survey of pyroconversion processes for biomass. Pp. 38–41 in: *Biochemical Engineering: Renewable Sources of Energy and Chemical Feedstocks*, edited by J. M. Nystrom and S. M. Barnett. American Institute of Chemical Engineers, New York, New York, USA.

Reesen, L. and R. Strube (1978) Complete utilization of whey for alcohol and methane production, *Process Biochemistry* 13(2):21–24.

Republic of Korea (1976) *Design Manual for Methane Gas Producing Facilities*. Institute of Agricultural Engineering and Utilization, Division of Agricultural Engineering, Republic of Korea.

Rice, F. C. (1980) Second generation technology for the production of pipeline quality gas from sanitary landfills. Pp. 267–274 in: *Symposium Papers, Energy from Biomass and Wastes IV*. January 21–25, 1980, Lake Buena Vista, Florida. Institute of Gas Technology, Chicago, Illinois, USA.

Rimkus, R. R., J. M. Ryan, and A. Michuda (1979) Digester Gas Utilization Program. Paper presented at a convention of the American Society of Civil Engineers, October 22–26, 1979, Atlanta, Georgia, USA.

Rogers, K. E., ed. (1979) *Technology and Economics of Wood Residue Gasification. Proceedings of the Tenth Texas Industrial Wood Seminar*. March 13, 1979, Lufkin, Texas, USA.

Ru-chen, C., X. Zhi-ping, and L. Nian-guo (1979) *Digesters for Developing Countries—Water Pressure Digesters*. Guangzhou Institute of Energy Conversion, Chinese Academy of Sciences, Guangzhou, PRC.

Schellenbach, S., W. Turnacliff, and F. Varani (1977) *Methane on the Move: A Discussion of Small Anaerobic Digesters*. Bio-Gas of Colorado, Inc., and Colorado Energy Research Institute, Loveland, Colorado, USA.

Simon, E. and P. Solis (1977) Economic stove that burns as fuel, *Appropriate Technology* 4(1).

Sintunawa, J. (1978) Power Alcohol from Agricultural and Agro-Industrial Products. An Appropriate Measure to Develop Energy Sources for Agricultural Developing Country [*sic*]. Thesis, Mahidol University, Bangkok, Thailand.

Skrinde, R. T. (1978) Biomass to Energy Technology, International Solar Energy Society Congress, New Delhi. Report to the U.S. Department of Energy, Washington, D.C., USA.

Skrinde, R. T. (1980) Preliminary Energy Sector Assessments of Jamaica. Vol. III of: *Renewable Energy: Biogas Applications*. U.S. Agency for International Development, Washington, D.C., USA.

Solar Energy Research Institute (1979) *Survey of Biomass Gasification*. SERI, Golden, Colorado. Available from NTIS, Order No. SERI/TR-33-239.

Solar Energy Research Institute (1980) *Fuel from Farms: A Guide to Small-Scale Ethanol Production*. SERI, Golden, Colorado. Available from NTIS, Order No. SERI/SP-451-519.

Spears, J. S. (1978) Wood as an Energy Source: The Situation in the Developing World. Paper presented at the 103rd Annual Meeting of the American Forestry Association, October 8, 1978, Hot Springs, Arkansas, USA.

Srinivasan, T. G. and U. Ghosh (1977) A social benefit-cost appraisal of community gobar-gas plant, *Energy Management* 1(4):243–248.

Staff, Compost Science and Land Utilization (1980) *Biogas and Alcohol Fuels Production: Proceedings of a Seminar on Biomass Energy for City, Farm, and Industry*. October 25–26, 1979, Chicago, Illinois. JG Press, Emmaus, Pennsylvania, USA.

Starr, P. J., D. W. Finn–Carlson, and C. J. Nachtsheim (1978) Crop residue as energy source, assuming the cost of energy feasibility in direct firing, *Energy Sources* 3(3/4): 353–373.

Stout, B. A. and T. L. London (1977) Energy from organic residues. Pp. 353–400 in: *Residue Utilization Management of Agricultural and Agro-Industrial Residues*. Vol. 1. Food and Agriculture Organization of the United Nations, Rome, Italy.

Sukreeyapongse, P. (1979) Biogas and its Impact on the Rural Development: A Socio-Economic Analysis of Thai Rural Households. Thesis, Mahidol University, Bangkok, Thailand.

Sukreeyapongse, P. (1979) The Findings of Social and Economic Evaluation of Biogas Technology in Rural Thailand. Paper presented at the International Development Research Centre Biogas Workshop, Bangalore, India.

Sussman, A. and R. Frazier (1978) *Handmade Hot Water Systems*. Available from Garcia River Press, P.O. Box 527, Point Arena, CA 95468, USA.

Sutton, D. L. and W. H. Ornes (1977) Growth of *Spirodela polyrhiza* in static sewage effluent, *Aquatic Botany* 3:231–237.

Tatom, J. W., *et al.* (1978) A Prototype Mobile System for Pyrolysis of Agricultural and/or Silvicultural Wastes. Final Report, Project B-446, Engineering Experiment Station, Georgia Institute of Technology, Atlanta, Georgia, USA.

Tatom, J. W., *et al.* (1978) Pyrolytic Conversion of Agriculture and Forestry Wastes to Alternate Energy Sources in Indonesia. Georgia Institute of Technology, Atlanta, Georgia, USA.

Tatom, J. W. (1979) Feasibility of Industrial Fuel Production from Pyrolysis of Wood Wastes in Papua New Guinea. Energy Planning Unit, Department of Minerals and Energy, Konedobu, Papua New Guinea.

Tatom, J. W., H. W. Wellborn, F. Harahap, and S. Sasmojo (1980) Third World applications of pyrolysis of agricultural and forestry wastes. Pp. 671–688 in: *Thermal Conversion of Solid Wastes and Biomass*, edited by J. L. Jones and S. B. Radding. ACS Symposium Series, Vol. 130. American Chemical Society, Washington, D.C., USA.

Toyama, N. and K. Ogawa (1977) Sugar production from rice straw by saccharification with *Trichoderma viride* cellulose, *Symposium on Bioconversion of Cellulosic Substances into Energy, Chemicals, and Microbial Protein*. February 21–23, 1977, New Delhi, India.

United Nations Economic and Social Commission for Asia and the Pacific (1979) *Proceedings of the Workshop on Biogas and Other Rural Energy Resources, Held at Suva, and the Roving Seminar on Rural Energy Development, Held at Bangkok, Manila, Tehran and Jakarta*. Energy Resources Development Series No. 19. United Nations, New York, New York, USA.

United Nations University (1979) Bioconversion of Organic Residues for Rural Communities. Food and Nutrition Bulletin Supplement 2. United Nations University, Tokyo, Japan.

U.S. Agency for International Development (1976) Pyrolytic Conversion of Agricultural and Forestry Wastes in Ghana: A Feasibility Study. AID/ta-C-1290. U.S. Agency for International Development, Washington, D.C., USA.

Van Buren, A., ed. (1979) *A Chinese Biogas Manual*. Intermediate Technology Publications, Ltd., London, England.

Vergara, W. and D. Pimentel (1978) Fuels from biomass. Comparative study of the potential in five countries: the United States, Brazil, India, Sudan, and Sweden. Pp. 125–173 in: *Advances in Energy Systems and Technology*. Vol. 1. Academic Press, New York, New York, USA.

Volunteers in Technical Assistance (1979) *Designs of Woodburning Stoves/Ovens.* VITA, Mt. Rainier, Maryland, USA.

Volunteers in Technical Assistance. (1979) *Wood Conserving Stoves: Two Stove Designs and Construction Techniques*. VITA, Mt. Rainer, Maryland, USA.

Volunteers in Technical Assistance (1979) *Woodstove Design Manual.* VITA, Mt. Rainier, Maryland, USA.

Volunteers in Technical Assistance (1980) Small-Scale Biogas Production for Cooking. Energy Fact Sheet Number 6. VITA, Mt. Rainier, Maryland, USA.

Wik, O. (n.d.) *How to Build an Oil Barrel Stove*. Available from Alaska Northwest Publishing Co., Box 4-EEE, Anchorage, AK 99509, USA.

Williams, R. D. and B. Horsfield (1977) Generation of low-BTU fuel gas from agricultural residues. Pp. 465–487 in: *Food, Fertilizer, and Agricultural Residues*, edited by R. Loehr, Ann Arbor Science Publishers, Ann Arbor, Michigan, USA.

Williams, R. D., *et al.* (1978) Development of pilot plant gasification systems for the conversion of crop and wood residues to thermal and electric energy. Pp. 142–161 in: *Solid Wastes and Residues*. ACS Symposium Series, Vol. 76. American Chemical Society, Washington, D.C., USA.

Wise, D. L. (1977) Fuel gas production from biomass: progress and plans. P. 151 in: *Proceedings, Fuels from Biomass Symposium*, edited by J. J. Pfeffer and J. J. Stukel. Department of Engineering, University of Illinois, Urbana, Illinois.

Wittwer, S. H. (1979) Future technological advances in agriculture and their impact on the regulatory environment, *Bio Science* 29(10):603–610.

Wortman, S. (1976) *Food and Agriculture*. W. H. Freeman, San Francisco, California, USA.

Wortman, S. and R. W. Cummings Jr., (1978) *To Feed This World*. Johns Hopkins University Press, Baltimore, Maryland, USA.

Yaroson, E. O. and J. D. Edewor (1980) Conversion of Nigerian cellulose (forestry and agricultural wastes) to oils and gases—a future potential, *Chemical Age of India* 31:182–184.

GEOTHERMAL

Addoms, J. F., B. Breindel, and C. M. Gracey (1978) Wellsite verification testing of an advanced geothermal primary heat exchanger, *Geothermal Resources Council Annual Meeting Transaction* 2.

Armstead, H. C. H. (1977) Geothermal energy development. Pp. 181–231 in: *Progress in Energy Combustion Science*. Vol. 12. Pergamon Press, Oxford, England.

Armstead, H. C. H. (1978) *Geothermal Energy*. E. and F. N. Spon, London, England.

Cerini, D. J. (1978) Geothermal rotary field tests, *Geothermal Resources Council Annual Meeting Transaction* 2.

Electric Power Research Institute (1978) Geothermal Energy Prospects for the Next 50 Years. Preliminary Report to the World Energy Conference, ER-611-SR. EPRI, Palo Alto, California, USA.

Geothermal Resources Council (1978) *Direct Utilization of Geothermal Energy: A Symposium*. January 31–February 2, 1978, San Diego, California, USA.

Geothermal Resources Council (1978) Geothermal energy: a novelty becomes a resource, *Geothermal Resources Council Annual Meeting Transaction* 2.

Geothermal Resources Council (1979) Expanding the geothermal frontier, *Geothermal Resources Council Annual Meeting Transaction* 3.

Goodman, L. J. and R. N. Love, eds. (1980) *Geothermal Energy Projects*. Pergamon Press, New York, New York, USA.

Kalenga, M. W. and I. K. Ng'Ise (1979) L'Energie Solaire, Sur Les Energies Nouvelles. Dakar University, Senegal (in French).

Lindal, B. (1977) Geothermal energy for space and process heating, *Energy Technology Handbook*. McGraw-Hill, New York, New York, USA.

Mubayi, V., P. F. Palmedo, and A. B. Doernberg (1979) Framework for Energy Policy and Technology Assessment in Developing Countries: A Case Study of Peru. BNL-27613. Brookhaven National Laboratory, Upton, New York, USA.

National Conference of State Legislatures (1978) Geothermal resources ease energy woes, *Energy Report to the States* 14(13).

Ratigan, J. L. and U. E. Lindblom (1978) The single-hole concept of heat extraction from hot dry rock masses, *Geothermal Resources Council Annual Meeting Transaction* 2.

Sandia Laboratories (1978) Geothermal well logging device under development at labs, *Sandia Science News* 13(1).

Schmitt, R. C., *et al*. (1978) Fluidized bed low-temperature geothermal drying, *Geothermal Resources Council Annual Meeting Transaction 2*.

Schultz, R. J., *et al*. (1976) Direct applications of geothermal energy, *Geothermal Environmental Seminar*. October.

Swanson, C. R. (1978) San Diego Gas and Electric Company's geothermal program, *Geothermal Resources Council Annual Meeting Transaction* 2.

Varnado, S. G. and H. M. Stoller (1978) Geothermal drilling and completion technology development, *Geothermal Resources Council Annual Meeting Transaction* 2.

ENERGY STORAGE

Barney, D. L. (1979) Lithium/Metal Sulfide Battery Program. Paper presented at the DOE Battery and Electrochemical Contractors' Conference. December 10–12, 1979, Arlington, Virginia, USA.

Carr, P. (1978) Zinc–Chlorine Hydrate Battery. Energy Development Associates paper presented at the Second Annual Battery and Electrochemical Storage Technology Conference, June 5–7, 1978, Arlington, Virginia, USA.

Clark, R. P. (1979) Batteries for Specific Solar Applications. Sandia Laboratories, Albuquerque, New Mexico. Available from NTIS, Order No. SAND-79-1428C.

Davis, D. (1977) Rocketdyne's High-Energy-Storage Flywheel Module for the U.S. Army, *Proceedings of the 1977 Flywheel Technology Symposium.* October 5–7, 1977, San Francisco, California, USA.

Energy Development Associates (1978) Development of High-Efficiency Cost-Effective, Zinc–Chlorine Batteries for Utility Peak-Sharing, 1976. Report prepared for the Electric Power Research Institute, EPRI-EM-711, Research Projects 226-2 and 226-7-2, Palo Alto, California, USA.

General Electric Company (1976) Demonstration of an Inductor Motor/Alternator/Flywheel Energy Storage System. Technical Quarterly Progress Reports Nos. 1 and 2, June 28 and September 28. General Electric Company, Schenectady, New York, USA.

Harlow, R. A. (1979) Status of DOE/Ford Sodium–Sulfur Battery Development Program. Ford Aerospace & Communications Corporation paper presented at the DOE Battery and Electrochemical Contractors' Conference, December 10–12, 1979, Arlington, Virginia, USA.

Henry, E. A., K. W. Johnson, F. E. McMurphy, and T. M. Quick (1977) Bibliography of Flywheel Energy Storage Systems. Lawrence Livermore Laboratory, Livermore, California, USA.

Lustanader, E. L., I. H. Edelfelt, D. W. Jones, A. Plunkett, and E. Richter (n.d.) Regenerative Flywheel Energy Storage System. Research and Development Center, General Electric Company, Schenectady, New York, USA.

Millner, A. R. and T. Dinwoodie (1980) System Design, Test Results, and Economic Analysis of a Flywheel Energy Storage and Conversion System for Photovoltaic Applications. Lincoln Laboratory, Massachusetts Institute of Technology, Lexington, Massachusetts. Available from NTIS, Order No. COO-4094-70.

Mitoff, D. P. and D. Chatterji (1978) Development of Sodium–Sulfur Batteries for Utility Applications. Paper prepared for the Electric Power Research Institute, EPRIEM-463 Project 128-4 Interim Report by the General Electric Company. EPRI, Palo Alto, California, USA.

Nelson, P. A., *et al.* (1978) Development of Lithium/Metal Sulfide Batteries at Argonne National Laboratory. Summary Report for 1978, ANL-79-64. Argonne National Laboratory, Argonne, Illinois, USA.

Pickrell, R. L., G. O'Sullivan, and W. C. Merrill (1978) Inverter/Controller Subsystem Optimized for Photovoltaic Applications. NASA–Lewis Research Center, Cleveland, Ohio. Available from NTIS, Order No. DOE/NASA/1022-78/31.

Redox Project Office (1978) Redox Flow Cells and Solar Applications. NASA–Lewis Research Center, paper presented at the Second Annual Battery and Electrochemical Storage Technology Conference, June 5–7, 1978, Arlington, Virginia, USA.

Sandia Laboratories (1979) Batteries for Specific Solar Applications: Project Description. Albuquerque, New Mexico. Available from NTIS, Order No. SAND-79-1782C.

Schueler, D. G. and G. J. Jones (1978) Energy Storage Considerations in Photovoltaic Central Station Utility Applications. Sandia Laboratories, Albuquerque, New Mexico. Available from NTIS, Order No. SAND-78-0956C.

William M. Brobeck and Associates (1979) Conceptual Design of a Flywheel Energy Storage System. William M. Brobeck and Associates, Berkeley, California, USA.

STIRLING ENGINES

Advanced Mechanical Technology, Inc. (1980) Design and Development of Stirling Engines for Stationary Power Generation Applications in the 500 to 3000 Horsepower Range. First Quarterly Report. Advanced Mechanical Technology, Inc., Newton, Massachusetts. Available from NTIS, Order No. DOE/ET/15207-T1.

Bell, G. C. (1979) Solar Powered Liquid Piston Stirling Cycle Irrigation Pump. Research Grant, May 1, 1978–December 31, 1978. University of New Mexico, Albuquerque, New Mexico. Available from NTIS, Order No. SAN-1894/1.

Dochat, G. R., H. S. Chen, S. Bhate, and T. Marusak (1979) Design Study of a 15 kw Free-Piston Stirling Engine–Linear Alternator for Dispersed Solar Electric Power Systems: Final Report, September 1978–August 1979. Mechanical Technology, Inc., Latham, New York. Available from NTIS, Order No. N80-22787/9.

Hamilton, N. I. and P. O. Jarvinen (1979) Conceptual Design of a Solar-Heated–Air Receiver Coupled to a Brayton or Stirling Engine. Lincoln Laboratory, Massachusetts Institute of Technology, Lexington, Massachusetts. Available from NTIS, Order No. COO-4878-7.

Marciniak, T. J., J. C. Bratis, A. Davis, and C. Lee (1979) Assessment of Stirling Engine Potential in Total and Integrated Energy Systems. Argonne National Laboratory, Argonne, Illinois. Available from NTIS, Order No. ANL/ES-76.

Martini, W. R. (1978) Stirling Engine Design Manual; Final Report. Joint Center for Graduate Study, University of Washington, Richland, Washington. Available from NTIS, Order No. N78-23999/3ST.

Mechanical Technology, Inc. (1978) Demonstration of a Free Piston Stirling Engine Driven Linear Alternator System. Annual Report. Mechanical Technology, Inc., Latham, New York. Available from NTIS, Order No. COO-2764-002.

Parker, M. D., M. L. Malik, R. B. Wright, J. F. Staley, and J. D. Mangus (n.d.) Investigation of a 3 kw Stirling Cycle Solar Power System. General Motors Corporation, Indianapolis, Indiana. Available from NTIS, Order No. AD-276 975.

Walker, G. (1978) *Stirling Engines*. University of Calgary, Alberta, Canada.

SOLID-STATE ELECTRIC INVERTERS

Applebaum, J., *et al.* (1977) Performance analysis of solar electrical system with a load and storage batteries, *Energy Conversion*. 16(3):105–110.

Dewan, S. B. and A. Straughen (1975) *Power Semiconductor Circuits*. John Wiley & Sons, New York, New York, USA.

Grafham, D. R. and J. C. Hey, eds. (1977) *SCR Manual*. Fifth edition. General Electric Company, Schenectady, New York, USA.

Meyer, H. (1976) Synchronous inversion—concept and applications, *Proceedings, Sharing the Sun! Solar Technology in the Seventies Conference*. Vol. 7. Winnipeg, Manitoba, Canada.

Ramakumar, R. G. (1977) A review of wind-electric conversion technology, *Proceedings, International Conference on Alternative Energy Sources*, December 5–7, 1977, Miami Beach, Florida, USA.

Ramakumar, R. G. (1978) Wind-electric conversion utilizing field modulated generator systems, *Solar Energy* 20(1):109–117.

Ramakumar, R. G. and K. Bahrami (1979) Application of Field Modulated Generator Systems to Dispersed Solar Thermal Electric Generation. Report No. DOE/JPL 1060-25. Jet Propulsion Laboratory, Pasadena, California, USA.

Index

Page numbers in **boldface** refer to the original volume of *Energy for Rural Development: Renewable Resources and Alternative Technologies for Developing Countries*. The appendixes to both volumes are not indexed here, but they contain extensive information on many of the subjects listed below.